# BIOTECHNOLOGY IN AGRICULTURE SERIES

**General Editor: Gabrielle J. Persley, Biotechnology Adviser, Environmentally Sustainable Development, The World Bank, Washington DC, USA.**

For a number of years, biotechnology has held out the prospect for major advances in agricultural production, but only recently have the results of this new revolution started to reach application in the field. The potential for further rapid developments is, however, immense.

The aim of this book series is to review advances and current knowledge in key areas of biotechnology as applied to crop and animal production, forestry and food science. Some titles focus on individual crop species, others on specific goals such as plant protection or animal health, with yet others addressing particular methodologies such as tissue culture, transformation or immunoassay. In some cases, relevant molecular and cell biology and genetics are also covered. Issues of relevance to both industrialized and developing countries are addressed and social, economic and legal implications are also considered. Most titles are written for research workers in the biological sciences and agriculture, but some are also useful as textbooks for senior-level students in these disciplines.

**Editorial Advisory Board:**
*P.J. Brumby*, formerly of the World Bank, Washington DC, USA.
*E.P. Cunningham*, Trinity College, University of Dublin, Ireland.
*P. Day*, Rutgers University, New Jersey, USA.
*J.H. Dodds,* International Center for Agricultural Research in the Dry Areas, Syria.
*J.J. Doyle*, formerly of the International Laboratory for Research on Animal Diseases, Nairobi, Kenya.
*S.L. Krugman*, United States Department of Agriculture, Forest Service.
*W.J. Peacock*, CSIRO, Division of Plant Industry, Australia.

# BIOTECHNOLOGY IN AGRICULTURE SERIES

**Titles Available:**
1: Beyond Mendel's Garden: Biotechnology in the Service of World Agriculture *
   G.J. Persley
2: Agricultural Biotechnology: Opportunities for International Development
   Edited by G.J. Persley
3: The Molecular and Cellular Biology of the Potato *
   Edited by M.E. Vayda and W.D. Park
4: Advanced Methods in Plant Breeding and Biotechnology
   Edited by D.R. Murray
5: Barley: Genetics, Biochemistry, Molecular Biology and Biotechnology
   Edited by P.R. Shewry
6: Rice Biotechnology
   Edited by G.S. Khush and G.H. Toenniessen
7: Plant Genetic Manipulation for Crop Protection *
   Edited by A. Gatehouse, V. Hilder and D. Boulter
8: Biotechnology of Perennial Fruit Crops
   Edited by F.A. Hammerschlag and R.E. Litz
9: Bioconversion of Forest and Agricultural Plant Residues
   Edited by J.N. Saddler
10: Peas: Genetics, Molecular Biology and Biotechnology
    Edited by R. Casey and D.R. Davies
11: Laboratory Production of Cattle Embryos
    I. Gordon
12: The Molecular and Cellular Biology of the Potato, 2nd edn
    Edited by W.R. Belknap, M.E. Vayda and W.D. Park
13: New Diagnostics in Crop Sciences
    Edited by J.H. Skerritt and R. Appels
14: Soybean: Genetics, Molecular Biology and Biotechnology
    Edited by D.P.S. Verma and R.C. Shoemaker
15: Biotechnology and Integrated Pest Management
    Edited by G.J. Persley
16: Biotechnology of Ornamental Plants
    Edited by R.L. Geneve, J.E. Preece and S.A. Merkle
17: Biotechnology and the Improvement of Forage Legumes
    Edited by B.D. McKersie and D.C.W. Brown
18: Milk Composition, Production and Biotechnology
    Edited by R.A.S. Welch, D.J.W. Burns, S.R. Davis, A.I. Popay and C.G. Prosser
19: Biotechnology and Plant Genetic Resources: Conservation and Use
    Edited by J.A. Callow, B.V. Ford-Lloyd and H.J. Newbury
20: Intellectual Property Rights in Agricultural Biotechnology
    Edited by F.H. Erbisch and K.M. Maredia
21: Agricultural Biotechnology in International Development
    Edited by C. Ives and B. Bedford
22: The Exploitation of Plant Genetic Information: Political Strategies in Crop Development
    R. Pistorius and J. van Wijk
23: Managing Agricultural Biotechnology: Addressing Research Programs Needs and Policy Implications
    Edited by J.I. Cohen

* Out of print

# MANAGING AGRICULTURAL BIOTECHNOLOGY

## Addressing Research Program Needs and Policy Implications

*Edited by*
**Joel I. Cohen**

*International Service for National Agricultural Research (ISNAR)*
*The Hague*
*The Netherlands*

CABI *Publishing*
*in association with the*
International Service for National Agricultural Research (ISNAR)

**CABI** *Publishing* **is a division of CAB** *International*

CABI Publishing
CAB International
Wallingford
Oxon OX10 8DE
UK

CABI Publishing
10 E 40th Street
Suite 3203
New York, NY 10016
USA

Tel: +44 (0)1491 832111
Fax: +44 (0)1491 833508
Email: cabi@cabi.org

Tel: +1 212 481 7018
Fax: +1 212 686 7993
Email: cabi-nao@cabi.org

©CAB *International* 1999. All rights reserved. No part of this publication may be reproduced in any form or by any means, electronically, mechanically, by photocopying, recording or otherwise, without the prior permission of the copyright owners.

A catalogue record for this book is available from the British Library, London, UK.
A catalogue record for this book is available from the Library of Congress, Washington DC, USA

*Published in association with:*
International Service for National Agricultural Research (ISNAR)
PO Box 93375
2509 AJ The Hague
The Netherlands

Tel: (31) 70 349 6100
Fax: (31) 70 381 9677
Email: isnar@cgiar.org
http://www.cgiar.org/isnar/

ISBN 0 85199 400 8

Printed and bound in the UK by Biddles Ltd, Guildford and King's Lynn

To
Roz, Rachel, Ari, and Edie,
and all of the friends and colleagues
who have made this work possible

# Contents

| | |
|---|---|
| **Contributors** | xi |
| **Foreword** | xiii |
| **Preface** | xv |
| **Acknowledgments** | xvii |
| **Acronyms** | xix |
| **Introduction and Overview**<br>*Joel I. Cohen* | 1 |

**SECTION I   Addressing Management and Policy Issues** — 5

1. Identifying Needs and Priorities: A Decision-Making Framework for Agricultural Biotechnology — 7
 *Joel I. Cohen, Cesar A. Falconi, and John Komen*

2. The Debate on Genetically Modified Organisms: Relevance for the South — 15
 *Robert Tripp*

3. Agricultural Biotechnology Research Indicators and Managerial Considerations in Four Developing Countries — 24
 *Cesar A. Falconi*

**SECTION II   Setting and Implementing Priorities** — 38

4. Methods for Priority Setting in Agricultural Biotechnology Research — 40
 *Cesar A. Falconi*

5. Setting Research Priorities for the Chilean Biotechnology Program — 53
 *Thomas Braunschweig, Willem Janssen, Carlos Muñoz, and Peter Rieder*

6. Managing Biotechnology in AARD, Indonesia: Priorities, Funding, and Implementation — 66
 *Sugiono Moeljopawiro*

## SECTION III  Maximizing Benefits from Resources    77

7   Issues in Human Resource Management and Development    80
    *Bruce W. Holloway*

8   Managing Bioprospecting and Biotechnology for Conservation and Sustainable Use of Biological Diversity    92
    *Ana Sittenfeld and Annie Lovejoy*

9   Managing Genetic Resources and Biotechnology at IRRI's Rice Genebank    102
    *Michael T. Jackson*

10  International Collaboration in Agricultural Biotechnology    110
    *John Komen*

11  Public- and Private-Sector Biotechnology Research and the Role of International Collaboration    128
    *Joel I. Cohen*

12  Indo-Swiss Collaboration in Biotechnology: Lessons Learned and Future Strategies    140
    *Katharina Jenny and Ernst Schaltegger*

## SECTION IV  Ensuring Environmental Responsibility    152

13  Biosafety Management: Key to the Environmentally Responsible Use of Biotechnology    155
    *Patricia L. Traynor*

14  Formulating Guidelines for Field-Testing in the Philippines    166
    *Emerenciana B. Duran*

15  Addressing Public Acceptance Issues for Biotechnology: Experiences from Japan    174
    *Yutaka Tabei*

16  Balancing Needs for Productivity and Sustainability: Genetic Engineering of Rice at IRRI    184
    *John Bennett*

17  Managing Target Pest Adaptation: The Case of Bt Transgenic Plant Deployment    194
    *Mark E. Whalon and Deborah L. Norris*

## SECTION V  Managing IPR, Proprietary Science, and Technology Transfer    206

18  Intellectual Property Rights and Agricultural Biotechnology    209
    *Michael Blakeney, Joel I. Cohen, and Stephen Crespi*

19  Agricultural Research and the Management of Intellectual Property    228
    *Michael Blakeney*

20  Managing Intellectual Property in Embrapa: A Question of Policy and a Change of Heart    240
    *Maria José Amstalden Sampaio and Elza A.B. Brito da Cunha*

21 Managing Proprietary Science and Institutional Inventories for
   Agricultural Biotechnology                                          249
   *Joel I. Cohen, Cesar Falconi, John Komen, Silvia Salazar, and Michael
   Blakeney*

22 International Collaboration: Intellectual Property Management and
   Partner-Country Perspectives                                        261
   *Catherine L. Ives, Karim M. Maredia, and Frederic H. Erbisch*

23 Industrial Research and Business Development: Experiences from the
   Singapore Institute of Molecular Agrobiology                        272
   *Tai-Sen Soong*

24 Introducing Transgenic Crops in India: A Joint Venture Approach     279
   *Ellora Mubashir*

**APPENDIX   Accessing Electronic Information**                        **286**
*John Komen and Patricia Traynor*

**Glossary**                                                           **302**

**Index**                                                              **311**

# Contributors

**John Bennett**, Senior Molecular Biology and Biotechnology Coordinator, Genetic Resources Center, IRRI, MCPO Box 3127, 1271 Makati City, The Philippines; or IRRI, P.O. Box 933, Los Baños, 1099 Manila, The Philippines.

**Michael Blakeney**, Director, Queen Mary Intellectual Property Research Institute, University of London, Queen Mary and Westfield College, 339 Mile End Road, London E1 4NS, United Kingdom.

**Thomas Braunschweig**, Research Fellow, Institutional Development and Governance Program, ISNAR, P.O. Box 93375, 2509 AJ The Hague, The Netherlands; and Department of Agricultural Economics, Swiss Federal Institute of Technology, ETH-Zentrum, 8092 Zürich, Switzerland.

**Elza A.B. Brito da Cunha**, Executive Director, Embrapa – SAIN, C.P. 04.0315, Parque Rural, Final AV. W3 Norte, Brasília – DF, CEP 70770-901, Brazil.

**Joel I. Cohen**, Director, Information and New Technologies Program, ISNAR, P.O. Box 93375, 2509 AJ The Hague, The Netherlands.

**Stephen R. Crespi**, Patent Consultant, 16 Kenlegh, Bognor Regis, West Sussex, PO21 3TS, United Kingdom.

**Emerenciana B. Duran**, Member, National Committee on Biosafety of The Philippines, 2nd Floor DOST Compound, General Santos Ave., Bicutan, Tagui, Metro Manila, The Philippines.

**Frederic H. Erbisch**, Director, Office of Intellectual Property, Michigan State University, 238 Administration Building, 1046 East Lansing, MI 48824, USA.

**Cesar Falconi**, Research Officer, ISNAR Biotechnology Service (IBS), ISNAR, P.O. Box 93375, 2509 AJ The Hague, The Netherlands.

**Bruce Holloway**, Project Coordinator, Crawford Fund for International Agricultural Research, 22 Reading Avenue, North Balwyn, 3104 Victoria, Australia.

**Catherine L. Ives**, Director, ABSP Project, Michigan State University, 324 Agriculture Hall, East Lansing, MI 48824, USA.

**Michael T. Jackson**, Head, Genetic Resources Center, IRRI, MCPO Box 3127, 1271 Makati City, The Philippines

**Willem Janssen**, Senior Research Officer, Institutional Development and Governance Program, ISNAR, P.O. Box 93375, 2509 AJ The Hague, The Netherlands.

**Katharina Jenny**, Program Manager, Indo-Swiss Collaboration in Biotechnology, Institute of Biotechnology, ETH-Hönggerberg, 8093 Zurich, Switzerland.

**John Komen**, Associate Research Officer, ISNAR Biotechnology Service (IBS), ISNAR, P.O. Box 93375, 2509 AJ The Hague, The Netherlands.

**Annie Lovejoy**, Outreach Officer, 3727 Reservoir Rd. NW, Washington DC 20007, USA.

**Karim M. Maredia**, Technology Transfer Coordinator, Michigan State University, ABSP, 416 Plant & Soil Science Bldg, 1325 East Lansing, MI 48824, USA.

**Sugiono Moeljopawiro**, Head, Molecular Biology Division, Research Institute for Food Crops Biotechnology (RIFCB), Jl. Tentara Pelajar 3A, 16114 Bogor, Indonesia.

**Ellora Mubashir**, Regulatory Officer, Proagro-PGS India Ltd, Dhumaspur Road, Badshahpur, Gurgaon – 122001, Haryana, India.

**Carlos Muñoz**, General Manager, Instituto de Investigaciones Agropecuarias, Casilla 16.077-9, Santiago, Chile.

**Deborah L. Norris**, Research Technician, Michigan State University, East Lansing, MI 49824, USA.

**Peter Rieder**, Professor, Department of Agricultural Economics, Swiss Federal Institute of Technology, ETH-Zentrum, 8092 Zürich, Switzerland.

**Silvia Salazar**, Legal Consultant, Apdo. 91-3100, Santo Domingo de Heredia, Costa Rica.

**Maria José Amstalden Sampaio**, Researcher, Embrapa–SAIN, C.P. 04.0315, Parque Rural, Final AV. W3 Norte, Brasília–DF, CEP 70770-901, Brazil. (Currently located at USDA-ARS/ Cornell University as a member of the USDA/EMBRAPA-LABEX project.)

**Ernst Schaltegger**, Task Force Moderator, Indo-Swiss Collaboration in Biotechnology Project, Tulum Ltd., 6987 Caslano, Switzerland.

**Ana Sittenfeld**, Associate Professor, Center for Research in Cellular and Molecular Biology, University of Costa Rica, San José, Costa Rica.

**Tai-Sen Soong**, First Vice President, Life Science Investment Unit, Overseas Investment Department, China Development Industrial Bank, 125 Nanking East Road, Section 5, Taipei, Taiwan.

**Yutaka Tabei**, Deputy Director, Innovative Technology Division, Ministry of Agriculture, Forestry & Fisheries, 1-2-1 Kasumigaseki, Chiyoda, Tokyo, 100-8950 Japan.

**Patricia L. Traynor**, Science Coordinator, Information Systems for Biotechnology, Virginia Polytechnic Institute & State University, 120 Engel Hall, Blackburg, VA 24061, USA.

**Robert Tripp**, Research Fellow, Overseas Development Institute (ODI), Portland House, Stag Place, London SW1E 5DP, United Kingdom.

**Mark Whalon**, Professor of Entomology, Director, Center for Integrated Plant Systems, Michigan State University, East Lansing, MI 48824, USA.

# Foreword

The judicious introduction of molecular techniques in agricultural research has built on and stimulated fundamental research, leading to rapid advancements in plant science. This has convinced many young scientists to devote their efforts to furthering our knowledge in this field. Plant scientists, particularly in the USA and Japan, were fast in grasping the opportunity offered. The result was a rapid growth of knowledge in stress physiology and plant defence mechanisms. Approaches based on molecular genetics that were successful in animal and microbial research are contributing to the correlation of the complete sequence and genetic map of the genome of *Arabidopsis thaliana*. Progress with rice and corn is well underway.

It is now documented that, where studied, substantial stretches of the genome of other plants show the same gene order as *Arabidopsis*. So we are learning that part of the synteny discovered among the grasses may well be extended to most plants. This genomic data is increasing the interest in plant biochemistry and will stimulate plant-science research in general.

Substantial efforts are required to integrate this emerging knowledge in agricultural research. Till recently, many agronomists, breeders, and ecologists remained sceptical that the new tools offered by molecular biologists could be of real value. Indeed, the large-scale introduction of the first-generation transgenic plants did not begin until 1996. Even now, molecular marker-assisted breeding is a complex process, requiring substantial investments in bioinformatics. This means that its utility in breeding programs is still limited.

I believe that we have to urgently develop the plants that can improve the productivity of developing-country agricultural ecosystems, from industrial agriculture to the family-size farm of the subsistence farmer. But who will do the research? What networks are needed to initiate the fundamental research that is still required? How will such work progress successfully? What are the priorities? And how do we build public acceptance? This book introduces and discusses all these and many more important questions in detail. The approach is very realistic and direct. Reflections on the available knowledge are followed by case studies from different regions of our planet. Such case studies help focus attention and advance our learning of the real challenges.

The task of implementing biotechnology in developing-country agriculture is tremendous. Where does one start? Insight must be gained into which traits in tropical crop plants should be altered and which plants should be treated with priority. In the third millennium, the majority of the world population will continue to live in developing countries. Achieving a global equilibrium requires that, in addition to ensuring food and feed supply, we need to build less-polluting industries and offer affordable health care. One contribution to achieve this will come from plant-based industries for fine chemicals and pharmaceuticals.

Will the CGIAR system, well acquainted with the economics and the sociology of tropical agriculture be able to convince the donor countries to help establish networks with competitive centers of fundamental research in the North? There, many young post-docs, well-trained in molecular biology techniques, are eager to provide assistance in programs for developing-country agriculture, but preferably under conditions where they can continue to participate in cutting-edge research. Is it possible to convince foundations of the importance of funding such post-docs?

Decision makers and government officials may ask whether it is advisable to support such efforts now, given the diversity of public opinion. This is a very important question. Agriculture is not only science and technology; it is a core activity of our societies. The sociology and economics of agriculture are therefore as important as the production skills. When large segments of the public claim that they are concerned, then it is essential to listen to them. But one should also understand the real meaning of this concern. The public wants to be assured that scientists' claims are accurate and objective, and that they are not influenced by personal gains. Until now, nobody has demonstrated that genetically modified organisms have any substantial, long-term health hazards for humans or animals. Neither is there any proof of danger to the environment. This book provides several important initiatives to further our understanding of these complex topics.

However, the careful use of our skills in biotechnology is only one way of achieving success. Political problems, such as injustice, discrimination, and malnutrition, also influence tropical agriculture. Significant effort is therefore required in considering the role of biotechnology in relation to food production in traditional agroecosystems. For this reason, the release of this book is very timely. Scientists and directors of research share a sense of urgency, as engineering new plants can take up to 10 years or more. While many research efforts have already begun, a thorough analysis of the problems and in-depth discussions of the technological needs are essential to define how research results can be achieved. The book provides a wealth of information in this regard, coupling scientific options with relevant sociological and economic functions. Those working in the agricultural research community should give careful consideration to the points presented in the various chapters. It helps them extend and apply useful findings and methodologies to their own research institute.

*Marc C.E. Van Montagu*
*Full Professor and Director*
*Laboratory of Genetics, Faculty of Science*
*University of Gent*
*Belgium*

*September 1999*

# Preface

As this book goes to press, debate continues about the contribution of modern biotechnology towards meeting the food security needs of developing countries. Almost 25 years ago, scientists from around the world met at the Asilomar Conference Center, Pacific Grove, California, USA, to take a first look at the remarkable developments making news in the world of molecular genetics and to think collectively regarding implications of their use. Since that meeting, applications of biotechnology have increasingly found their way into global and local food production. Recent developments in agriculture are poised to transcend initial applications for herbicide and insect resistance, offering products that can improve human nutrition and health while contributing to ameliorate damage to the environment.

Over the past 15 years, additional efforts were taken to ensure that the safe application of new biotechnologies becomes part of international agricultural research, with potential benefits reaching farmers and consumers in developing countries. Remarkable advances have occurred, building capacity and competency for biotechnology in countries where a few years ago none existed. International collaboration provided opportunities not only to advance the technical dimensions of biotechnology, but also to learn of the related managerial and policy implications that accompany emerging technologies. However, the time available for learning has been too short. Scientists and directors of research institutions in developing countries rush to catch up, seeking knowledge of technologies and their implications needed to ensure safe and reliable sources of food and feed.

When considering biotechnology as applied to food and agriculture, it is essential that potential benefits and risks to society be made clear. Increasing productivity alone is clouded by concerns of the technologies used to produce such gains, with consumers having a closer eye on safety than productivity. For this reason, those undertaking research in biotechnology are always one step away from the myriad debates and issues that surround its application to agriculture. Scientists, managers, and policymakers are being urged again to participate in debates, dialogs, and opportunities that help the public understand the benefits from biotechnology and how biotechnology can be effectively and responsibly regulated.

Therefore, the need for sharing and providing information does not diminish; rather, its urgency increases. To meet this need, this book provides the combined experiences of individuals and specific managerial decisions taken to implement biotechnology research priorities and programs. Special attention is given to those implementing biotechnology in the context of international agricultural research. The methodology adopted for the book and the associated course, "Managing Biotechnology in a Time of Transition," advances learning needed by those involved in decision making and management of biotechnology, whether they be from national research institutions, nongovernmental organizations, or the private sector.

Contributors to this book have been instrumental in helping others to understand many of the transformations that are occurring in agricultural research. To capture and extend their knowledge, I have relied on a highly interactive process, distilling key management challenges faced by the authors and deriving recommendations to address them. I have drawn on the experience of 16 managers and directors of research, with experiences based on accomplishments in developing countries, international agricultural research centers, nongovernmental organizations, universities, and industry. The remaining chapters summarize methodologies, frameworks, and theory regarding policy implications for biotechnology, including chapters from those undertaking research on intellectual property and environmental responsibility, international collaboration, and for setting priorities.

The combination of synthesis, recommendations, personal and institutional experiences, methodologies, and tools provides a close look at the challenges confronting research managers with responsibilities for biotechnology. Meeting these challenges requires a willingness to facilitate discussion on the benefits as well as the potential difficulties associated with biotechnology. Such challenges will not be settled by easy answers or technological fixes, but rather by patient efforts that stimulate learning, advancing our abilities to find agricultural solutions adapted to the needs and benefits of its users.

*Joel I. Cohen*
*The Hague, the Netherlands*

# Acknowledgments

This book originated from the combined efforts of many people working with ISNAR's Biotechnology Service (IBS) over the past three years, moving from conception to review, consultation, synthesis, and editing. I express my deep gratitude to all those who have devoted their valuable time and attention to the chapters and concepts of the book. Without such dedication and commitment, this book would never have been possible.

For many of these individuals, I offer not one but two expressions of gratitude. I thank many of the authors not only for their papers, but also, if not more importantly, for their continued support to the ISNAR management course "Managing Biotechnology in a Time of Transition," which builds on the lessons presented in the book. Many of these authors join our annual migration to Asia to present their chapters to participating scientists, managers, and directors. We have all benefited from the high degree of professionalism provided by them as they embody the message conveyed in their chapters, bringing a sense of purpose to this book and to the course.

I specifically thank the following contributors to this book and the course: John Bennett, Michael Blakeney, Thomas Braunschweig, Emmy Duran, Cesar Falconi, Bruce Holloway, Catherine Ives, Mike Jackson, Katharina Jenny, John Komen, Sugiono Moeljopawiro, Ellora Mubashir, Maria José Sampaio, Ana Sittenfeld, Tai-Sen Soong, Yutaka Tabei, Patricia Traynor, Robert Tripp, and Mark Whalon.

Special acknowledgement is due for John Komen and Cesar Falconi. Their ceaseless energy, during all stages of the process, began when there was nothing more to show than a three-page concept paper. Their support of the work of other contributors has been of great importance. Special thanks is extended to Patricia Traynor as well, who did much to improve the book and the course, working diligently with particular attention to chapters on environmental responsibility. I could always count on her for critical and constructive reviews.

My appreciation is also extended to the Government of Japan for its support to the course and the book. Colleagues in Japan include Ken-Ichi Hayashi, member of ISNAR's Board of Trustees; Nobuyoshi Maeno, Director General of the Japan International Research Center for Agricultural Sciences; and colleagues at the Ministry of Agriculture, Forestry and Fisheries.

Additional support for these combined efforts has come from the Crawford Fund for International Agricultural Research. Specifically, I would like to thank Bruce Holloway, Project Coordinator of the Crawford Fund's Biotechnology Master Classes, and Alex Buchanan, former director. The Fund supported a consultation and review meeting of the book's authors at ISNAR in March 1999. This meeting facilitated in-depth analysis of each chapter and the first work on a synthesis for each section. During this meeting, we gained much from the attendance of Klaas Tamminga of the Directorate General for International Cooperation of the Netherlands' Ministry of Foreign Affairs, Niels Louwaars and Jan-Peter Nap of the Dutch Agricultural Research Service (Wageningen), Colm Lawler of Michigan State University, Reynaldo dela Cruz of BIOTECH, The Philippines, Manuel

Ruiz of Peru's Society for Environmental Law, Byron Mook of ISNAR, Mao Yeding from the Chinese Academy of Tropical Agriculture, and Tim Hardwick of CABI Publishing.

Initial review of the workplan and mechanisms for the course and book came from members of the Steering Committee for ISNAR's Biotechnology Service. Most valuable to this process has been the continuing support from Paul Egger of the Swiss Agency for Development and Cooperation; Stein Bie, Director General of ISNAR; Teresa Fogelberg of the Netherlands' Ministry of Foreign Affairs, and my long-standing collaboration with Gabrielle Persley of the World Bank.

The course itself and the development of these chapters gained much from interactions and collaboration with ISNAR's Training Unit, with extensive help in concept development from Zenete Franca and further support from Helen Hambly, Richard Claase, Mirela Zoita, and Albetine Huybrechts. I have also benefited from the contributions of ISNAR's Publications Unit, including especially close collaboration with Jan van Dongen. Additional support was provided by Michelle Luijben and Kathleen Sheridan. IBS secretaries Karin Felix-Faure and Maaike Vergeer were essential for both the book and the authors' meeting.

I owe a great debt and gratitude to professors at the Harvard Institute of International Development and Kennedy School of Government for inspiring concepts and methodologies used in this book and the ISNAR course. Among those I especially wish to thank are John Thomas, Merilee Grindle, Michael Watkins, Ronald Heifetz, and Kathy Eckroad.

Substantial improvements in the chapters were made through formal reviews by Howard Elliott and Douglas Horton of ISNAR, Youngyuth Yuthavong of Thailand's National Science and Technology Development Agency, and an anonymous fourth reviewer.

Finally, a special note of thanks to Marc Van Montagu for making the time to prepare the book's foreword and reviewing its contents.

It has been a sincere privilege, honor, and professional achievement to work continuously with these people while steadily advancing this book to its readers.

*Joel I. Cohen*
*September 1999*

# Acronyms

| | |
|---|---|
| AARD | Agency for Agricultural Research and Development, Indonesia |
| AARP | Applied Agricultural Research Project, Indonesia |
| ABSP | Agricultural Biotechnology for Sustainable Productivity, Michigan State University, USA |
| ACIAR | Australian Centre for International Agricultural Research |
| ACNFP | Advisory Committee on Novel Foods and Processes, UK |
| AGERI | Agricultural Genetic Engineering Research Institute, Egypt |
| AHP | analytic hierarchy process |
| ARBN | Asian Rice Biotechnology Network |
| ARI | advanced research institute |
| ARM project | Agricultural Research Management project, Indonesia |
| Bappenas | National Development Planning Agency, Indonesia |
| BORIF | Bogor Research Institute for Food Crops, Indonesia |
| BRC | biosafety review committee |
| BRI | Biotechnology Research Institute, Zimbabwe |
| Bt | *Bacillus thuringiensis* |
| CBD | Convention on Biological Diversity |
| CENARGEN | National Research Center for Genetic Resources and Biotechnology, Brazil |
| CGIAR | Consultative Group on International Agricultural Research |
| CICY | Scientific Research Center of Yucatan, Mexico |
| CINVESTAV | Center for Research and Advanced Studies, Mexico |
| CRDB | Center for Research and Development of Biotechnology, Indonesia |
| CRIFC | Central Research Institute for Food Crops, Indonesia |
| CRIH | Central Research Institute for Horticulture, Indonesia |
| Embrapa | Brazilian Agricultural Research Corporation |
| ETH | Federal Institute of Technology, Switzerland |
| FAO | Food and Agriculture Organization of the United Nations |
| GATT | General Agreement on Tariffs and Trade |
| GDP | gross domestic product |
| HRD | human resource development |
| GMO | genetically modified organism |
| IBC | institutional biosafety committee |
| IBS | ISNAR Biotechnology Service |
| IBT-UNAM | Biotechnology Institute of the Autonomous National University of Mexico |
| ICARDA | International Center for Agricultural Research in the Dry Areas |
| ICGEB | International Centre for Genetic Engineering and Biotechnology |
| ICRISAT | International Crops Research Institute for the Semi-Arid Tropics |
| INBio | National Biodiversity Institute, Costa Rica |

| | |
|---|---|
| INIA | National Agricultural Research Institute, Chile |
| INPI | National Industrial Property Institute, Brazil |
| IPM | integrated pest management |
| IPR | intellectual property right(s) |
| IRRI | International Rice Research Institute |
| ISAAA | International Service for the Acquisition of Agri-biotech Applications |
| ISCB | Indo-Swiss Collaboration in Biotechnology |
| ISNAR | International Service for National Agricultural Research |
| KABP | Kenya Agricultural Biotechnology Platform |
| KARI | Kenya Agricultural Research Institute |
| MAFF | Ministry of Agriculture, Forestry and Fisheries, Japan |
| MINAE | Ministry of the Environment and Energy, Costa Rica |
| MINAGRI | Ministry of Agriculture, Chile |
| MSU | Michigan State University, USA |
| MTA | material transfer agreement |
| NACBAA | National Advisory Committee on Biotechnology Advances and Their Applications, Kenya |
| NAFTA | North American Free Trade Agreement |
| NARO | national agricultural research organization |
| NARS | national agricultural research system(s) |
| NBC | national biosafety committee |
| NBP | National Biotechnology Program, Chile |
| NCBP | National Committee on Biosafety of the Philippines |
| NGO | nongovernmental organization |
| NRC | National Research Council, USA |
| ODI | Overseas Development Institute, UK |
| OECD | Organisation for Economic Co-operation and Development |
| PBG | Philippine Biosafety Guidelines |
| PBR | plant breeders' right |
| PCT | Patent Cooperation Treaty |
| PPP | purchasing power parity |
| PRONDETYC | National Program for Technological and Scientific Development, Mexico |
| PVR | plant variety rights |
| R&D | research and development |
| RIFCB | Research Institute for Food Crops Biotechnology, Indonesia |
| RLO | research liaison office |
| SDC | Swiss Agency for Development and Cooperation |
| SINAC | National System of Conservation Areas, Costa Rica |
| SIRDC | Scientific and Industrial Research and Development Centre, Zimbabwe |
| STAFF | Society for Techno-Innovation of Agriculture, Forestry and Fisheries, Japan |
| TAC | Technical Advisory Committee of the CGIAR |
| TCCP | Tissue Culture for Crops Project |
| TRF | Tea Research Foundation, Kenya |
| TRIPs | (Agreement on) Trade-Related Aspects of Intellectual Property Rights |
| TRR | traditional resource rights |
| UNEP | United Nations Environment Programme |

| | |
|---|---|
| UNIDO | United Nations Industrial Development Organization |
| UPM | Universiti Putra Malaysia |
| UPOV | International Union for the Protection of New Varieties of Plants |
| USAID | United States Agency for International Development |
| USDA | United States Department of Agriculture |
| WIPO | World Intellectual Property Organization |
| WTO | World Trade Organization |
| WWW | World Wide Web |

# Introduction and Overview

*Joel I. Cohen*

Innovation is essential for sustaining and enhancing agricultural productivity in developing countries. Agricultural innovation has always involved new, science-based products and processes that have contributed reliable methods for increasing productivity and environmental sustainability. The set of techniques commonly referred to as biotechnology has introduced a new dimension to such innovation. For most chapters of this book, biotechnology research focuses on cellular and molecular biology and the resulting techniques coming from these disciplines for improving the genetic makeup and agronomic management of crops and animals. Thus, the definition of biotechnology as used here includes less-advanced techniques such as tissue culture, as well as genetic engineering and emerging genomic-based research methods.

Effective development and safe deployment of new agricultural products, particularly those derived from biotechnology, entails decisions on priorities and resource allocations. It also requires research managers, policymakers, and scientists to address new responsibilities, such as participation in dialogues regarding public awareness and acceptance of biotechnology. These decisions and responsibilities form the strategic context in which developing countries consider their options for undertaking research in biotechnology.

The ability to make strategic decisions regarding agricultural biotechnology is critical in determining if present and future products from such research will address the needs of developing-country farmers. Such decisions have greater importance now than even five years ago because of the increase in investments (human, financial, and managerial) as well as the range of biotechnology partnerships occurring in many developing countries. This increase in investments and collaboration means that more effort is needed to manage resources, address research priorities, understand the complexities of partnerships, and attend to the policy implications of intellectual property rights and environmental responsibility.

New advances in biotechnology are already on the horizon. Recent examples of new and expanded applications include a method to control plant gene expression and germination, commonly referred to as "terminator" technology, bioinformatics, and genomic research. Making informed managerial decisions on these advances requires a constant updating of knowledge and increased insight into the policy implications of such developments.

---

© CAB *International*. 1999. *Managing Agricultural Biotechnology—Addressing Research Program Needs and Policy Implications* (ed. J.I. Cohen)

Recognizing the importance of these strategic decisions and responsibilities for developing-country agricultural research organizations, the International Service for National Agricultural Research (ISNAR), in partnership with developing countries and international expertise, has conducted a research and advisory program on agricultural biotechnology policy, organization, and management. This book is a product of that program; it shares information and advice emanating from the program and aims to improve access to knowledge among its readers, particularly those in developing countries

## Goal and purpose of the book and context for management

The goal of this book is to provide agricultural research managers and directors with information that will help them achieve more effective management of agricultural research programs involving biotechnology. Effective management is defined here as the judicious use of means and resources, which together help achieve a strategy for agricultural biotechnology research. The information will enable managers to

- think strategically about priorities and research programs, and the management implications inherent in them
- address concerns regarding environmental responsibility, including biosafety and public acceptance
- better manage human, biological, genetic, and information resources, which serve to integrate biotechnology with agricultural research
- strengthen their understanding and management of various forms of intellectual property protection and international technology transfer.

The book targets individuals with responsibilities for managing research programs, organizations, or institutions where agricultural biotechnology research and policy analysis is assuming increased importance. Within each section, most chapters identify challenges, expectations, and responsibilities that confront managers, and provide the author's first-hand experiences in addressing these matters. Their experiences are then analyzed in relation to one of the themes of the book in order to, where possible, provide supportive decision-making tools and recommendations for managers. The individual chapters were synthesized and reviewed during a meeting at ISNAR of authors, external reviewers, and experts in the application of agricultural biotechnology. The meeting focused on the lessons learned for each topic.

## Thematic sections

The book's five thematic sections highlight research and case studies that were identified as priorities for ISNAR's work regarding policy and management issues for biotechnology, with papers undertaken in collaboration with national and international partners. Each section begins with a synthesis of lessons learned taken from the following chapter and identifies specific managerial recommendations. The sections cover the following topics:

- Section I ("Addressing Management and Policy Issues") orients the reader to a management and decision-making perspective.
- Section II ("Setting and Implementing Priorities") highlights decisions with regard to priority setting that take into account special features of biotechnology.

*Introduction and Overview*  3

- Section III ("Maximizing Benefits from Resources") focuses on the management of various resource needs, including human and genetic resources and biodiversity, and it describes opportunities for international collaboration and the resources to be gained through this collaboration. Together, these help build a national capacity in agricultural biotechnology.
- Section IV ("Ensuring Environmental Responsibility") covers managerial and policy issues such as biosafety, integrating into research projects the diverse goals of productivity and sustainability, public acceptance, and the management of pest resistance in relation to transgenic plant deployment.
- Section V ("Managing Intellectual Property, Proprietary Science, and Technology Transfer") focuses on intellectual property issues and the new managerial responsibilities placed on directors of research.
- The appendix provides up-to-date information on electronic biotechnology databases and resources available. Giving the reader access to a wide range of supplemental information, the specific examples are selected and screened for their utility and accuracy.

## Relation to technical and scientific content

Many chapters provide examples of how the most recent developments in agricultural biotechnology are being used to address research needs of developing countries. To some extent, a discussion of scientific methodologies is essential when speaking of research program management and policy implications for biotechnology. The focus of this book, however, is not on technologies per se but rather their implications for managing agricultural programs, policies, and their various institutional settings. Chapters featuring examples of biotechnology research and methods are related to specific managerial choices and implications. For this reason, and in keeping with the book's focus on management, detailed accounts of scientific and technical procedures are covered in annexes and suggested references.

## Using this book

Readers will enhance the analyses in this book by comparing and contrasting their own experiences with those presented in the individual chapters. In comparing experiences, however, readers should keep in mind that management and policy implications for biotechnology may be nationally or culturally specific. As each country is unique, biotechnology initiatives have evolved differently in different parts of the world. Cultural specificity means that it is not always possible to overlay practices from one national system onto a very different one. Some adaptation is often required.

National organizations involved in promoting agricultural biotechnology can use this book as a resource for developing research management programs that more closely fit their own circumstances. To facilitate the design of these programs, particular chapters from this book have been incorporated into a comprehensive training module for national biotechnology researchers and potential national trainers.

The training module and this book together comprise a thorough plan for organizing and structuring a two-week course. The training module contains clearly identified and planned assignments that guide participants in reading the book and applying the contents to their own organization. The module consists of the curriculum, including learning objectives for each day's activities; descriptions of the training approach, methods and techniques; and master copies of all exercises, handouts, worksheets, overhead transparencies, registration and evaluation forms, and a recommended bibliography for use by trainers and program participants.

# SECTION I

# Addressing Management and Policy Issues

Deciding how to best make use of the rapid advances in biotechnology requires careful judgment and experience. Agricultural research managers must weigh potential productivity increases alongside uncertain factors such as environmental risks, potential returns on investment, and alternative approaches to address food security. In addition, policymakers and the general public increasingly call upon research managers to inform them about the latest developments in biotechnology. This responsibility increases with the growing debate on the safety and public acceptance of biotechnology used for food production and consumption.

Managers responsible for agricultural biotechnology research find that they need technical, managerial, and policy skills that go beyond the scope of scientific education. In the first section of this book, the main issues facing these managers are presented from three different perspectives. Chapter 1 ("Identifying Needs and Priorities: A Decision-Making Framework for Agricultural Biotechnology") analyzes identified needs and findings from regional seminars on biotechnology that were organized by ISNAR.

Findings are presented in relation to a decision-making framework applied to biotechnology involving four steps: identifying research priorities for which biotechnology offers a comparative advantage, determining relevant national policies, formulating an appropriate research agenda, and providing for delivery of products to end users. Seminar findings and decision-making steps are also related to broader strategic thinking for agricultural research, as undertaken by the International Food Policy Research Institute. The policy seminars helped participants to

- explore each country's context for agricultural biotechnology
- identify factors unique to formulating strategies and policies for agricultural biotechnology
- think systematically about decision making and implementing biotechnology
- understand the high level of uncertainty faced by decision makers when making policy on agricultural biotechnology.

Identifying, understanding, and addressing these needs is the starting point for the material presented in this book, as well as for the ongoing ISNAR course "Managing Biotechnology in a Time of Transition."

Taking a different perspective, chapter 2 ("The Debate on Genetically Modified Organisms: Relevance for the South") focuses on the public debate and emerging controversial issues of biotechnology, specifically regarding the introduction of genetically modified organisms (GMOs). In the UK, for example, GMO foods are available on grocery shelves, but their use is so politically sensitive that they are banned from the restaurants of parliament. Research managers in developing countries should be prepared to respond to concerns raised by policymakers and the public regarding the environmental safety and food safety of GMOs and the perceived trend towards corporate control and market concentration in the life sciences.

The final chapter of this section ("Agricultural Biotechnology Research Indicators and Managerial Considerations in Four Developing Countries") analyzes key considerations for national policymakers regarding the mobilization and use of human, financial, and physical resources for agricultural biotechnology and the institutional setting in which the agricultural biotechnology research system operates. Information on the size, structure, and content of public research is needed to improve policy decisions, clarify the roles of the public and private sectors, and support public-sector implementation of biotechnology research. There is, however, a lack of structured data on resources available for agricultural biotechnology in developing countries.

ISNAR initiated this study to collect and analyze baseline data. Initial findings provide new information that strengthens the basis for making decisions on agricultural biotechnology and lead to a set of policy recommendations. The information provides valuable background material for other themes covered by this book, such as strategic planning and priority setting. It facilitates greater understanding of the institutional developments required for agricultural biotechnology in developing countries, and it provides the means for comparison of information between selected countries.

The main recommendations from the chapters in this first section to policymakers, donors, and managers of national agricultural research systems are the following:

- Support future policy dialogues that identify needs and mechanisms for follow-up regarding policy and managerial dimensions of biotechnology.
- Raise awareness of the potential benefits and costs of using biotechnology to achieve national goals.
- Assure relevant stakeholder and end-user participation in policy dialogues for identifying needs regarding biotechnology policy.
- Develop mechanisms that help developing countries find funding for their own research by addressing issues of sustainability and user orientation.
- Institute policy analysis on socioeconomic aspects of biotechnology, necessary legal reforms, and build regulatory capacity to deal with biotechnology and related agricultural policies.
- Conduct regular studies to analyze trends in public and private investments and capacity development in biotechnology.
- Initiate policies and programs to encourage partnerships with the private sector that complement investments made in the public sector.

# 1 Identifying Needs and Priorities: A Decision-Making Framework for Agricultural Biotechnology

*Joel I. Cohen, Cesar A. Falconi, and John Komen*

## Abstract

*This chapter presents findings of four regional policy seminars held by ISNAR's Biotechnology Service (IBS). The regional seminars dealt with needs related to policy and management aspects for agricultural biotechnology as identified by stakeholders in developing countries. The chapter reviews a framework for decision making in agricultural biotechnology that provided the basis for the seminars. Identified needs and recommendations of the seminars are discussed as related to specific research papers and case studies presented in this book.*

## A framework for decision making

An increasing number of developing countries are investing in human resources and physical infrastructure for agricultural biotechnology. Analyzing this trend, ISNAR's Biotechnology Service (IBS) conducted a comparative review in 1993 of 10 developing countries and their experiences in stimulating the use of biotechnology in agricultural research (Komen and Persley 1993). The review found that the institutional arrangement adopted in a country for biotechnology-based agricultural research depends on three main factors: the size of the country, its scientific capacity, and its existing research infrastructure. Indeed, some developing countries may be unable to use biotechnology due to a combination of (1) pressing needs that do not require biotechnology, (2) an inability to maintain a strong base for research, and (3) lack of funding.

In 1994, IBS introduced a decision-making framework that takes account of all key policy, scientific, and economic considerations (Cohen 1994). It starts with the national policy environment and incorporates institutional, financial, and program issues involved in priority setting and in determining needs and objectives for biotechnology-based research. The framework helps decision makers ensure that resources allocated to biotechnology—including the activities and contributions of international research centers and donors—are consistent with agreed national objectives. The framework has four phases, which are analogous to those used in planning conventional agricultural research programs:

© CAB *International*. 1999. *Managing Agricultural Biotechnology—Addressing Research Program Needs and Policy Implications* (ed. J.I. Cohen)

1. identify research priorities for which biotechnology offers a comparative advantage
2. determine relevant national policies
3. formulate and implement an appropriate research agenda based on agreed priorities
4. provide for the transfer and delivery of research products to users

The framework also takes into account characteristics particular to biotechnology, such as high development costs, integration with conventional research programs, opportunities for international collaboration, and challenges of public perception, biosafety, as well as intellectual property rights (IPR). It suggests that an overview of existing national policies, priorities, and research activities for agricultural biotechnology be prepared as an introduction to the decision-making process.

Each phase requires interaction between technical, financial, and policy specialists. In addition, periodic meetings with stakeholders are essential for selecting appropriate research activities. Decision makers can foster these interactions by encouraging scientists to focus their research on agreed priorities and supporting policies that facilitate the development and delivery of research products.

## Four regional seminars on agricultural biotechnology policy

ISNAR refined and tested the decision-making framework through a series of agricultural biotechnology policy seminars. The seminars addressed issues of strategic planning, priorities, and policies for agricultural biotechnology by doing the following:

1. examining planning, investments, and priorities for biotechnology research related to agriculture
2. analyzing case studies on future needs in planning and managing biotechnology-based research in agriculture
3. identifying follow-up initiatives and suitable collaboration tailored to the needs of each participating country

The overall goal was to enhance biotechnology's integration with broader national priorities and to strengthen the management of demand-driven, user-oriented biotechnology programs in agriculture.

Between September 1994 and October 1996 four regional policy seminars were held—in Southeast Asia (Komen, Cohen, and Lee 1995), in Eastern and Southern Africa (Komen, Cohen, and Ofir 1996), in West Asia/North Africa (Brenner and Komen 1997), and in Latin America and the Caribbean (Komen, Falconi, and Hernández 1998). Each seminar was attended by country delegations that included a mix of government planners, agricultural policymakers, research managers from public- and private-sector organizations, and end-user representatives. This broad mix of participants emphasizes the interactive and multilevel nature of the decision-making process and ensured that practicing scientists and managers conversed with end users and policymakers, thus broadening the perspectives of each regarding research priorities and focus.

Each seminar was seen as a first step in planning a process of decision making for agricultural biotechnology. In-depth analysis of the country-specific findings was therefore left for subsequent national or institutional meetings. Plenary sessions were based on case studies or examples (technical, policy, or managerial) that illustrated each of the different phases of the decision-making framework. In working groups, the national delegations discussed issues raised in plenary, identifying gaps and needs in their own country and suggesting follow-up activities for overcoming them. The working-group exercises were an essential part of the seminars. They served not only to record possibilities for national country follow-up activities, but also to guide ISNAR on future agricultural biotechnology issues that it should address.

The agenda included six general themes: (1) introduction and overview, (2) needs and priorities for biotechnology research, (3) identifying national policy issues, (4) mobilizing and allocating resources, (5) delivering benefits through product development and technology transfer, and (6) follow-up planning at the national level. Specific topics to be addressed under each theme were identified during preparatory visits to the region and participating countries. Sessions covering socioeconomic issues and international initiatives in agricultural biotechnology were based on research at ISNAR.

## Analysis of seminar findings

At each seminar, the country delegations used standardized response sheets to record the "needs identified" and "suggestions for follow-up actions." The response sheets were analyzed and categorized by ISNAR. In total, 227 distinct needs were identified by delegations representing 17 countries. The most important categories are discussed below, following the four phases of the decision-making framework.[1]

### Phase 1: Identifying needs and priorities

Priority setting is the complex process of choosing between alternative sets of research activities. A formal priority setting exercise aims to make the most effective use of available resources by selecting the best portfolio of projects for a research system, institution, or program.

The priority setting needs reported by the seminar participants reflected a strong interest in establishing formal priorities for biotechnology research programs. They also reflect more general issues faced by developing-country research institutes in initiating and implementing a priority setting exercise. For example, participants raised concern about the continuity of the priority setting process, as the ability to initiate concrete proposals after priorities are defined is not always assured. Discussions confirmed the need for participatory processes involving end users, such as farmers and private companies.

A range of approaches to priority setting for biotechnology, including their costs and benefits, were presented as case studies during the seminars. These ranged from top-down priorities set by government to bottom-up priority setting methodologies used by donor-supported initiatives, scoring methods for national and international research, and, finally, application of the analytic hierarchy process (AHP). Chapters 4 ("Methods for Priority Setting for Agricultural Biotechnology Research") and 5 ("Setting Research Priorities for the Chilean Biotechnology Program") discuss issues related to priority setting in depth.

---

[1] For further details on the synthesis of this information, please refer to ISNAR Briefing Paper No. 38, *Strategic Decisions for Agricultural Biotechnology: Synthesis of Four Policy Seminars*, (ISSN 1021-2310) by the same authors.

*Incorporating socioeconomic considerations*

Results of the seminars indicate that all country delegations attending the seminars consider socioeconomic analysis to be a necessary tool for decision making and priority setting. Scientists and research managers are increasingly asked to demonstrate the expected or actual impact of agricultural research projects. A strong need was recognized for institutionalizing both ex ante and ex post socioeconomic analysis of (planned) biotechnology research. However, obtaining funding for research on the socioeconomic implications of biotechnology remains a challenge. The involvement of social scientists in natural-science projects is not widespread. Success stories are needed to promote such multidisciplinarity. In addition, the analytical methods available for assessing the impact of biotechnology products—particularly ex ante assessment—have not been extensively employed (Qaim 1998).

## Phase 2: Formulating national policies

Seminar participants agreed that appropriate policies must be put into place to deal with biosafety and intellectual property.

*Biosafety*

Biosafety means the policies and procedures adopted to ensure the environmentally safe application of biotechnology. The countries that participated in the seminars mentioned biosafety as one of their top concerns. Their concern is reinforced by the attention given to biosafety in international forums such as the negotiations on an internationally binding "biosafety protocol" under the Convention on Biological Diversity.

Case studies discussed at the seminars, especially those presented by representatives of international agricultural research centers, emphasized that biotechnology products are being introduced in a growing number of developing countries. Moreover, a range of products can be expected from national and international collaborative programs (chapter 10 of this volume; Moffat 1999; Qaim 1998, 1999). Some of these products, for example, transgenic plants and recombinant livestock vaccines, require biosafety reviews. Donors supporting international collaborative research programs often request that a formal biosafety review be done before technology transfer can take place. While much experience has been gained in genetically improving crops in developing countries through conventional breeding, there is still concern about the limited knowledge available on transgenic products.

The three most important biosafety-related needs identified at the seminars were to develop biosafety guidelines, establish a responsive national system, and increase capabilities to perform risk analysis of genetically modified organisms (GMOs). These issues and tasks are discussed in more detail in section IV ("Ensuring Environmental Responsibility"). However, it was also acknowledged that discussions on potential benefits and environmental risks are hampered by a lack of data. This is particularly true regarding the introduction of transgenic organisms into tropical ecosystems and centers of diversity. The concerns can be summarized as follows:

- risks related to the introduction of transgenic plants in centers of diversity, such as crops becoming weeds, geneflow to wild plants, or erosion of genetic diversity
- development of new viruses as a result of genetic recombination between virus genes

- accelerated emergence of pest resistance to genetically engineered insect-control proteins
- deleterious effects on nontarget organisms, such as beneficial microorganisms, insects, animals, and humans

## Intellectual property

IPR is a broad term for the various rights granted by law for the protection of economic investment in creative effort. The main categories of intellectual property relevant to agricultural research are patents, plant variety rights, and trademarks. In most developing countries, policies for the application of IPR to biotechnology products are still under formulation. Some countries have never explicitly excluded living material from patent protection. Others have recently adopted IPR for biotechnology or are discussing IPR legislation in which the inclusion of living material is envisaged. The seminars highlighted the need to explore options and implications for agricultural research of national IPR policy decisions on biotechnology.

Given the increasing importance of IPR in relation to agricultural innovation, research managers must carefully consider which type of protection is appropriate for each innovation, whose needs are being served, and how to weigh expected costs and benefits. Their decisions must reconcile various factors: scientists' "perceived need" for IPR protection, institutional goals, the interests of end users of the innovation, and national policy objectives. Intellectual property issues are discussed in section V ("Managing Intellectual Property, Proprietary Science, and Technology Transfer").

## Phase 3: Developing and implementing a research agenda

The national delegations highlighted a number of program management issues. Among the most important were the need to enhance general management capabilities for biotechnology and improve infrastructure, access to information, and international collaboration. The consensus was that, ultimately, good research depends on the quality and critical mass of human resources available for biotechnology. At the same time, it was acknowledged that there is a lack of qualified human resources in agricultural biotechnology, as well as a range of factors inhibiting human resource management and a lack of information on opportunities for human resource development in this area.

Strategic planning for human resource development and management can integrate needs for biotechnology with financial and infrastructure needs necessary for research. Chapter 7 describes steps that enhance the human skills needed for research and development in agricultural biotechnology. The first step is to identify the skills needed in relation to the overall objectives of the research institution with respect to agricultural biotechnology. Thereafter, three basic components are necessary for strategic planning:

- identifying skills needed for success on a continuing basis
- auditing skills already present and knowing what to change to reach strategic objectives
- merging total resource management of people, equipment, consumables, and financial provision into a comprehensive plan

*Managerial constraints*

The seminars also focused on managerial challenges and responsibilities for biotechnology in addition to human resources. Case studies were developed to illustrate particular management challenges faced by agricultural research directors in developing and developed countries. Particular chapters in this book focus on these challenges as they pertain to research program management, policy implications, biological and genetic resources, and international development and collaboration.

*Financial constraints*

Most of the country delegations indicated that biotechnology institutes or programs have suffered from funding limitations or simply that the amount of public and private investment in biotechnology is too small. Research in agricultural biotechnology is affected by this limited funding, due to the long-term and continuous nature of the research. Reasons reported for funding constraints include (1) implementation of fiscal austerity policies, (2) lack of understanding of biotechnology among decision makers, (3) lack of research impact, (4) dependence on a single source of funds, particularly government and donors, and (5) the lack of political and financial support from farmers and their organizations and agribusiness.

The country delegations suggested several alternatives to reverse the situation. There were three major suggestions to explore: (1) demonstrate impact, (2) build strategic alliances, and (3) set up competitive financing mechanisms to promote interaction between the public sector and private sector. In particular, the introduction or promotion of mechanisms that do not require new institutions (such as joint ventures, collaborative research, research levies, and contract research) was strongly encouraged.

## Phase 4: Delivering the products

Product delivery and technology transfer were discussed most intensely during the seminars in Southeast Asia and Latin America, reflecting the more advanced stage of development that agricultural biotechnology has reached in those regions. In particular, participants at those seminars discussed the products expected from international collaboration and technology transfer. ISNAR's analysis of data on international initiatives in agricultural biotechnology, recorded in its BioServe database, was presented and reviewed at the policy seminars. The findings show that most collaborative research initiatives are now in the experimental, laboratory phase (see chapter 10).

Some of these have already resulted in products that are ready for wider diffusion (Ives and Bedford 1998). The following are some examples:

- **Diagnostics and vaccines for livestock diseases.** This appears to be the most significant area for product development to date, with diagnostic tests and rDNA vaccines for rinderpest, cowdriosis (heartwater), theileriosis (East Coast fever), and foot-and-mouth disease.
- **Disease-free planting material.** Various tissue-culture techniques are applied for the micropropagation of disease-free planting material. They involve mainly export crops such as coffee, cocoa, banana, oil palm, and sugarcane.
- **Biocontrol agents.** Products include biopesticides based on *Bacillus thuringiensis* and *B.sphaericus* and pheromone-based attractant decoy for tick vector control.

- **Transgenic plant varieties.** Virus-resistant transgenic potatoes have been planted as part of international programs. The range of crops is expected to expand.

Decisions about the production and delivery of products to users must be considered at an early stage of a research program. An important finding of the seminars regarding product delivery is that collaboration or joint ventures between the private, commercial sector and public institutes or universities is essential. In some cases, national or international intermediary organizations have facilitated technology transfer from public- to private-sector organizations. The strong relationship between the public and private sectors in product development was emphasized, specifically in the areas of product price regulation and registration; offering on-farm demonstrations, pilot production facilities, and science parks for startup companies; and procuring and distributing planting material.

## Placing biotechnology in the context of the 2020 Vision

Findings from these seminars can be related to broader strategic planning needs for biotechnology, as recommended by the International Food Policy Research Institute's study, *A 2020 Vision for Food, Agriculture, and the Environment.* In their summary of recommendations, the following points were made pertaining to biotechnology: First, developing countries should develop a clear policy on and an agenda for biotechnology research. Second, partnerships should be forged between developing countries, international research institutions, and public and private research institutes in industrialized countries. Third, incentives should be provided to the private sector for undertaking biotechnology research targeted at the problems of poor farmers (IFPRI 1995). Specific chapters and introductions to the sections in the book contribute towards these three recommendations, as well as presenting ways in which managers help achieve a strategy for biotechnology. Readers will find information pertaining to developing clear policies and agenda for biotechnology research in sections II, IV, and V. Case studies and examples of partnerships and private-sector research are found in sections III and V.

## Conclusion and recommendations

This chapter summarized needs identified by stakeholders in developing countries with regard to management and policy implications arising from agricultural biotechnology. The policy seminars confirmed a number of significant findings, many of which are addressed in this book, and relate to broader agricultural strategies as put forward by the International Food Policy Research Institute (IFPRI 1995), as well as to ongoing work by many other sources of capacity building and collaboration (see Ives and Bedford 1998). Some of the most important findings reinforced the following:

- Building capacity and competency in biotechnology poses complex and difficult investment decisions for many developing countries.
- Enhancing the management of biotechnology research must be given high importance. However, training opportunities are limited, with most support provided instead for fundamental research.
- Using biotechnology to address agricultural objectives necessitates building on a strong foundation of conventional research.

- Time frames for development of biotechnology products, capabilities, and policies are long term, often taking 10–15 years.
- Socioeconomic considerations, defining appropriate priorities, and addressing end-user needs have gained renewed importance, as these are means to ensure that broader segments of the agricultural sector benefit from biotechnology research.
- Biosafety competence is of highest national importance, regardless of the technological level of the country.
- New institutional partnerships and collaboration are crucial to building competency. However, they are also complicated by the nature of the private sector, proprietary technologies, and regulatory concerns.

## References

Brenner, C. and J. Komen. (eds). 1997. Integrating Biotechnology in Agriculture: Incentives, Constraints and Country Experiences. Report of a Policy Seminar for West Asia and North Africa, Morocco, 22-24 April, 1996. The Hague / Paris: Intermediary Biotechnology Service / OECD Development Centre.

Cohen, J. 1994. Biotechnology Priorities, Planning, and Policies: A Framework for Decision Making. ISNAR Research Report No. 6. The Hague: International Service for National Agricultural Research.

Cohen, J.I., C. Falconi, and J. Komen. 1998. Strategic Decisions for Agricultural Biotechnology: Synthesis of Four Policy Seminars. ISNAR Briefing Paper 38. The Hague: International Service for National Agricultural Research.

IFPRI. 1995. A 2020 Vision for Food, Agriculture, and the Environment. Washington D.C.: International Food Policy Research Institute.

Ives, C.L. and B.M. Bedford (eds). 1998. Agricultural Biotechnology in International Development. Oxon: CABI Publishing.

Komen, J. and G. Persley. 1993. Agricultural Biotechnology in Developing Countries: A Cross-Country Review. ISNAR Research Report No. 2. The Hague: International Service for National Agricultural Research.

Komen, J., J. Cohen, and S-K Lee (eds). 1995. Turning Priorities into Feasible Programs. Proceedings of a Regional Seminar on Planning, Priorities and Policies for Agricultural Biotechnology in Southeast Asia. Singapore, 25-29 September 1994. The Hague / Singapore: Intermediary Biotechnology Service / Nanyang Technological University.

Komen, J., J. Cohen, and Z. Ofir. (eds). 1996. Turning Priorities into Feasible Programs. Proceedings of a Policy Seminar on Agricultural Biotechnology for East and Southern Africa. South Africa, 23-27 April 1995. The Hague / Pretoria: Intermediary Biotechnology Service / Foundation for Research Development.

Komen, J., C. Falconi, and H. Hernández. (eds). 1998. Transformación de las Prioridades en Programas Viables. Actas del Seminario de Política Biotecnólogica Agrícola para América Latina. Perú, 6-10 Octubre 1996. The Hague / Mexico: Intermediary Biotechnology Service / CamBioTec.

Moffat, A.S. 1999. Crop engineering goes south. *Science* 285:370-371.

Qaim, M. 1998. Transgenic virus resistant potatoes in Mexico: potential socioeconomic implications of North-South biotechnology transfer. ISAAA Brief No. 7-1998. Ithaca, NY: International Service for the Acquisition of Agri-biotech Applications.

Qaim, M. 1999. Assessing the impact of banana biotechnology in Kenya. ISAAA Brief No. 10-1999. Ithaca, NY: International Service for the Acquisition of Agri-biotech Applications.

## 2 The Debate on Genetically Modified Organisms: Relevance for the South

*Robert Tripp*

### Abstract

*With the growing range of genetically modified crops being used in food products, a public debate has evolved regarding their implications for the environment, human health, and the overall future directions for agriculture. Although the debate largely takes place in Europe and the USA, it is likely to take place in developing countries as well. This chapter introduces some of the most controversial issues and concludes with important considerations and recommendations for agricultural and regulatory policy.*

### Introduction

The debate over the future of genetically modified organisms (GMOs) in agriculture has captured a wide audience. GMO foods are widely available on grocery shelves in many countries, but their use is so politically sensitive that they are banned from the restaurants of the House of Commons in the UK. The majority of agricultural scientists have welcomed the advent of the new technology. But the public remains skeptical. Corporate advertising promises bountiful harvests, but environmental activists have destroyed field trials of GMOs in several countries.

Although the debate has thus far been conducted largely in the North, there are significant implications for policy in the South. This chapter begins with a review of the nature of GMO technology, the major concerns with its development and utilization. It concludes with some observations on regulatory and research policy.

### The technology

A wide range of GMOs has been developed in the past few years. The first GMO used on a large scale was a virus-resistant tobacco variety developed and now widely grown in China. The first GMO to achieve major commercial use in the North was a tomato with lengthened shelf life. In the past two years GMOs have accounted for a significant proportion of the maize, soybean, cotton, and rapeseed grown in North America. Most of these are varieties

---

This chapter is a slightly amended version of a Briefing Paper (ISSN 01-40-8682) by the same title, published by the UK's Overseas Development Institute (ODI).

© CAB *International*. 1999. *Managing Agricultural Biotechnology—Addressing Research Program Needs and Policy Implications* (ed. J.I. Cohen)

with pest resistance or herbicide tolerance. Only a few other countries have begun to grow GMOs (table 2.1).

GMOs are just one product of the rapidly growing field of biotechnology. New techniques have been developed that make it easier for plant breeders to monitor the outcomes of conventional crossing and selection, allow useful genes to be identified and cloned, and make possible for genes from the same species to be utilized more quickly and precisely than traditional plant breeding. But most attention has been focused on new techniques that allow genetic transfers between species. The products of these transfers, "transgenic crops," are those referred to as GMOs.

GMOs incorporate genes from another plant species, an animal, a bacterium, or a virus. Although this represents a significant change from conventional plant breeding, excessive emphasis in the debate on "alien genes" is not particularly useful. The genetic code is universal, and even distantly related species share a large proportion of their genes. When farmers or plant breeders select a new variety it usually includes novel genes. These may have been transferred from a related variety or species. Or they may be mutations that occurred "naturally" in the field or were induced artificially in the laboratory. For many years, plant breeders also have had access to techniques that allow the production of interspecific and intergeneric crosses, so that a wheat variety may benefit from the disease resistance genes of a wild grass, for instance (Tudge 1988). The new genes, regardless of their source, are responsible for certain changes in the new variety's metabolism. The unique characteristic of a transgenic crop is that its genetic makeup allows it to mimic an aspect of the metabolism of an unrelated species and often introduces this capacity to a new environment.

GMOs have attracted significant investments from multinational companies, but they are also the focus of attention for public research organizations and universities in both the North and the South. Although most current GMOs offer herbicide tolerance or insect resistance, the technology offers possibilities for a wide range of products (table 2.2). Many plant breeders believe that GMOs offer significant opportunities for agriculture in the South, including reducing the reliance on dangerous pesticides, promoting soil conservation through the rational use of herbicides, and eventually, the development of varieties that can withstand environmental stresses such as drought.

**Table 2.1.** Area under Genetically Modified Organisms by Country and Year (in Millions of Hectares)

| Country[1] | 1996 | 1997 | 1998 |
|---|---|---|---|
| USA | 1.5 | 8.1 | 20.5 |
| Argentina | 0.1 | 1.4 | 4.3 |
| Canada | 0.1 | 1.3 | 2.8 |
| Australia | <0.1 | 0.1 | 0.1 |
| Mexico | <0.1 | <0.1 | 0.1 |
| Spain | – | – | <0.1 |
| France | – | – | <0.1 |
| South Africa | – | – | <0.1 |
| **Total** | **1.7** | **10.9** | **27.8** |

*Source:* James 1997, 1998.
[1] The table does not include China.

**Table 2.2.** Types of GMO

| Characteristic | Examples | Rationale |
| --- | --- | --- |
| Consumer or industrial qualities | Long shelf-life tomato, high-starch maize | Development of new foods or sources of industrial products |
| Herbicide tolerance | Various crops tolerant to specific herbicides | More efficient herbicide use and/or use of safer herbicides |
| Disease or insect resistance | Bollworm-resistant cotton, virus-resistant tobacco | Reduction in pesticide use |
| Tolerance to abiotic stresses | Research on drought-tolerant maize | Improved production in marginal areas (But involves polygenic modifications; more difficult than other GMOs) |

*Source:* Adapted from Farmer's Link (1998).

However, although GMOs are products of the evolution of plant breeding, they are not necessarily a vanguard of the benign progress of scientific research. The development and introduction of GMOs raises a number of legitimate technical, political, and ethical concerns. The following section reviews some of the most important of these, including environmental protection, food safety, corporate control of agriculture, and the future direction of farming.

## The Concerns

### *Environmental protection*

Cultivated crops are capable of crossing with related wild species growing in the same environment. One of the principal concerns about the release (or even the restricted field testing) of GMOs is the possibility of this type of genetic interchange. Interbreeding between new crop varieties and their wild relatives is certainly not unique to GMOs. But it is usually assumed that such crosses confer no competitive advantage and are subsequently eliminated by selection pressure in the wild population. Some GMOs, on the other hand, might provide traits (such as herbicide tolerance) that allow the recipient plants to compete successfully and to displace related species. This could have serious consequences for wild plant populations and for biodiversity. In addition, some GMOs themselves might become weeds (Rissler and Mellon 1996).

The excessive use of disease- or pest-resistant crop varieties (or pesticides) always involves the danger of promoting the evolution of resistant strains of pests or pathogens. Over-reliance on GMOs raises particular concerns. Many insect-resistant GMO varieties take advantage of a gene from *Bacillus thuringiensis* (Bt) that produces a specific toxin. As these Bt varieties become widespread it is only a matter of time until the affected insects develop resistance (Whalon and Norris 1996). Many Bt GMOs are sold with advice to farmers on maintaining refuge areas of a non-Bt variety, but this is difficult to control or regulate. There are also concerns about the effects of the diffusion of the toxin on other insects and vertebrates in the food chain of the target insects.

## Food safety

There has been an exceptional amount of controversy about the food safety of GMOs. Several countries have blocked or severely restricted the importation or use of such crops. Some of these actions are related to fears that the GMOs might contain allergens or other harmful substances, although the same arguments could be made for many conventional crop varieties.

A distinguishing characteristic of many GMOs that causes particular alarm is the presence of an antibiotic-resistance marker gene. GMOs are developed by linking the target gene (e.g., for insect resistance) to a gene of an easily identifiable ("marker") trait. Indeed, the only exotic gene in some GMOs, such as the extended shelf-life tomato, is a marker gene. The most widely used marker gene confers resistance to a particular class of antibiotics. Although current research is moving towards less controversial markers, most GMOs currently available carry an antibiotic resistance gene. The possibility of the incorporation of antibiotic resistance in humans, or in the animals that consume GMOs as feed, causes understandable concern. The principal fear is not that the resistance would be transferred directly to humans, but rather that it could be incorporated by bacteria in the gut. There is sufficient uncertainty that the UK Advisory Committee on Novel Foods and Processes (ACNFP) requires a case-by-case review of GMOs containing antibiotic resistance genes.

## Corporate control

In addition to these biological questions, the advent of GMOs has elicited strong concerns about the corporate control of agriculture. The new technology is responsible for a confluence in the interests of chemical companies and seed companies. GMOs may substitute for agricultural chemicals (in the case of pest or disease resistance), be linked to chemicals (herbicide tolerance), or be engineered to produce industrial products (such as oils or pharmaceuticals). These links have motivated a number of mergers and takeovers between seed and chemical companies. Because biotechnology requires considerable investment, the companies have attempted to exercise exceptional control over the processes, genes, and chemicals. In many instances one company's protected gene is made available through licensing arrangements with other companies.

The field of biotechnology has produced a remarkable scramble for patent rights and other forms of protection. Both the genes and the various techniques used for their incorporation are the subject of patent protection so that a single GMO may have several "owners." Some companies have tried to establish exceptionally broad patent protection. One company was initially successful in applying for rights to all genetically modified soybean and cotton, although these rulings are now being challenged.

A particularly controversial aspect of the battle for control of biotechnology is the attempt to limit farmers' saving seed. This issue is not confined to GMOs, however. For instance, hybrid crop varieties have been available for more than 50 years. Seed companies have long favored hybrids because the loss of hybrid vigor in the second generation motivates farmers to purchase fresh seed each year. A recent development affecting seed saving is the Agreement on Trade-Related Aspects of Intellectual Property Rights (TRIPs) as part of the establishment of the World Trade Organization (WTO). Member countries must establish plant variety protection legislation. This will limit farmers' capacities to save or trade seed of protected varieties. The legislation's major current application is for conventional varieties, but the advent of GMOs has provided additional impetus.

In addition to these legal mechanisms, a recent innovation has caused widespread concern. The so-called "terminator technology" is a genetic mechanism that renders the seed's progeny infertile. The technology was developed by a private company and the United States Department of Agriculture. This mechanism is obviously different from legal protection, which can be rescinded or modified as the situation warrants. It is a biological alteration that, unlike hybridization or gene transfer, promises no productive advantage but merely provides the company additional control over its variety. Although the technology is still in the developmental phase, it has attracted worldwide apprehension and protest.

## Future directions for agriculture

These controversies over the environmental implications and legal control of GMOs are often embedded in larger debates about the future of agriculture. Science is moving so fast that the public's fear of a technology out of control is easy to understand. Past experience with damaged ecologies and pesticide-tainted crops, combined with the prospects of cloned animals and "Frankenfoods," makes many people wish for a simpler agriculture.

The debates over conflicting visions of the future of agriculture in the North are easily transferred to discussions about agricultural development in the South. The lack of focus in these debates is exemplified by the wildly differing interpretations of the term "Green Revolution." Depending on the context and the speaker, the term can represent anything from the triumph of agricultural science to the destruction of traditional agriculture (Tripp 1996). The fact that discussions about a set of technological changes initiated over 30 years ago are still characterized by an astounding lack of clarity bodes ill for our capacity to come to terms with the complex issue of GMOs. The South needs a strategy for dealing with biotechnology. This section discusses implications for regulation and research policy.

## Regulation

Because many of the problems raised by GMOs involve environmental protection or food safety, they imply an obvious appeal to regulatory mechanisms. But a regulatory response is not straightforward. Problems in regulating GMOs have emerged in the North and the lessons of these experiences are of direct relevance to the South. These involve the complexity and control of the regulatory process.

### *Complexity*

Many countries have regulatory procedures for testing and releasing conventional crop varieties, but protocols for testing GMOs are still being developed (Krattiger and Rosemarin 1994). Field testing GMOs in the US requires an analysis of the molecular biology of the organisms (donor, recipient, vector) used in the gene transfer, the demonstration of safeguards to prevent contamination, and an environmental analysis of potential dangers.

The requirements for testing GMOs are the subject of continuing debate. One question is whether regulation should focus on GMOs per se or on the characteristics of the particular variety. Herbicide tolerance, for instance, is a major focus of GMO research, but it could be developed through conventional plant breeding as well as through biotechnology. The appropriate level of regulation for GMOs also depends on the environment where they will be used. In the South, considerable information is required for adequate assessment of GMOs' risks because less is understood about the ecologies there into which GMOs might be introduced.

Regulating the food safety of GMOs also presents challenges. The novelty of GMOs is not accommodated by most food safety regulations, which focus on food additives. There is uncertainty about how much testing is required and to what extent the testing should focus on the new genes themselves or on the substances they produce in the plant.

If there is sufficient consumer concern it is possible to label foods that contain (or that do not contain) GMOs. An example of "negative" labeling is provided by the certification of organic produce, to guarantee the absence of agricultural chemicals (and more recently, of GMOs). In the UK, organic certification is administered by the Soil Association, a private body. A "positive" alternative, where the presence of GMOs must be acknowledged, may be the subject of public regulation. However, labeling of GMOs requires consensus on what constitutes genetic modification, and this is still a problem. A recent European Parliament law requires the labeling of foods containing GMOs, but critics contend that the law's focus on detectable genetic material or proteins allows many GMO food products to continue unlabeled.

### *Control*

Besides the technical complexities of deciding how to regulate the cultivation and consumption of GMOs, there are serious questions about the control of the regulatory process. In theory, regulation assumes that decisions are made in the public interest by an independent authority. But the process is rarely as simple as this. The recent BSE experience in the UK illustrates how the management of regulation designed to protect consumers can easily conflict with the temptation to protect the regulated industry. This type of "regulatory capture" is a common occurrence as competing interests battle for control of the regulatory process (Tripp 1997).

In the case of GMOs, there is considerable pressure from the multinationals to simplify the regulatory process. Although this is obviously motivated by commercial interest, similar pressures might well come from public-sector researchers attempting to move their own GMOs through national regulatory procedures. On the other hand, those urging a more cautious approach may include not only environmental activists but also commercial agricultural interests that stand to lose from competition with the cultivation or importation of GMOs.

In summary, regulation of something as complex as GMOs can never be done on a purely objective, technical basis. The assessment of risk and the interpretation of data will always be affected by the values of the regulators and the political and economic pressures exerted on the regulatory process. However, progress towards more satisfactory regulation of GMOs can be made with access to adequate technical and environmental data, and regulatory procedures that are as transparent as possible.

## Agricultural research policy

There is little doubt that GMOs will become increasingly prevalent in developing-country agriculture. A number of developing countries, including those that decry the predations of multinationals, have their own public- and sometimes private-sector biotechnology programs. Through these efforts, they are beginning to produce their own GMOs, which will be put forward for regulatory approval and public acceptance. In addition, multinationals have established joint programs with several private companies and public research organizations in the South.

The majority of commercial GMO crops are currently grown in the North. It is not likely that many of these varieties, which have been bred for the environments and crop management practices of industrial agriculture, will be grown on a large scale in the South in the near future. Thus, the immediate concern is not that subsistence farmers will become dependent on Monsanto maize seed and herbicide, given the cost and complexity of the technology. But the multinationals are actively developing opportunities to sell their technology in the South. The most likely targets are commercial crops such as cotton that are already subject to relatively high levels of management, or large commercial farms that can afford the technology. Appropriate policies must be in place to make sure that GMOs are not imposed upon a compliant regulatory structure but rather are judged and utilized in the interests of national agricultural development goals.

The relation between GMOs and agriculture in the South has come under further scrutiny in the light of corporate advertising that attempts to present GMOs as an essential step towards eliminating world hunger. Such public relations strategies can be quite misleading. It is true that any increase in food output may potentially lead to lowering global food prices. But it is deceitful to argue that a technology currently aimed at US soybean farmers is part of a strategy to address poverty and hunger in the South. National policies need to ensure that the poor have the resources to acquire their food—be it imported or domestically produced—and that new technology is used to promote equitable agricultural development.

## Conclusions

GMOs are one product of a remarkable expansion in agricultural biotechnology. They offer the possibility of addressing some difficult problems. But they also present uncertainties. Their development has sparked debates about the direction of agriculture and the control of technology. These debates are partially grounded in differing values, so there is little prospect of simple resolution. This review has urged consideration of the separate aspects of the GMO debate. Examination of GMOs' safety should not be linked to opinions about their most visible corporate sponsors. Similarly, concerns about industrial agriculture or input use may be articulated independently of judgements about the potential of biotechnology. No matter how the debate is conducted, there are several areas that can benefit from policy analysis. These include legal reform, regulatory capacity, and agricultural policy.

The intellectual property protection mechanisms established for new varieties require careful examination. Companies have the right to protect their products, but the current bout of predatory patenting and legal maneuvering threatens to deliver excessive privileges to a handful of companies. An unprepared legal system may be surrendering potential for fostering further competition and innovation in a rapidly growing field. The current legal climate also affects the willingness of farmers and public-sector plant breeders to exchange germplasm.

A nation's ability to deal with GMOs depends on its regulatory capacity. Regulation involves the careful interpretation of adequate technical data. But it is also an inherently political process. A raft of conflicting interests will affect any decision regarding the approval of GMOs. In the South, pressures will be applied from both domestic and foreign sources; these will include the interests of plant breeders, agricultural input and commodity firms, and a range of political and advocacy groups. Establishing regulatory procedures that allow transparent and representative debate is a tall order for any country.

Regulatory decisions about GMOs require access to high-quality technical information about environmental interactions. This information is costly to acquire, and most developing countries do not have adequate resources for this purpose. External funding is required to support environmental studies, as well as for the broader concerns of biodiversity conservation. New techniques and insights can be shared among regulatory agencies in the South. But because of the location-specific character of environmental management, each country must be prepared to take responsibility for much of its own research. On the other hand, there are opportunities for linking regulatory agencies concerned with food safety. Any information developed about the food safety of GMOs should be of universal relevance.

GMOs are a challenge to several aspects of agricultural policy in the South. These include identifying the roles and complementarities of public- and private-sector participation in agricultural development; formulating policies that promote the contribution of commercial agricultural firms; and strengthening public-sector research and extension to meet agriculture's responsibilities for poverty reduction, productivity improvement, and environmental protection.

The most pressing need is for good information. These are complex issues that cannot be debated using formulae, slogans, or slick advertising. The majority of the reporting and analysis on both sides of the GMO issue has not been accompanied by adequate technical information. Until both the public and the commentators in the North are better informed, it is best to be modest in giving advice to other countries. Poorly informed arguments between the supporters of high science and low inputs do little to further the development of responsible policies in the South.

## References

ACNFP. 1996. The use of antibiotic resistance markers in genetically modified plants for human food: Clarification of principles for decision-making. London: Advisory Committee on Novel Foods and Processes.

Farmers' Link. 1998. Smart Plants. A Farmer's Guide to Genetically Modified Organisms (GMOs) in Arable Agriculture. Norfolk: Farmers' Link.

James, C. 1997. Global Status of Transgenic Crops in 1997. ISAAA Briefs No. 5. Ithaca, NY: International Service for the Acquisition of Agri-biotech Applications.

James, C. 1998. Global Review of Commercialized Transgenic Crops: 1998. ISAAA Briefs No.8. Ithaca, NY: International Service for the Acquisition of Agri-biotech Applications.

Krattiger, A. and A. Rosemarin. 1994. Biosafety for Sustainable Agriculture. Sharing Biotechnology Regulatory Experiences of the Western Hemisphere. Ithaca, NY / Stockholm: International Service for the Acquisition of Agri-biotech Applications / Stockholm Environment Institute.

Rissler, J. and M. Mellon. 1996. The Ecological Risks of Engineered Crops. Cambridge, MA: MIT Press.

Tripp, R. 1996. Biodiversity and modern crop varieties: Sharpening the debate. *Agriculture and Human Values* 13(4): 48-63.

Tripp, R. 1997. New Seed and Old Laws: Regulatory Reform and the Diversification of National Seed Systems. London: Intermediate Technology Publications.

Tudge, C. 1988. Food Crops for the Future. London: Basil Blackwell.

Whalon, M. and D. Norris. 1996. Resistance management for transgenic Bacillus thuringiensis plants. *Biotechnology and Development Monitor* 29:8–12.

# 3 Agricultural Biotechnology Research Indicators and Managerial Considerations in Four Developing Countries

*Cesar A. Falconi*

## Abstract

*In 1998 ISNAR conducted a survey on agricultural biotechnology in the national agricultural research systems of Mexico, Kenya, Indonesia, and Zimbabwe. The findings show that only a few public-sector research organizations use advanced biotechnology techniques, and that most organizations are still in the first stages of developing biotechnology research capacity. Although expenditures on agricultural biotechnology research in these countries have grown every year, the percentage of biotechnology expenditures to total agricultural research expenditures has been small, with the public sector accounting for almost all of it. Some of the policy recommendations proposed in this chapter to overcome the identified constraints in biotechnology research include (1) increasing investment levels in agricultural biotechnology research, (2) promoting the involvement of the private sector in this research, (3) fostering partnerships between the public sector and the private sector, and (4) developing a comprehensive strategy for biotechnology research.*

## Introduction

Urgent attention needs to be paid to the impact of biotechnology on medium- and long-term agricultural production in developing countries. Countries that wish to participate in the global biotechnology arena call on their national research organizations to incorporate basic capabilities and tools of modern biotechnology into productive processes. This is particularly important for agriculture, which in developing countries is often the largest economic sector in terms of income, employment, and foreign exchange earnings. Biotechnology can help by promoting sustainable development and creating and maintaining competitive positions in international agricultural markets.

The level of resources available for investment in agricultural biotechnology is one indicator of a country's efforts to strengthen or create these capabilities. Information on the size, structure, and content of public research is needed to improve policy decisions, clarify the roles of the public and private sectors, and support public-sector implementation of biotechnology research. There is, however, a lack of structured data on resources for developing-country agricultural biotechnology. An exhaustive review of the literature reveals that few studies collect and analyze this kind of information at the national level.

© CAB *International*. 1999. *Managing Agricultural Biotechnology—Addressing Research Program Needs and Policy Implications* (ed. J.I. Cohen)

Some of these are Cabral (1993) in Brazil and Jaffe (1992) in Argentina, Venezuela, and Costa Rica. But Jaffe gathered and analyzed investment and human resources data for public and private organizations involved in agricultural biotechnology activities for only one year (1989).

The scarcity of information in this area prompted ISNAR's Biotechnology Service (IBS) to launch a survey of research indicators in agricultural biotechnology to learn how resources are mobilized and used to implement agricultural biotechnology.

This chapter provides statistical and institutional information that was collected through the survey, as well as information gleaned about the development of the most relevant public and nonpublic organizations involved in agricultural biotechnology research activities in four developing countries: Mexico, Kenya, Indonesia, and Zimbabwe.

## Definitions and methodology

Public-sector organizations that were surveyed consisted of national institutes for agricultural research, universities, and parastatals. Private-sector entities included commercial firms (input companies, farm sector, and food-processing companies) as well as private noncommercial organizations (foundations and nongovernmental organizations).

Agricultural biotechnology research, as defined here, focuses on cellular and molecular biology and new techniques coming from these disciplines for improving the genetic makeup and agronomic management of crops and animals. Tissue culture applications, biofertilizer, and bioinsecticides are included in this definition.

Research expenditures include personnel, operational, and capital expenses. The research expenditures are presented in real local currency that is deflated by the gross domestic product (GDP) index (base year 1985). These figures are also presented in real dollars but converted to international standards by the purchasing power parity (PPP) index (base year 1985) (Heston and Summers 1991). The expenditures presented in PPP dollars can be used to compare to other country expenditure levels.

The countries were selected to permit comparison on the basis of size of population and level of development of agricultural biotechnology research. Of the four selected countries, Mexico is the most advanced in biotechnology. Indonesia's level of biotechnology is intermediate. Kenya has an intermediate-low level of biotechnology, as does Zimbabwe, the smallest country.

The surveys begin with a section on the nature of the biotechnology programs or institutions. Next, they focus on information on human, physical, and financial resources available for agricultural biotechnology. The last part deals with the agricultural biotechnology research projects that the institutions currently conduct.

The survey covered the countries' most relevant public and private organizations doing agricultural biotechnology research. The sample represents 70–80% of the total expenditures on agricultural biotechnology research in each country. Of the 34 organizations surveyed, 13 were public research institutes, 11 were public universities, six were private noncommercial organizations, and four were private commercial entities (see the annex to this chapter for an overview of these organizations). The period of analysis was 1985–97 for Mexico, 1989–96 for Kenya, 1989–97 for Indonesia, and 1989–98 for Zimbabwe.

The surveys were completed by the directors or research leaders of the biotechnology research institutes, programs, or units. When data seemed inconsistent, one or more follow-up interviews were conducted to clarify the information provided.

## Institutional and policy development

Each country followed a different path in developing its biotechnology capacity, reflecting not only differences in the implementation of strategies, when in place, but also the influence of donors.

**Indonesia** first formulated a strategy and policy document, the National Program of Development of Biotechnology, in 1983. The program did not identify any priority areas, however. In order to implement the strategy, in 1985 the Government of Indonesia established the Inter-University Center for Agricultural Biotechnology at Bogor Agricultural University to train university faculty in biotechnology. (The implementation of the strategy was severely affected by the 1997 financial crisis, however; the funding of biotechnology research activities showed a sharp decline in that year.) It also created the Research and Development Center for Biotechnology at the Indonesian Institute of Sciences to enhance national capability in biotechnology (Sasson 1993).

The Biotechnology Division of the Central Research Institute for Food Crops of the Agency for Agricultural Research and Development (AARD) was established in 1989. The Central Research Institute for Food Crops and the Research and Development Center for Biotechnology were selected as centers of excellence for agricultural biotechnology in 1990, after an assessment of all institutions involved in biotechnology. These are now the two leading institutes in agricultural biotechnology in Indonesia, applying several advanced biotechnology techniques.

Even though the government regards biotechnology as a priority, it realized that significant resources are required to achieve the goals set out in the plans. It therefore launched a special grant program in 1992 to commit funds for research in biotechnology and other sciences.

In 1995, in accordance with the national development plan, the Biotechnology Division of the Central Institute for Food Crops and the Bogor Research Institute for Food Crops were merged to become the Research Institute for Food Crops Biotechnology, which is now AARD's main biotechnology arm.

The Biosafety Regulations and the Biosafety Commission were put in place in 1997. Indonesia approved a patent law in 1989 and revised it in 1997 to include protection to biotechnology products. However, plant breeders' rights are not in place in Indonesia.

**Mexico** established its first tissue culture laboratory in 1970. After an assessment of the biotechnology situation in the country in the early 1980s, the following national biotechnology research units were established: the Biotechnology Institute of the Autonomous National University of Mexico, the Center for Research and Advanced Studies–Irapuato Unit (CINVESTAV-I), and the Scientific Research Center of Yucatan. Since the 1980s these three research centers have applied advanced biotechnology techniques and have been instrumental in making Mexico one of the more advanced countries in biotechnology in the developing world (Pedraza et al. 1998).

In 1989 the Mexican Biosafety Committee was established, and in 1991 the Industrial Property Act was approved, making Mexico the first developing country that explicitly patents biotechnological inventions. Mexico also legislates plant breeders' rights.

In the early 1990s the government initiated a program to promote international standards of scientific excellence in the country. It also obtained a World Bank loan for a program to build scientific capacity and infrastructure. While both programs focused on the scientific quality of research projects, no priority areas were identified and biotechnology was not explicitly mentioned. The same holds true for the 1995–2000 national development plan.

Although Mexico has no biotechnology policy and strategy in place, it has been active in developing biotechnology during the last years. Because several developments indicated that some elements of a national strategy were pursued, the pressure to compete in the North American market, and the membership in the North American Free Trade Agreement. It is close to generating the first transgenic product to be released by a national organization (CINVESTAV-I, in collaboration with Monsanto) in Latin America (Qaim 1998).

**Kenya** and **Zimbabwe** are in very similar stages of biotechnology development. Kenya's leading institution in agricultural biotechnology is the Kenya Agricultural Research Institute (KARI), which first began biotechnology research in 1982. Zimbabwe's lead institute is the Biotechnology Research Institute (BRI) of the Scientific and Industrial Research and Development Centre, established in 1992 (Woodend and Muza 1993).

Biotechnology research in both countries benefited from a Special Program on Biotechnology implemented by the Government of the Netherlands in 1992. Under this program, priority setting exercises were conducted in both countries. These included the participation of farmers, researchers, extensionists, and policymakers. One of the outcomes of the exercise was to establish an Agricultural Biotechnology Platform in Kenya and a Biotechnology Advisory Committee in Zimbabwe. Both were created in the mid 1990s. They advise the Dutch-supported special programs and the respective governments on the development of agricultural biotechnology in the countries. The Dutch government supports agricultural biotechnology in both countries through these two entities.

Kenya's Biosafety Committee was formed in 1996. The Industrial Property Act was implemented in 1993, and the Plant Varieties Act was approved in 1994 (Olembo 1998). Zimbabwe's Biosafety Regulations were put in place in 1998, and the National Biosafety Committee was established in 1999. Zimbabwe has had an industrial patent law and plant breeders' rights since the 1970s, but these need to be amended to cover biotechnology products.

It is expected that the Dutch, Kenyan, and Zimbabwean governments' recognition of the significance of biotechnology in agricultural development will strengthen agricultural biotechnology research in Kenya and Zimbabwe.

The biosafety committees and the revised property rights legislation are also meant to encourage the private sector to participate in agricultural biotechnology research. Private-sector participation has been limited in the four countries, however. Only a few commercial companies are engaged in more advanced biotechnology research, such as Empresa La Moderna in Mexico; most private companies are specialized in tissue culture of fruits and ornamentals.

## Agricultural biotechnology research indicators

The following paragraphs analyze the evolution of agricultural biotechnology research in the four countries.

### Structure and organization

A research organization is considered specialized in biotechnology if its core activity is biotechnology. Organizations that use biotechnology as a tool to support other research activities are nonspecialized. Following this classification, biotechnology is a core activity in nearly 25% of the research organizations surveyed (table 3.1). Note that four of the eight specialized research organizations began their biotechnology research only recently.

Table 3.2 shows the percentages of research expenditures in the various sectors of agricultural biotechnology research. Public-sector organizations accounted for almost 92% of research expenditures during the period of analysis. The average 8% for the private sector, however, showed higher annual growth than did the public universities (except in Indonesia). Moreover, the universities showed a significant decline in research expenditures, which is probably due to economic recession and a drop in donor funding. Public research institutes showed not only the highest share of financial resources but also the highest annual growth rate (Mexico and Kenya 9%, Indonesia 30%, Zimbabwe 70%).

**Table 3.1.** Number of Organizations in Agricultural Biotechnology Research in 1997

| Country | Core activity | Support activity |
|---|---|---|
| Indonesia | 3 | 5 |
| Kenya | 1 | 5 |
| Mexico | 3 | 11 |
| Zimbabwe | 1 | 5 |
| Total | 8 | 26 |

Source: Wafula and Falconi 1998; Qaim and Falconi 1998; Moeljopawiro and Falconi 1999; Gopo and Falconi 1999.

**Table 3.2.** Research Expenditures in Agricultural Biotechnology (% of 1985 Research Expenditures)

| | Indonesia | | Kenya | | Mexico | | Zimbabwe | |
|---|---|---|---|---|---|---|---|---|
| Sector | 1989 | 1997 | 1989 | 1996 | 1985 | 1997 | 1989 | 1998 |
| Public research institute | 66 | 85 | 47 | 72 | 50 | 60 | 1 | 81 |
| Public university | 14 | 11 | 49 | 24 | 50 | 28 | 98 | 3 |
| Private noncommercial | 0 | 1 | 4 | 4 | 0 | 4 | 0 | 16 |
| Private commercial | 20 | 3 | 0 | 0 | 0 | 8 | 0 | 0 |
| Total | 100 | 100 | 100 | 100 | 100 | 100 | 100 | 100 |

Source: Wafula and Falconi 1998; Qaim and Falconi 1998; Moeljopawiro and Falconi 1999; Gopo and Falconi 1999.

This explains why financial resources are concentrated in only a few public research institutes: KARI in Kenya (70% of total expenditures in 1996), BRI in Zimbabwe (80% in 1998), three research organizations in Indonesia (70% in 1997), and three Mexican research organizations (55% in 1997). This is in sharp contrast to most developed countries; in the USA, in 1992 70% of the financial resources in agricultural biotechnology research came from the private commercial sector (Fuglie et al. 1996; Caswell et al. 1994).

## *Personnel*

The number of researchers in biotechnology at least doubled, while the number of PhDs has at least tripled (see chapter 7, table 2). This growth may be explained by the significant increase of the number of postgraduate programs in biotechnology, the establishment of specialized research organizations that required more scientists trained in biotechnology, and special grant programs that encourage scientists to become involved in biotechnology research. The scientists are concentrated in only a few research organizations: in Kenya some 45% in KARI, in Mexico 60% in only four research organizations, in Indonesia 60% in three research organizations, and in Zimbabwe 70% are in three research organizations.

Downer et al. (1990) has suggested a minimum efficient size for research groups in agricultural biotechnology. For genetic engineering and tissue culture a ratio of one researcher to two support personnel[1] (technicians) was recommended. In the four countries there's an average of one technician for every two researchers. Most of the research organizations show a low technical support – researcher ratio, which could affect the potential development of research outputs. It is interesting to note that only one Indonesian and two Mexican private commercial entities had a greater than one technical support staff – researcher ratio (five technicians per researcher on average).

## *Expenditures*

During the period of analysis, the number of researchers in agricultural biotechnology research in Kenya, Mexico, and Zimbabwe outgrew research expenditures (table 3.3). This resulted in a 7% annual decrease in expenditures per researcher. The Dutch and the World Bank's support to Kenya could reverse the trend in that country, and the Dutch support to Zimbabwe could do the same there. The recovery of the Mexican economy will no doubt help Mexico. It is interesting to note that only two Mexican private commercial entities showed a positive growth and higher level of expenditures per researcher than in the public sector.

Indonesia was the only country that showed a significant annual growth in the expenditures per researcher during the period of analysis. However, expenditures per researcher dropped in 1997 due to the beginning of the financial crisis in the country. In that year all the research organizations showed a negative annual growth. The private commercial sector showed the most significant decline.

Expenditures per researcher were higher in Mexico and Indonesia than in Kenya and Zimbabwe. This implies that Mexican and Indonesian researchers have more resources and are more likely to generate biotechnology research results.

---

[1] Technical support staff assist in designing and conducting agricultural research. They include laboratory technicians and biometricians and usually have post-secondary professional education.

Table 3.3. Expenditures in Agricultural Biotechnology Research

| Expenditures | Indonesia[1] | | | Kenya | | | Mexico | | | Zimbabwe | | |
|---|---|---|---|---|---|---|---|---|---|---|---|---|
| | 1989 | 1997 | Annual growth | 1989 | 1996 | Annual growth | 1985 | 1997 | Annual growth | 1989 | 1998 | Annual growth |
| In millions 1985 PPP US $ | 2.4 | 18.7 | 29.3% | 2.5 | 3.0 | 2.6% | 9.7 | 20.4 | 8.5% | 1.8 | 3.5 | 7.5% |
| In nominal US$ millions | 0.7 | 6.0 | 30.8% | 1.0 | 1.2 | 2.5% | 4.3 | 11.5 | 6.3% | 1.0 | 1.4 | 3.8% |
| Per researcher ('000 1985 PPP US $) | 19.1 | 53.6 | 13.7% | 77.2 | 45.5 | -7.2% | 187.5 | 85.1 | -6.4% | 92 | 43 | -8.0% |
| In % of agric. GDP | 0.003 | 0.018 | 25.1% | 0.046 | 0.048 | 0.6% | 0.026 | 0.052 | 5.9% | 0.12 | 0.23 | 7.5% |
| In % of total agric. research | 1.7 | 9.6 | 24.1% | 3.3 | 2.8 | -2.3% | 3.1 | 9.6 | 9.8% | 4.6 | 10.0 | 9.0% |

*Source:* Wafula and Falconi 1998; Qaim and Falconi 1998; Moeljopawiro and Falconi 1999; Gopo and Falconi 1999.
[1] Total agricultural research expenditures include only those for the Agency for Agricultural Research and Development (AARD).

Even though the research intensity ratio has grown annually, the percentage of agricultural biotechnology research expenditure in relation to agricultural gross domestic product is quite minimal: 0.014% in Indonesia, 0.04% in Mexico and Kenya, and 0.12% in Zimbabwe, on average. The percentage of agricultural biotechnology research expenditures to total agricultural research expenditures was around 2.3% for Kenya, 6.5% for Mexico, 7.0% for Indonesia, and 5.0% for Zimbabwe, on average.

There is no rule for which percentage of the agricultural research budget should be allocated to biotechnology, but for comparison, the Consultative Group on International Agricultural Research (CGIAR) spent about 8% of its budget on biotechnology research in 1997. The USA allocated 13% of its agricultural research expenditures to biotechnology in 1992 (Fuglie et al. 1996; Caswell et al. 1994). And to make a comparison with an industrialized country, of the total amount that the USA invested in biotechnology, Kenya and Indonesia each invested only 4%, Mexico 9%, and Zimbabwe 12%. Research leaders and decision makers should take this enormous gap in the resources committed to conducting biotechnology into consideration in planning the development of biotechnology in their countries.

## Financing

Table 3.4 presents the sources of funding for agricultural biotechnology research for the four countries. Public research institutes have been the main recipients of donors funding; KARI in Kenya accounted for nearly 85% of total donor support in 1996 and BRI in Zimbabwe for almost 90% in 1998. The sustainability of these levels of funding will be compromised in the medium term if there is no effort to obtain funding from local sources.

Some public research institutes and universities fund their biotechnology research activities from nontraditional sources of funding, such as sales of products and services and contractual arrangements. Although these sources of funding are still minimal, they increased during the period of analysis. Contracts and levies fund biotechnology research done by private noncommercial organizations, while private commercial organizations are financed by the sales of their products.

The limited funding from nontraditional sources for public research institutes and universities indicates a minimal interaction between public entities and the private sector. In a study of the poor interaction between these sectors in Mexico, Wagner (1998) concluded that (1) the private sector can import technology more cheaply, (2) the

Table 3.4. Sources of Funding for Agricultural Biotechnology (in %)

| Funding source | Indonesia (1997) | Mexico (1997) | Kenya (1996) | Zimbabwe (1998) |
|---|---|---|---|---|
| Government | 93 | 60 | 28 | 34 |
| Sales of products | 4 | 12 | 3 | 16 |
| Contracts | 1 | 4 | 0 | 0 |
| Donors | 2 | 24 | 67 | 50 |
| Levies | 0 | 0 | 2 | 0 |
| Total | 100 | 100 | 100 | 100 |

Source: Wafula and Falconi (1998); Qaim and Falconi (1998); Moeljopawiro and Falconi (1999); Gopo and Falconi (1999).

government neglects the use of science to foster economic development, (3) the regulatory framework confuses foreign and local companies in introducing biotechnology products, (4) the basic research orientation of scientists impedes the collaboration between scientists and businessmen, and (5) there is a lack of funding mechanisms to bring the two sectors closer to each other.

## Focus of research

In the institutes surveyed, most researchers involved in biotechnology focus on crop research (almost 75% in Kenya, 100% in Mexico, 90% in Indonesia, and 80% in Zimbabwe), while the remainder work in livestock research programs. The allocation of biotechnology resources should perhaps be more in line with the contribution of livestock to the value of total agricultural production—livestock contributed 50% in Kenya in 1996, 45% in Mexico in 1997, 30% in Indonesia in 1997, and 35% in Zimbabwe in 1998.

The distribution of researchers on the gradient of biotechnology techniques is an indicator of a country's level of biotechnology development (Jones 1990). In Mexico and Indonesia, 50% of the researchers are applying advanced techniques, such as genetic engineering. The other half use less sophisticated techniques, such as tissue culture. In Kenya and Zimbabwe, 30% of the researchers use more advanced techniques, and 70% use less sophisticated biotechnology techniques.

Only three public research organizations in Mexico, one in Kenya, three in Indonesia, and two in Zimbabwe use more advanced biotechnology techniques. Other research organizations are in the first stages of developing a biotechnology capacity.

Tissue culture and genetic engineering techniques were chosen to indicate the research focus of the institutions in the various countries (table 3.5). In all countries, private-sector organizations use mainly less advanced techniques such as tissue culture, which are less costly, less risky, and closer to the market than more advanced techniques such as genetic engineering. As more advanced techniques are expensive and the payoffs uncertain, they are used mainly by public-sector organizations. However, a significant proportion of public-organization researchers frequently use less advanced techniques to complement advanced techniques. The public sector also applies biotechnology to "orphan" commodities and to solve problems facing marginal farmers. Research leaders and decision makers should take these differences in research between the public and the private sector into account when planning the development of biotechnology.

**Table 3.5.** Agricultural Biotechnology Research Focus (% of Researchers)

| Techniques | Kenya (1996) Public | Kenya (1996) Private | Mexico (1997) Public | Mexico (1997) Private | Indonesia (1997) Public | Indonesia (1997) Private | Zimbabwe (1998) Public | Zimbabwe (1998) Private |
|---|---|---|---|---|---|---|---|---|
| Tissue culture | 76 | 100 | 68 | 90 | 40 | 90 | 66 | 67 |
| Genetic engineering | 24 | 0 | 32 | 10 | 60 | 10 | 34 | 33 |
| **Total** | **100** | **100** | **100** | **100** | **100** | **100** | **100** | **100** |

*Source:* Wafula and Falconi 1998; Qaim and Falconi 1998; Moeljopawiro and Falconi 1999; Gopo and Falconi 1999.

## Findings and policy recommendations

The main findings from the survey include the following:

- Most agricultural research institutions in the four countries are in the first stages of developing biotechnology research capacity.
- Although expenditures on agricultural biotechnology research have grown in the four countries, the proportion of biotechnology expenditures to total agricultural research expenditures is still small.
- As the number of researchers has grown much faster than have expenditures, there's been a significant decline in the operating expenditures per researcher. This indicates a trend towards unsustainability or low performance.
- Funding and execution of biotechnology research in the four countries is highly dependent on the public sector. The participation of the private sector is limited.
- Donor contributions constituted the largest source of funding for agricultural biotechnology research in Kenya and Zimbabwe. The high degree of donor dependency raises concerns regarding the sustainability or expansion of agricultural biotechnology.
- Most agricultural biotechnology research is focused on crops. There's limited attention to livestock.
- The private sector focuses on the near-market and low-technology end of biotechnology and on horticultural crops, such as ornamentals and fruits. These are high-value crops with a faster payback time.
- The four countries' governments have supported agricultural biotechnology research in their institutes by establishing biotechnology research centers, creating post-graduate programs, and formulating a regulatory framework for biosafety and intellectual property rights. These efforts provide a good basis for future developments of biotechnology. But a comprehensive strategy is lacking.

These findings lead to the following policy recommendations:

- The countries should design a development strategy to foster biotechnology. Strategic decisions need to incorporate a clear understanding of the costs of biotechnology and its potential to meet national goals and the impact to potential beneficiaries.
- Planning and priority setting tools should be developed to help policymakers make informed decisions on commodities and research objectives.
- Countries that have earmarked agricultural biotechnology as a priority area for research should step up their commitment in funding this research.
- Developing countries should find alternative ways to fund their research to ensure that their biotechnology programs remain funded and that they are user oriented.
- Policies and incentive mechanisms should be developed to encourage the private sector to invest and participate in agricultural biotechnology. Public- and private-sector research should be complementary and not in competition with each other.
- While the policy framework should promote the safe use of biotechnology, it should not discourage investments by the private sector and collaboration in the global agricultural research system.

## References

Cabral, R. 1993. The Financing of Biotechnology in Developing Countries: A Brazilian Case in Point. *Technology in Society*, Vol. 15, pp.311-325.

Caswell, M., K. Fuglie and C. Kotz. 1994. Agricultural Biotechnology: An Economic Perspective. Agricultural Economic Report No. 687. Washington, D.C.: USDA Economic Research Service.

Downer, R., E. Dumbroff, B. Glick, J. Pasternack and K. Winter. 1990. Guidelines for the Implementation and Introduction of Agrobiotechnology into Latin America and the Caribbean. San José: Interamerican Institute for Cooperation in Agriculture.

Fuglie, K., N. Ballenger, K. Day, C. Kotz, M. Ollinger, J. Reilly, U. Vasavada and J. Yee. 1996. Agricultural Research and Development: Public and Private Investments under Alternative Markets and Institutions. Agricultural Economic Report No. 735. Washington, D.C.: USDA Economic Research Service.

Gopo, J. and C. Falconi. 1999. Agricultural Biotechnology Research Indicators: Zimbabwe. Discussion Paper No. 99-8. The Hague: International Service for National Agricultural Research.

Jaffe, W.R. 1992. Agricultural Biotechnology Research and Development Investment in Some Latin American Countries. *Science and Policy*, Vol. 19, No. 4, pp. 229-240.

Jones, K.A., 1990, Classifying Biotechnologies. In *Agricultural Biotechnology: Opportunities for International Development*, edited by G.J. Persley, CABI.

Heston, A. and R. Summers. 1991. The Penn World Table (Mark 5): An Expanded Set of International Comparisons, 1950-1988. *Quarterly Journal of Economics*, May 1991, pp.327-368.

Moeljopawiro, S. and C. Falconi. 1999. Agricultural Biotechnology Research Indicators: Indonesia. Discussion Paper No. 99-7. The Hague: International Service for National Agricultural Research.

Olembo, N. 1998. Intellectual Property Rights in the Context of International Collaborations: Case Study Kenya. Paper presented at the International Seminar "Biotechnologies for Dryland Agriculture: Prospects and Constraints", Hyderabad, India, July 16-18, 1998.

Qaim, M. 1998. Transgenic Virus Resistant Potatoes in Mexico: Potential Socioeconomic Implications of North-South Biotechnology Transfer. ISAAA Briefs No. 7. Ithaca, NY: International Service for the Acquisition of Agri-biotech Applications.

Qaim, M. and C. Falconi. 1998. Agricultural Biotechnology Research Indicators: Mexico. Discussion Paper No. 98-20. The Hague: International Service for National Agricultural Research.

Pedraza, L., E. Brockmann, V. Hernández, R.G. de Alba, V. Peñaloza, and R. Quintero. 1998. La Biotecnología en México: Una Reflexión Retrospectiva 1982-1997. Unpublished Paper. Centro de Investigación en Biotecnología, Universidad Autónoma del Estado de Morelos, Cuernavaca, Mexico.

Sasson, A. 1993. *Biotechnologies in Developing Countries: Present and Future. Vol. 1: Regional and National Surveys*. Paris: United Nations Educational, Scientific and Cultural Organization.

Wafula, J. and C. Falconi. 1998. Agricultural Biotechnology Research Indicators: Kenya. Discussion Paper No. 98-14. The Hague: International Service for National Agricultural Research.

Wagner, C. 1998. Biotechnology in Mexico: Placing science in the service of business. *Technology in Society*, Vol. 20, pp. 61-73.

Woodend, J.J. and F. Muza. 1993. Biotechnology for Resource Poor Farmers in Zimbabwe. Study commissioned by the Biotechnology Forum of Zimbabwe and ENDA-Zimbabwe.

## Annex 3.1

Table A.3.1. Agricultural Biotechnology Research Institutes in Mexico (as of 1997)

| Institutional category | Name of institute | Focus of biotech. research | First year of biotech. res. |
|---|---|---|---|
| Public research institute | National Research Institute on Forestry, Plants and Livestock (INIFAP) | Crops<br>Tissue culture<br>Molecular biology | 1992 |
| | Scientific Research Center of Yucatan (CICY) | Crops<br>Tissue Culture | 1979 |
| | Center for Research and Advanced Studies – Irapuato (CINVESTAV-I) | Crops and fruits<br>Tissue culture<br>Molecular biology<br>Plant microbe interact.<br>Genetic engineering | 1981 |
| | Center for Research and Advanced Studies-Mexico City – Biotechnology and Bioengineering Department (CINVESTAV-DF) | Medicinal plants<br>Bioinsecticides<br>Tissue culture | 1974[1] |
| | Center for Development of Biotic Products (CEPROBI) | Fruits<br>Tissue culture<br>Molecular biology | 1992 |
| Public university | Postgraduate School in Agricultural Sciences (CP) | Fruits and crops<br>Tissue culture | 1970 |
| | Metropolitan Autonomous University–Natural Products Area and Biochemistry Area (UAM) | Fruits<br>Food processing<br>Tissue culture<br>Fermentation | 1974 |
| | Autonomous University of Morelos State–Center for Biotechnology Research (CIB-UAEM) | Medicinal plants<br>Tissue culture | 1995 |
| | Autonomous National University of Mexico–Biotechnology Institute (IBT-UNAM) | Crops<br>Tissue culture<br>Molecular biology<br>Genetic engineering | 1982[2] |
| | Autonomous University of Chapingo –Department of Plant Breeding (UACH) | Crops<br>Tissue culture<br>Gene characterization | 1979/1995[3] |
| Private non-commercial | Institute for Technology and Advanced Studies of Monterrey (ITESM) | Crops and fruits<br>Tissue culture | 1990 |
| | International Center for Agricultural Research and Training (CIICA) | Crops<br>Tissue culture | 1996 |
| Private commercial | Biogenetica Mexicana (BioMex) | Ornamentals<br>Bioreactors<br>Tissue Culture | 1986 |
| | Grupo Bioquimico Mexicano (GBM) | Biopesticides | 1990 |

*Source:* Qaim and Falconi (1998).
[1] Agricultural biotechnology research was intensified in 1995.
[2] Biotechnology activities began in 1991.
[3] The tissue culture laboratory and the molecular biology laboratory were established in 1979 and 1995 respectively.

**Table A.3.2.** Agricultural Biotechnology Research Institutes in Kenya (as of 1996)

| Institutional category | Name of institute | Focus of biotech. research | First year of biotech. res. |
|---|---|---|---|
| Public research institute | Kenya Agricultural Research Institute (KARI) | Crops and livestock<br>Tissue culture<br>Genetic engineering | 1982 |
| Public university | University of Nairobi–Biochemistry Department (UN Biochem) | Crops<br>Molecular biology | 1990 |
| | University of Nairobi–Crop Science Department (UNCS) | Crops<br>Tissue Culture | 1983 |
| | Jomo Kenyatta University of Agriculture & Technology–Institute for Biotechnology Research (JKUAT) | Crops<br>Tissue culture | 1995 |
| Private non-commercial | Coffee Research Foundation (CRF) | Coffee<br>Tissue culture | 1975 |
| | Tea Research Foundation (TRF) | Tea<br>Tissue culture<br>Genetic markers | 1990 |

*Source:* Wafula and Falconi (1998).

**Table A.3.3.** Agricultural Biotechnology Research Institutes in Zimbabwe (as of 1998)

| Institutional category | Name of institute | Focus of biotech. research | First year of biotech. res. |
|---|---|---|---|
| Public research institute | Department of Research and Specialist Service (DRSS) | Cotton<br>Horticulture<br>Tissue culture | 1995[1] |
| | Veterinary Research Laboratories (VRL) | Livestock<br>Tissue culture<br>Genetic engineering | 1989 |
| | Biotechnology Research Unit Scientific & Industrial Research & Development Centre (SIRDC) (BRI) | Crops<br>Tissue culture<br>Genetic engineering | 1992 |
| Public university | Faculties of Agriculture, Science and Veterinary Studies University of Zimbabwe (UZ) | Crops<br>Livestock<br>Tissue culture<br>Molecular biology | 1983 |
| | National University of Science and Technology (NUST) | Crops<br>Tissue culture | 1992 |
| Private non-commercial | Tobacco Research Board (TRB) | Tobacco<br>Tissue culture<br>Genetic markers | 1991 |

*Source:* Gopo and Falconi (1999).
[1] The Horticultural Research Institute and the Cotton Research Institute began in 1995 and 1997 respectively

Table A.3.4. Agricultural Biotechnology Research Institutes in Indonesia (as of 1997)

| Institutional category | Name of institute | Focus of biotech. research | First year of biotech. res. |
|---|---|---|---|
| Public research institute | Research Institute for Food Crops Biotechnology (RIFCB) | Crops<br>Tissue culture | 1995[1] |
| | Agency for Agricultural Research and Development (AARD) | Genetic engineering | |
| | Research and Development Center for Biotechnology (RDCB) | Crops<br>Livestock | 1986 |
| | Indonesian Institute of Sciences (LIPI) | Tissue culture<br>Genetic engineering | |
| | Biotechnology Research Unit for Estate Crops (BRUEC) | Crops<br>Tissue culture<br>Molecular biology | 1993 |
| | Indonesian Sugar Research Institute (ISRI) | Crops<br>Tissue culture<br>Genetic markers | 1992 |
| Public university | Inter-University Center for Biotechnology (IUC) | Crops<br>Livestock | 1985 |
| | Bogor Agricultural University | Tissue culture<br>Molecular biology | |
| Private non-commercial | Veterinary Biologic Center (VBC) | Livestock<br>Vaccines<br>Tissue culture | 1991 |
| Private commercial | PT Fitotek Unggul (Fitotek) | Crops<br>Tissue culture | 1987 |
| | Inagro | Biofertilizers<br>Tissue culture | 1994 |

Source: Moeljopawiro and Falconi (1999).
[1] A biotechnology division was established in 1989.

# SECTION II

# Setting and Implementing Priorities

Integrating biotechnology with national agricultural research must take into account the scarcity of resources available to research organizations. Preparing a strategy for biotechnology provides a starting point for developing research agendas.

The first step in preparing a strategy, and part of the decision-making framework introduced in chapter one, is defining the priorities of a country or institute in developing its capacity in biotechnology. Priority setting helps decision makers plan research and allocate scarce resources. As part of the decision-making process, formal and systematic priority setting requires a method suitable for the specific decision problem. The first chapter in this section ("Methods for Priority Setting in Agricultural Biotechnology Research") reviews the most common formal priority setting methods in agricultural research and the factors influencing the choice. A necessary input for priority setting is collecting, processing, and analyzing information on the mobilization and use of resources and the institutional setting of the agricultural research system.

The case study presented in chapter 5 ("Setting Research Priorities for the Chilean Biotechnology Program") concludes that participation and the unique features of biotechnology research have to be addressed explicitly in priority setting. The main special characteristics of biotechnology research are substantial uncertainty, lack of data, public acceptance, high investments, integration to conventional research, and legal regulations.

The case study presented in chapter 6 ("Managing Biotechnology in AARD, Indonesia: Priorities, Funding, and Implementation") addresses the issue of implementing the results of a priority setting exercise. Among the issues to be considered are the identification of priority projects, the development of human resources, the provision of facilities and equipment, and the identification of appropriate collaborators.

The choice of a priority setting method depends on the nature of the decision that has to be made and on factors such as the degree of detail in the analysis, the availability of skills and resources, and the participants' or clients' understanding of the process. There is *a priori* no right or wrong method. However, as explained in chapter 4 and further confirmed in chapter 5, the analytic hierarchy process (AHP) has proven to be a powerful decision-support tool for setting priorities for agricultural biotechnology research.

Obtaining information in advance is a critical input for any priority setting exercise. In the case of biotechnology, this information must compare advantages and performances among organizations undertaking agricultural biotechnology research. The case study from Chile highlights the importance of gathering information which takes into account unique features of biotechnology, including limited institutional experience and scanty data on potential impacts. Lack of critical information can mean that substantial uncertainty results from a priority-setting process. To address this uncertainty, the authors select a methodology allowing for broad participation that improves the quality of information available for the priority setting process, and accounts for the chance of success both in terms of research and adoption by the end users. Other features to consider are public acceptance, changing legal frameworks, risks to biodiversity, and the importance of enhancing scientific capacity. The Chilean case also brings out the need for different stakeholder groups to participate to obtain relevant information for the priority setting exercise and to increase ownership of the results.

The Indonesia case demonstrates how setting priorities becomes the first step in establishing a strategy for biotechnology research. The results of a priority-setting exercise can only be implemented if they match available resources and take into account human resources and financial sustainability. Finally, a legal and institutional framework has to be established to provide an environment that is conducive to innovative research to support national development goals.

## Recommendations

The main recommendations for research managers in this section are summarized as follows:

- Undertake priority setting as an integral part of the strategic planning process.
- Begin with an analysis of what is already being done in agricultural biotechnology and which resources are used.
- Determine the questions that need to be answered and then choose an appropriate priority setting method. AHP has been found to be a flexible approach that can incorporate the unique management concerns of biotechnology research.
- In the course of the priority setting exercise, focus on particularly biotechnology-related issues such as intellectual property, uncertainty, and public acceptance.
- Design participatory processes, ensuring end user and beneficiary participation, for priority setting that maximize the two-way flow of information, improve the quality of decision making, and ensure a commitment to implementation.
- Plan and reserve resources to implement priority setting results.

# 4 Methods for Priority Setting in Agricultural Biotechnology Research

*Cesar A. Falconi*

## Abstract

*Priority setting is a complex process of choosing among different sets of research activities to make the most effective use of available resources. Priority setting helps plan for research and resource allocation. Participation by those who are responsible for the decision making and those who provide the information is essential for any successful priority setting exercise. The best method for priority setting depends on the purpose of the analysis, the availability of resources, the inclusion of unique features of biotechnology research, and the participants' or clients' understanding of the process. The most useful priority setting methods for agricultural biotechnology are the analytic hierarchy process (AHP) and a combination of AHP and the economic surplus approach to facilitate consistency with the economic framework. Even though the selection of the method is critical for the priority setting, research managers should attach the same importance to identifying the final user(s) of the research results and the political support required and available for conducting and implementing the outputs of the priority setting exercise.*

## Introduction

Decision making in developing-country national agricultural research systems (NARS) is becoming increasingly complex. There is increasing need to introduce new techniques to deal with new issues such as distribution, financing, multiple goals, and imputing returns to basic versus applied research activities. There is also growing pressure to improve the efficiency and effectiveness of agricultural research. At the same time, research budgets are being tightened and even cut. Also, donor organizations, which contribute financially or provide new technologies, require NARS to make sound decisions on resource allocation. Consequently, they press for appropriate decision-making processes to be put in place.

As a result of these pressures, research managers are seeking new methods and tools to help them reconcile the sometimes conflicting demands of producers, farm-input suppliers, consumers, scientists, and politicians. To be helpful, these methods must be cost-effective and capable of incorporating technical and economic information, as well as key value judgments on goals and objectives for the research system.

Participants attending the four agricultural biotechnology policy seminars organized by ISNAR's Biotechnology Service (IBS) acknowledged that more formal (or more rigorous) priority setting is necessary for better decision making. Some mentioned that

© CAB *International*. 1999. *Managing Agricultural Biotechnology—Addressing Research Program Needs and Policy Implications* (ed. J.I. Cohen)

setting research priorities is crucial in enhancing the productivity of biotechnology research.

The first section of this chapter presents the basic priority setting process. The second section discusses issues for research priority setting in biotechnology. In the third section methods for priority setting are defined and analyzed. The fourth section compares priority setting methods.

## An overview of priority setting

Priority setting in research is the process of choosing between different sets of research alternatives. The aim of priority setting is to make the most effective use of available resources; its main objective is to select the best portfolio of research activities for a certain research system, institution, or program (Janssen 1995). In biotechnology, it is an essential first step in developing a national biotechnology research program.[1]

Priority setting is best done in a participatory mode. This (1) facilitates a review of existing resource allocations, (2) helps achieve a consensus on objectives as differences of opinion are clarified, (3) helps update the research agenda as new alternatives are considered, (4) makes decisions on resource allocation more transparent and unambiguous, giving clearer guidance, (5) helps consolidate programs by allowing different staff levels to participate in the process, and (6) strengthens the credibility of an institution or program and helps it to take a proactive role in soliciting government and donor support for crucial areas to agricultural research.

A formal, systematic, transparent, and participatory priority setting process facilitates the implementation of the resulting priorities and makes it less costly.

### Levels of and structure for priority setting

Priority setting exercises support planning and resource allocation decisions and should be conducted at the same organizational levels where those activities occur. Within a research system, priorities are set at several levels: national, institutional, and the research program.

At the national level, the government decides on the level of funding for research, the type of research, and the funding for the individual research institutes. Policy considerations and negotiations dominate the decision making at this level. The outcome depends strongly on decision-makers' expectations of the impact of research on national policy goals, such as increased rural welfare, poverty alleviation, export growth, and food security.

At the institutional level, the ministry and top management of the NARS decide on the relative importance of the different research programs (commodity or noncommodity) or major research zones as they contribute to the NARS' mandate and national agricultural development objectives. The potential impacts of research in each program are compared. Political considerations and technical judgments are critical at this level. Examples of the priority decisions made at this level are choosing between research on biotechnology versus natural resource management or sugarcane versus banana.

---

[1] Cohen (1994) described the most important steps in developing a national biotechnology research program: identification of priorities and setting national policies, program formulation, program implementation and monitoring, and technology transfer and delivery.

At the program level, research leaders and scientists prioritize the allocation of resources to specific research activities based on the likelihood of the research to generate adaptive technologies to help clients solve major problems. Technical assessment dominates this level, and projects within a program are compared for their technical soundness. For example, a breeding program that wishes to use biotechnology may have to decide between the genetic transformation of an export commodity and the use of molecular markers to understand the genetic diversity of a native crop.

To ensure consistency at all levels, political considerations based on clients' needs should be defined at the top level. These must be translated into clear objectives at the lower levels (top-down approach) (Dagg 1992). Technical considerations, also according to clients' needs, are mostly made at the bottom of the hierarchy (bottom-up approach). Senior managers or directors of research institutes should participate at both these levels to ensure good communication and to link both levels.

Participation is a crucial component of any priority setting exercise. The two main groups of participants are those who are responsible for the decision making and those who provide the information. These groups can be subdivided into the following categories:

- decision makers, including managers and senior government officials responsible for agricultural research
- economists, other social scientists, and management experts, who provide analytical tools and data required for the exercise
- researchers, who provide information about the technical feasibility of the research
- clients, including farmers, who provide information on technology needs

Participants can also be grouped according to national, institutional, and program levels. According to Mills and Mbabu (1998), the decision makers at the national level are senior government officials, while those providing the information include directors of research organizations, ministers of finance, and representatives of clients. Similarly, at the institute level, senior managers, institute directors, and research boards are involved in the decision making, while the information for the decision making is provided by researchers, economists, and clients. At the program level, decision makers are also the providers of information: researchers, program heads, economists, and clients.

Priority setting is a formal, structured decision process that reviews, analyzes, and shows which parties are involved, how long the process will take, what types of information are needed, what resources are required, what the basic principles and steps are, what outputs are produced, and how the results will be used. Although no rules exist about when to conduct a priority setting exercise, it is advisable to do so before any planning exercise. As priority setting exercises can be expensive, the frequency of holding them must be weighed against the costs of completing them.

### *Basic steps*

While there is no single, "right" way of doing priority setting, a series of basic steps have been tested and found useful (Janssen 1994; Chan-Halbrendt et al. 1995):

*Step 1. Identification of research objectives*

It is essential that research objectives be defined according to the national or agricultural development goals. Translating or relating research objectives to these national goals helps determine the potential contribution of research to the achievement of each agricultural development goal. These goals can usually be grouped into four categories: efficiency (economic growth), equity (income distribution across society), food security, and sustainability (environment). The categories are the main criteria for the priority setting process.

In this first step, decision makers, economists, and clients play a major role. For example, suppose that the objectives of a biotechnology center are (1) to increase net benefits to all producers and consumers through increased productivity and efficiency by developing or adapting biotechnology techniques and (2) to increase environmental sustainability by generating methods of biological control. The first objective can contribute to the national objective of increasing the welfare of a country's population (efficiency criterion), while the second contributes to protecting the natural resource base for future generations (sustainability criterion).

## Step 2. Defining feasible technology alternatives

Defining feasible technology alternatives is the second, crucial step in priority setting. A fast evolving technology, biotechnology can be applied across many commodities and programs. Tissue culture and micropropagation, for example, can be applied to both traditional and nontraditional crops. Researchers are the main actors in this step.

## Step 3. Deriving criteria and method

In this step we choose or derive the criteria and method to evaluate the technologies or research alternatives. Economists play a major role in this systematic assessment and comparison. Evaluation criteria normally correspond with measurable indicators, such as net present value, cost-benefit ratio, market standards, and resource capacity. These indicators are related to the research objectives identified earlier. A means of measurement is needed to estimate how the research or technology alternatives perform according to the criteria. There are many methods, ranging from simple, qualitative methods to complex quantitative approaches. It is therefore important to choose the right level of complexity for the problem at hand. In emerging technologies like biotechnology, where impacts are still unknown and the adoption of results also depends on public acceptance and health and environmental risks, it is difficult to generate measurable indicators. For example, indicators for assessing research alternatives can be the net present value for the efficiency criterion and small farmers' share of production for the equity criterion.

## Step 4. Performance assessment and comparison of alternatives

Once criteria and method are decided, the expected performance of the technologies is assessed. The assessments of different alternatives are then compared with measurement indicators to determine the best alternatives. Researchers and economists are the main actors in this step.

*Step 5. Approval and implementation*

The final step in priority setting is approving and implementing the results. The final results are discussed with decision makers and clients to reach consensus. Implementation will, of course, be constrained by the budget available.

## Issues specific to biotechnology

While the basic steps for priority setting can also be applied to biotechnology research, some issues specific to this emerging technology require special attention because they obscure the assessment of the positive and negative impact of biotechnology research. These include high developmental costs, uncertain risks, integration of biotechnology with conventional programs, and limited experience with biotechnology.

### Costs and investment

The potentially high cost of and investment in biotechnology require a close look at issues of efficiency in priority setting. First, fixed costs of laboratories and scientific personnel across biotechnology projects can be shared among commodity programs and biotech programs. For example, teams can shift between research on maize stemborer and cassava mosaic virus, thus sharing the fixed costs involved. But this undifferentiated biotechnology capacity is insufficient to develop all the knowledge and expertise needed for working on a particular crop. A decision must be made on whether a biotechnology program should pursue a specialization and, if so, along what lines.

Second, given the shrinking budgets and high fixed costs of many laboratories, centralization ensures a higher rate of use capacity. A major drawback of centralization, however, is that it directs biotechnology research away from the focus on more locally adapted cultivars.

Third is the issue of the investment to be made in the development of tools versus the application of existing techniques. The cost of tool development is quite high, even if such tools would solve a critical problem for a high-priority commodity in a country. Since existing techniques are usually less costly, it may be more efficient to acquire these techniques for lower-priority commodities. In addition, strategic alliances with public, private, and international institutions help research groups share the risk and uncertainty of high and lump-sum investments.

Fourth is the issue of timing. In most priority setting exercises, the question is whether or not to undertake a certain project. But in biotechnology, the question is also *when* to undertake a project—now or within the next two to five years. There is an advantage to a "late start," as scientific discovery in related areas may allow researchers to borrow results achieved by others. The danger of such a late start, however, is that others may already have obtained the patents.

Priority setting can help research managers solve cost issues before high and risky investments are made. However, problems in estimating the real cost of a particular biotechnology project or activity remain.

## Uncertainty

Agricultural research typically has a high degree of uncertainty. Biotechnology, a relatively new science where little experience has been acquired and a broad information base is still lacking, is particularly uncertain. This makes completing one of the most important steps in priority setting—assessing and predicting the performance of feasible technologies—especially difficult because of subjectivity involved. It is therefore crucial to apply a priority setting method that reduces individual biases in judgements and that incorporates technical and product knowledge.

## Qualitative impact

Quantitative data, such as expected increases in yield, can help researchers estimate the impact of a new technology quite accurately. However, the impact on human health of reduced use of pesticide on a pest-resistant variety is typically qualitative and thus much more difficult to measure. Public acceptance of the production and diffusion of genetically engineered products is another qualitative issue that is difficult to estimate. Complex and intangible though they may be, these impacts should be considered in the priority setting process.

## Other factors

Another complicating factor in assessing the performance of a biotechnology program is its complementarity to conventional research. Biotechnology research projects are usually not aimed at final users. Rather, they generate intermediate outputs for use as inputs in conventional research programs. A good example is the molecular markers that speed up conventional plant breeding. The benefits of such research-oriented biotechnology depend on to what extent it accelerates the plant breeding (which is difficult to assess) and if the final outcome of that research is accomplished with the release of a new variety or hybrid. The legal framework (biosafety regulations and intellectual property rights) also influences the impact of biotechnology research.

Biotechnology is surrounded by uncertainty and a broad range of potential impacts that are difficult to capture. That is why priority setting in biotechnology research tends to be based more on informed judgments than on quantifiable information, as in conventional research programs. The use of individual judgments should be minimized to avoid errors in the outcomes of the priority setting process. A participatory framework can contribute to improving the quality of information used for the priority setting exercise.

# Priority setting methods

To assess the costs and benefits of agricultural research, quantitative methods are needed, either before the research is undertaken (ex ante evaluation) or after it is completed (ex post evaluation). In the case of new technologies, where historical information is limited, the focus will be on ex ante evaluation methods.

Several publications exist on agricultural research evaluation and priority setting. Contant and Bottomley (1988) discuss some formal methods for priority setting. Alston, Norton, and Pardey (1995) present a very comprehensive and rigorous analytical economic framework for agricultural research evaluation and priority setting, as do, for example, Norton and Davis (1981) and Falconi (1993).

Several quantitative methods are available to assist agricultural research priority setting. The two most common methods are scoring and economic surplus. Two other methods, mathematical programming and simulation, have been used for selecting research projects. A more recent method is the analytic hierarchy process (AHP).

## *Scoring methods*

Scoring is a multicriteria approach that helps (1) select a broad set of research objectives (e.g., efficiency, equity, food security, or environment) and (2) establish indicators of research contributions to achieve these objectives. Examples of such indicators are the value of production, probability of research success, cost of research, and expected adoption. Relative weights are attached to the objectives or criteria, and weighted average scores are calculated for each research area. Next, commodities or research programs are ranked according to each objective, and these rankings are multiplied by the weights to derive a final composite ranking. Applications in agricultural research of this method are found in many studies all over the world (Norton 1993).

The advantage of scoring models is that they can be conducted in a relatively short period of time, they are relatively transparent, they allow extensive active participation, and they do not require advanced quantitative skills. They can be used to rank a long list of commodities as well as research areas, including nonproduction-oriented research and qualitative as well as quantitative information.

A major disadvantage of scoring models is that they seem simple to apply. Thus, users often overlap objectives, duplicate criteria (indicators), and add criteria at random. Other shortcomings may be the inaccurate accounting of research spillovers and ignoring the effects of domestic and trade policies. The precision of the components of scoring can be improved, however, by combining them with economic surplus calculations. A final criticism, which indeed applies to all priority setting procedures, is that the weights that are assigned to the objectives are highly subjective.

## *Economic surplus*

The economic surplus method is a single-criterion approach that estimates returns to investment (generally an average rate of return) by (1) estimating the benefits from research in terms of the change in consumer and producer surpluses that results from technological change and (2) using the estimated economic surplus together with research costs to estimate an internal rate of return.

Ex ante analysis of future research benefits requires information on expected values of production, expected yield increases, reduction of unit costs, probabilities of research success, market conditions, adoption rates, supply and demand elasticities, and the appropriate discount rate for converting future benefits and costs into present values. Benefits and costs are projected over several years and internal rates of return to research, benefit-cost ratios, or net present values are calculated. These values are used to help rank commodities, research programs, or projects. There are many examples of this approach in the economic literature on specific research commodities or production constraints (Falconi 1993).

The economic surplus approach can be used (1) to estimate the distribution of research benefits among producers and consumers, (2) to assess the spillover of research benefits among different technologies, commodities, regions, or countries, and (3) to estimate the effects of agricultural policies on the benefits arising from research. An important advantage of this approach is that it allows for more accurate and thus more credible calculations of efficiency and distributional effects of research, and it helps allocate resources to each commodity program or type of research.

The approach has several limitations. One is that the method requires substantial expenditures for collecting, processing, and interpreting economic and technical data. There is no place for group discussions during the priority setting process (low active participation). It is also not well suited to ranking noncommodity research areas, such as basic research, socioeconomic, or interdisciplinary research. And in order to incorporate multiple objectives, the economic surplus approach needs to be combined with scoring or mathematical programming.

## *Mathematical programming*

Mathematical programming is a problem-solving technique for maximizing the expected benefits of a research program. Instead of using ranking, it identifies an optimal research portfolio based on multiple goals and resource constraints (such as human capital, funding, and environmental factors). In theory, mathematical programming could help a research director allocate research resources among commodities, research areas, and programs. It can incorporate multiple goals as well as human resource and financial constraints on research. The information required is more stringent than that needed in the scoring and economic surplus approaches. Mathematical programming uses weight to reflect the political and economic importance of a set of research objectives.

Mathematical programming models have the advantage of explicitly incorporating human resource, budget, and other constraints in the system. They can also include multiple objectives. If constructed in a multiperiod format, they can identify how the research portfolio should change over time. This approach requires a great deal of time and analytical ability, however, and it is not an easy approach for managers or researchers to include in priority setting. It is also difficult to incorporate qualitative information. Used mainly by economists, this approach is low in active participation.

## *Simulation models*

In simulation models, mathematical relationships among variables are exposed to different scenarios to assess the best outcome. They can incorporate many factors that affect research priorities, such as multiple goals, research constraints, socioeconomic variables, risk, and uncertainty. The output of a simulation model is a set of research priorities that best meet the objectives of a system. A typical simulation procedure involves four steps: (1) formulating a mathematical relationship between, for example, research expenditures and agricultural productivity, (2) developing a set of scenarios to project growth in agricultural productivity, (3) evaluating the impact of research-derived productivity on total benefits, and (4) estimating the ex ante rate of return to research investments. Simulation models provide a flexible tool that can incorporate nonquantitative data and build up from relatively simple to complex models.

The main disadvantage of simulation models is the large investment of resources (in data and the time of a skilled analyst) required to implement them. Data requirements are more extensive than for other economic methods, and there are few practical applications for this approach (Falconi 1993).

## *Analytic hierarchy process*

The analytic hierarchy process (AHP) is a multiobjective, multicriterion decision-making tool that employs multiple paired comparisons to rank alternative solutions to a problem, formulated in hierarchic terms (Ramanujam and Saaty 1981) (see also chapter 5). The literature discusses a variety of applications, including planning, resource allocation, and priority setting examples in business, marketing, transportation, resource management, and many other fields (Saaty and Vargas 1994). AHP uses the following three-step procedure:

1. Creating a hierarchy of a minimum of three levels to structure the decision problem. The overall goal of the priority setting exercise (e.g., setting priorities among a set of research projects) is at the first (top) level, followed by a second (intermediate) level that consists of the decision criteria (e.g., research objectives) by which the alternatives (e.g., research projects), located in the third (bottom) level, will be evaluated.
2. Weighting the criteria and evaluating the alternatives. Criteria are compared in pairs with respect to their importance to the goal, while alternatives are compared in pairs with respect to the criteria.
3. Determining the overall priority of each alternative and getting the final ranking of the alternatives.

This approach can be seen as an extension of the scoring method, addressing some of the major shortcomings of that approach. While scoring models often use poorly measured or overlapping criteria for determining research contributions, AHP minimizes this through careful structuring of the decision hierarchy.

AHP is particularly suitable for situations in which much of the necessary data is subjective (such as biotechnology). It can be consistently introduced into the priority setting and it can deal with decision problems involving multiple criteria dimensions. Unique to AHP is that it recognizes bias and inconsistencies in subjective judgments. These inconsistencies can be tested and improved, resulting in a more consistent final ranking. Other advantages of the approach are that it allows group decision making by the different stakeholders in agricultural research (active participation in structuring the hierarchy and eliciting the judgments), and it is a transparent process. The method can be combined with other methods such as economic surplus and mathematical programming in order to improve the allocation of resources.

A problem with AHP is the pairwise comparison. If there are many alternatives (e.g., projects) to evaluate, then the number of comparisons will make the comparison process tedious. For example, to evaluate 10 projects, 45 comparisons will be required for each criterion.

## Choosing the right method: determining factors

As we have seen, each of the above methods has its advantages and disadvantages. Mathematical programming and simulation methods are less frequently used for priority setting than are economic surplus and scoring approaches. Simulation and mathematical programming are used in optimization exercises where there are many similar options for achieving the same output and the parameters are reasonably assumed. AHP, scoring, and economic surplus, however, use a partial equilibrium approach to assess the value of a singular activity. The more sophisticated methods are better suited for priority setting in commodity or other major research programs than for subprograms and projects.

The most appropriate method for a particular priority setting situation depends on (1) time available for the study, (2) data availability in relation to degree of analysis, (3) analytical capacity, (4) participation in the process, and (5) transparency in the process (adapted from Norton 1989).

### Time

Using a scoring method to rank commodities and program areas at the national and center levels typically takes two to four months. Compared with the economic surplus, mathematical programming, and simulation approaches, it asks fewer detailed questions of scientists, less secondary data is needed, and calculations are simpler. Research managers using AHP need four to six months to complete the ranking for their institute, because AHP involves more discussion and group decision than does the scoring method. Combining economic surplus analysis with scoring or AHP can introduce market situations, discounting, etc., into the analysis. This combined approach will usually take at least six months. Mathematical programming and simulations require more time and resources.

### Data required and level of analysis

Each priority setting method requires data. How much data is needed depends on the level of analysis and the nature of the question to be answered. The scoring and AHP methods collect data and information through interviews with scientists, economists, etc. These two approaches do not need any detailed information because the results are more general (e.g., list of ranking of projects). The economic surplus method requires more detailed data because the approach provides more information about research benefits, effects of pricing policies and trade on research benefits, and more accurate rate of returns than the other methods. Mathematical programming and simulation methods require extensive amounts of data.

### Analytical capacity

It helps if the individuals leading the priority setting process have a basic knowledge of economic theory. Scoring and AHP require fewer quantitative skills than do the other methods. However, in AHP it is important to have some knowledge of linear algebra and measurement theory.

## Participation

Economic surplus, mathematical programming, and simulation methods have a low degree of participation. Analysts dominate the priority setting process (i.e., modeling the process and eliciting the subjective judgments). The scoring and AHP approaches require extensive participation (e.g., eliciting information, defining criteria, assessing the different research activities, and setting the priorities). AHP also provides a consistent framework to formally incorporate subjective judgments, which are elicited and discussed in groups.

## Transparency

The process of how the final results or outcomes were obtained must be clear. The more transparent the priority setting process is, the greater the chance that the organization or institution will accept the results and the easier it will be to implement them. The scoring and AHP methods are more transparent than the other approaches.

Figure 4.1, which summarizes the above factors, shows that priority setting methods such as scoring and AHP are more transparent and participatory, while mathematical programming, simulation, and economic surplus require more time, resources, and data analysis. However, the latter approaches, in particular the economic surplus, provide rigor and finer analysis of trade-offs at the cost of requiring more data and analytical skills.

Based on the above five factors, the most useful methods for priority setting in agricultural biotechnology research are AHP, which handles subjective judgments and allows multiple objectives, or a combination of AHP and the economic surplus approach to facilitate consistency with the economic framework.

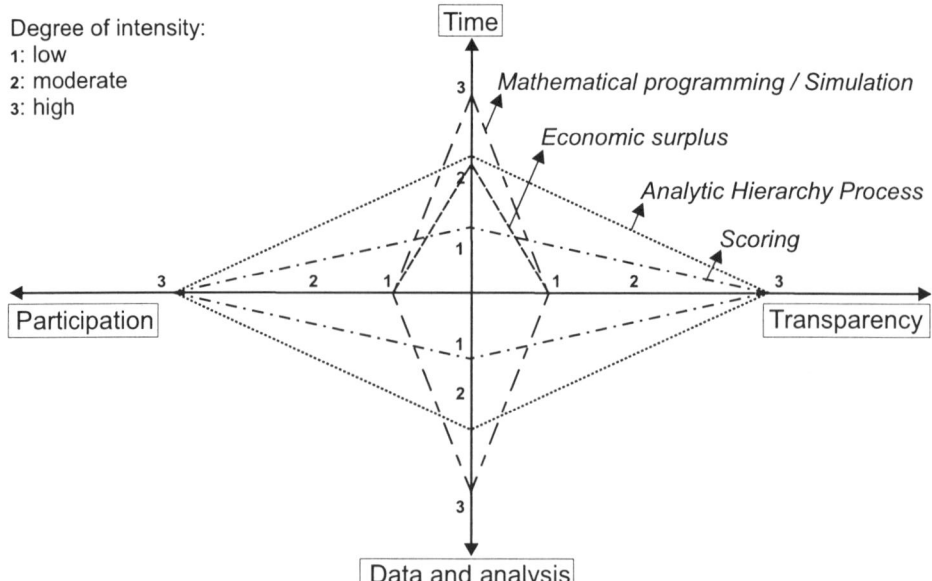

**Figure 4.1.** Priority setting methods compared

## Conclusions

Given the situation of decreasing research budgets, the demands for more accountability, and the high expectations of emerging technologies such as biotechnology, priority setting has become an important task in agricultural biotechnology research planning.

The particular characteristics of biotechnology require special attention in setting priorities. Little experience has been acquired in this field, and information about it is limited. Performance assessments of biotechnology projects are therefore often quite subjective. It is crucial to apply a priority setting method that reduces individual biases as well as the risks of arriving at the wrong choices.

The high investment into and cost of biotechnology also raises important issues with regard to priority setting, such as the possibility of sharing fixed costs, the question of centralization, and how much the biotechnology programs should invest in developing tools versus applying the existing techniques. All these issues should be taken into account in selecting a method or approach for setting priorities.

The appropriate method depends of the purpose of the analysis, the availability of resources, how the unique issues of biotechnology agricultural research are handled, and the participants' or clients' understanding of the process. The most useful methods for agricultural biotechnology research priority setting are AHP, which handles subjective judgments formally and allows multiple objectives, or a combination of AHP and the economic surplus method to facilitate consistency with the economic framework.

Even though selecting the right method is critical for the priority setting process, research managers should also attach the same importance to defining the final users of the results and establishing the kind of political support that is necessary and available in conducting and implementing the outputs of the priority setting exercise. Whatever the method chosen, the results of the analysis must be useful for NARS leaders to make decisions on people, budgets, and facilities.

## References

Alston, J.M., G. Norton, and P. Pardey. 1995. Science Under Scarcity: Principles and Practice for Agricultural Research Evaluation and Priority Setting. Ithaca and London: Cornell University Press.

Chan-Halbrendt, C., J.I. Cohen, W. Janssen, and T. Braunschweig. 1995. Improving Biotechnology Research Decision Making with Better Procedures and Information. *In* Assessing the Impacts of Agricultural Biotechnologies: Canadian-Latin American Perspectives, edited by B. Herbert-Copley. Proceedings of a workshop organized by IDRC, Ottawa, Canada, May 16-17, 1995.

Contant, R. and A. Bottomley. 1988. Priority Setting in Agricultural Research. Working Paper No. 10. The Hague: International Service for National Agricultural Research.

Cohen, J.I. 1994. Biotechnology Priorities, Planning, and Policies: A Framework for Decision Making. ISNAR Research Report No. 6. The Hague: International Service for National Agricultural Research.

Dagg, M.A. 1992. General Model for Research Program Planning. *In* Summary of Papers at the International Seminar on Agricultural Research Management, China, May 25-27 1992. The Hague: International Service for National Agricultural Research.

Falconi, C. 1993. Economic Evaluation. *In* Monitoring and Evaluating Agricultural Research: A Sourcebook, edited by D. Horton, P. Ballantyne, W. Peterson, B. Uribe, D. Gapasin and K. Sheridan. Oxon: CAB International.

Janssen, W. 1994. Biotechnology Priority Setting in the Context of National Objectives: State of the Art. *In* Turning Priorities into Feasible Programs: Proceedings of a Regional Seminar on Planning, Priorities and Policies for Agricultural Biotechnology in Southeast Asia, edited by J. Komen, J.I. Cohen, and S.K. Lee. The Hague / Singapore: Intermediary Biotechnology Service / Nanyang Technological University.

Janssen, W. 1995. Priority Setting as a Practical Tool for Research Management. *In* Management Issues in National Agricultural Research Systems: Concepts, Instruments, Experiences, edited by M. Bosch and H-J.A. Preuss. Hamburg: LIT Verlag.

Lynam, J.K. 1995. Building Biotechnology Research Capacity in African NARS. *In* Turning Priorities into Feasible Programs: Proceedings of a Regional Seminar on Planning, Priorities and Policies for Agricultural Biotechnology for East and Southern Africa, edited by J. Komen, J.I. Cohen and Z. Ofir. The Hague / Pretoria: Intermediary Biotechnology Service / Foundation for Research Development.

Mills, B. and A. Mbabu. 1998. The Role of and Levels for Agricultural Research Priority Setting. *In* Agricultural Research Priority Setting: Information Investments for the Improved Use of Research Resources, edited by B. Mills. The Hague: International Service for National Agricultural Research.

Norton, G. 1989. Methods to Assist with Agricultural Research Priority Setting. Paper presented at the International Agricultural Research Management Workshop, ISNAR, The Hague, November 7, 1989.

Norton, G. 1993. Scoring Methods. *In* Monitoring and Evaluating Agricultural Research: A Sourcebook, edited by D. Horton, P. Ballantyne, W. Peterson, B. Uribe, D. Gapasin and K. Sheridan. Oxon: CAB International.

Norton, G., and J. Davis. 1981. Evaluating Returns to Agricultural Research: A Review. *American Journal of Agricultural Economics*, Vol. 63:685-699.

Ramanujam, V. and T. Saaty. 1981. Technological Choice in the Less Developed Countries: An Analytic Hierarchy Approach. *Technological Forecasting and Social Change*, Vol. 19:81-98.

Saaty, T. and L. Vargas. 1994. Decision Making in Economic, Political, Social and Technological Environments with the AHP. University of Pittsburgh.

# 5 Setting Research Priorities for the Chilean Biotechnology Program

*Thomas Braunschweig, Willem Janssen, Carlos Muñoz, and Peter Rieder*

## Abstract

*This chapter reports on a priority setting exercise for biotechnology research projects in Chile. The primary objective of the exercise was to test a conceptual framework that can provide a list of prioritized biotechnology projects. The methodology applied is the analytic hierarchy process (AHP), a multicriteria decision-making approach that uses pairwise comparisons to determine preferences among a set of projects. AHP systematically structures a complex decision problem and analyzes only two elements at a time. The approach incorporates subjective judgments and combines them with hard data. It (1) decomposes an unstructured decision problem into a hierarchy, (2) makes a preference ranking among pairs of alternatives by making comparative judgments, and (3) synthesizes these judgments into the priority ranking of the projects. It is well-suited for group decision making and it is supported by a software package.*

## Introduction

The main objective of priority setting is to allocate scarce research resources efficiently, that is, to maximize the benefits that society expects from public expenditures for research. With new technologies such as biotechnology, it is even more pertinent to allocate resources efficiently. In setting priorities for new technologies, there is a strong need for a good decision-making tool, because the priorities influence future strategies for applying technologies. Agricultural biotechnology has some special features that should be taken into account in priority setting (see also chapter 4):

- **Uncertainty.** Substantial uncertainty exists in agricultural biotechnology, because many of the research techniques are still under development. The adoption process is an additional source of uncertainty.
- **Lack of data.** Little experience has been gained in this area, so there is little historical data.
- **Public acceptance.** Public perception can play an important role in introducing new biotechnology products, but it is difficult to assess how important this role is.
- **Investments.** The investments to initiate research activities are often high, particularly in biotechnology.

© CAB *International.* 1999. *Managing Agricultural Biotechnology—Addressing Research Program Needs and Policy Implications* (ed. J.I. Cohen)

- **Link with other research.** The extent to which a biotechnology project is integrated into the existing research system is an important determinant of the project's success.
- **Legal obligations.** The legal framework considerably influences the costs and outcome of research.
- **Institutional development.** Policymakers may pursue objectives such as human and institutional capacity building. The contributions to these objectives are difficult to quantify.

These issues led to the development of a collaborative project between ISNAR, the Department of Agricultural Economics of the Swiss Federal Institute of Technology (ETH), and the National Institute of Agricultural Research (INIA) of Chile. The project aimed to develop and test an analytical framework to support decision makers in public agricultural biotechnology research.

## Chile's national biotechnology program

Contributing 25% to total exports in 1994 (Gil et al. 1996), Chile's agricultural sector exported mainly fruits and vegetables, with most of the varieties cultivated in Chile developed abroad. In order to strengthen "technological self-sufficiency," the Ministry of Agriculture (MINAGRI) initiated a national biotechnology program (NBP) (Villalobos 1995). MINAGRI requested INIA to coordinate the NBP and commissioned the Food and Agriculture Organization of the United Nations (FAO) to develop the new biotechnology program. A delegation of national and international experts formulated a concrete proposal for the NBP (Vio 1995). Several national workshops were held to determine priorities for the program—a process taking one and a half years (Muñoz 1997). The priority setting exercise resulted in a list of potential research activities but failed to identify specific projects to be financed through the competitive fund that was planned for the program. The exercise did not result in well-defined decision criteria (Muñoz 1998).

Thus, the objective of the INIA-ETH-ISNAR study was to design and test a framework for choosing biotechnology research projects. The study was not intended to evaluate an exhaustive set of projects, and the resulting priorities were not directly translated into funding decisions.

## Methodological approach

At the heart of the conceptual framework to assist the priority setting process is a modified version of the analytic hierarchy process (AHP). The method is described as a

> "multiobjective multicriteria decision-making approach which employs a pairwise comparison procedure to arrive at a scale of preferences among a set of alternatives. To apply this approach, it is necessary to break down a complex unstructured problem into its component parts and arrange these parts, or variables, into a hierarchic order."
>
> (Saaty and Vargas 1991, p.14)

AHP was initially developed by Saaty (1980) and has been applied to a wide range of complex decision problems (Zahedi 1986; Golden, Wasil, and Levy 1989). The private sector has also used AHP in selecting research portfolios (Lockett et al. 1986; Liberatore 1989).

Decision making is a group process that involves discussions, learning, and checking results. AHP provides a consistent framework to formally incorporate subjective judgments in group decision making (Dyer and Forman 1992). The Expert Choice© software package facilitates the process considerably.

## Procedure

AHP follows three steps: (1) decomposition of a complex unstructured problem, (2) preference ranking among pairs of alternatives by making comparative judgments, and (3) synthesis of the judgments into priorities.

### Step 1: Stating the problem in a hierarchical structure

Figure 5.1 presents a basic hierarchy, which consists of three levels. The top level represents the main goal of the exercise, for example "prioritizing a given set of research projects." The second level contains the criteria (e.g., the research objectives) that are relevant for this goal, while the bottom level lists the alternatives (research projects). To introduce more precision into the evaluation process, criteria can be divided into subcriteria, which would insert an additional level into the hierarchy.

### Step 2: Weighting the criteria and evaluating the projects

Criteria are compared in pairs to define their importance with respect to the goal, and projects are compared in pairs with respect to each of the criteria. These relative comparisons are based on collected data (statistics, surveys, reports, etc.) as well as the intuition, experience, and expertise of the decision makers. AHP explicitly recognizes the legitimacy of subjective judgments in priority setting, which is concealed in many other methodological approaches (Shumway 1981). The fundamental scale presented in table 5.1 elicits the relative comparisons and translates verbal judgments into numerical values. These values are then placed in a matrix. Using the eigenvector method described in Saaty (1977), the weights of the criteria and the scores of the projects can be calculated.

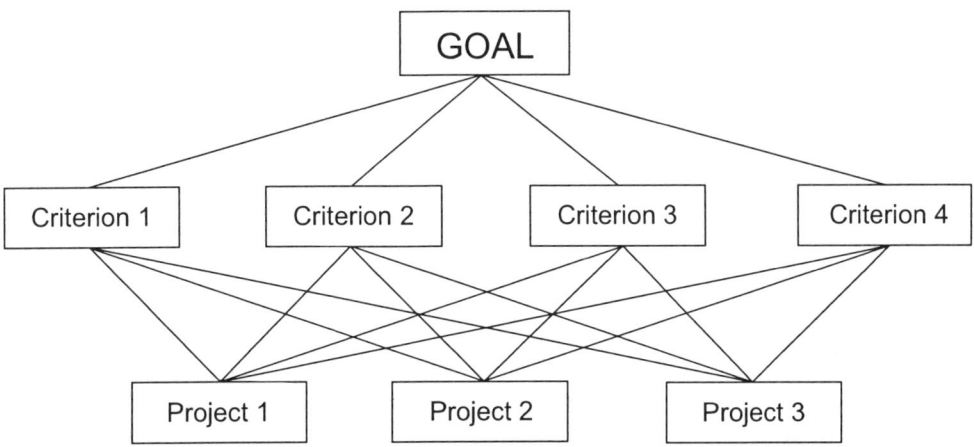

**Figure 5.1.** The basic structure of a hierarchy

**Table 5.1.** The Fundamental Scale for the Comparative Judgments

| Numerical value | Meaning |
|---|---|
| 1 | Equally important, likely, or preferred |
| 3 | Moderately more important, likely, or preferred |
| 5 | Seriously more important, likely, or preferred |
| 7 | Very seriously more important, likely, or preferred |
| 9 | Extremely more important, likely, or preferred |
| 2, 4, 6, 8 | Intermediate values to reflect compromises |

*Step 3: Judgments are synthesized into priorities*

For each project, the scores are multiplied by the corresponding criterion weight, and the results are summed up to get the degree of priority of a project with respect to the overall goal.

### Extensions

If there are many projects to be prioritized, then the comparison process may become very difficult to manage. For example, for $n=12$ projects there are already $N=66$ comparisons to complete for each criterion $[N=n(n-1)/2]$. To prevent such time-consuming comparisons, ratings may be applied to evaluate projects, as is customary in more traditional multicriteria approaches such as scoring methods. Criteria are still weighted by comparative judgments, however. In the Chilean study both pairwise comparison and ratings were used.

## The priority setting process

The priority setting procedure for the Chilean biotechnology program consists of 10 steps (see figure 5.2), structured around four meetings. The first meeting lasted one day, the second and third meeting two days, and the final meeting a half day. It is estimated that it will take INIA four to five months to repeat the exercise. To keep the exercise manageable, it focused strictly on projects by INIA. The private sector was not included because it did not have a specific interest in biotechnology.

Researchers of several regional INIA institutes were asked to formulate biotechnology projects on specially designed project proposal forms. The resulting ideas were presented and discussed at the initial meeting. To keep the pilot exercise manageable, the coordinator of INIA's biotechnology program and two planning specialists preselected seven projects (see table 5.2). They applied the following criteria: (1) the maturity of the project idea, (2) the scientific relevance for the NBP, and (3) the existence of an important research component as compared to the development.

Based on a review of policy documents from MINAGRI, INIA, and the NBP, four decision criteria were derived: "economic," "social," "environmental," and "institutional." The last one was included to assess to what extent research projects contributed to building capacity at the institutional and human resource levels. It can be used to estimate the contribution towards increased "technological self-sufficiency," an important objective for Chile's government. Criteria to capture the probability of research success and the rate of

# Setting Research Priorities for the Chilean Biotechnology Program

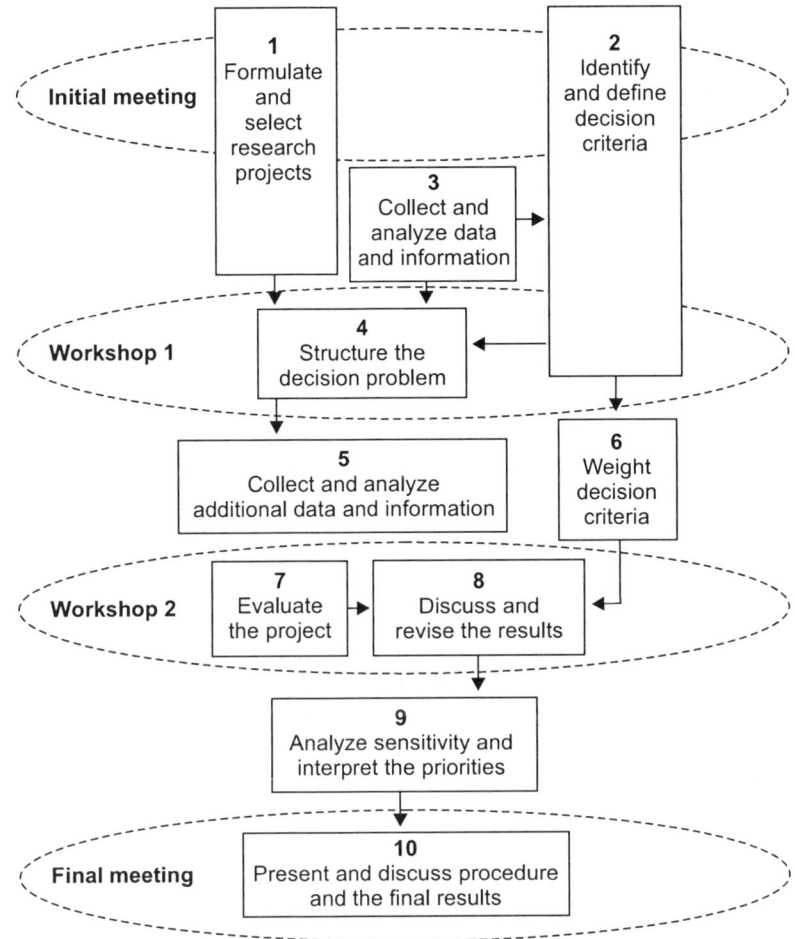

**Figure 5.2.** Priority setting procedure

**Table 5.2.** Selected Projects

| Commodity | Research focus |
|---|---|
| chirimoya | genetic transformation to obtain fruits with delayed ripening |
| grapes | genetic transformation to induce resistance to phytopato-gene fungi |
| potato | use of ligament maps of RFLP with the gene h1 marker to improve resistance to cyst nematodes in Chile |
| tomato | use of molecular markers to study the diversity of native germplasm |
| wheat | implementing genetic engineering to manipulate fungus-based diseases |
| nothofagus | biochemical, molecular, and dasometrical characterization of six species of the genus *nothofagus* |
| flowers | characterization and selection of native flowers with export potential |

adoption were also developed. For each criterion, subcriteria were formulated and indicators were determined to describe them.

Two working groups were formed: a *strategic* group consisting of research managers and experts in particular fields and a *technical* group consisting mainly of researchers who formulated a project. The first group selects and weights the criteria and subcriteria. The second evaluates the projects.

The list of criteria was modified according to the availability of data. Appropriate and unambiguous indicators were identified. Crop profiles and project information were collected.

In a two-day workshop, researchers, biotechnology and planning specialists, and an international expert designed a basic decision model with one hierarchy for the potential impacts of the projects (H1, see figure 5.3), another to evaluate the chances of research success (H2, see figure 5.4), and one to estimate adoption of the results by the end users (H3, see figure 5.5). The explicit treatment of uncertainty in two separate hierarchies (H2 and H3) is based on ISNAR's experience that potential project impact, research success, and adoption rates can best be estimated separately and multiplied afterwards (Janssen and Kissi 1997).

The participants of the second workshop were given a document to help them evaluate the projects. The document defined each criterion and described the performance of the projects with respect to all the indicators. In this phase projects were submitted for external peer review.

The members of the strategic group (half of whom were from INIA, the other half from other public institutions) weighted the relevant criteria and subcriteria of the three hierarchies in individual interviews. Calculating the average for each criterion and subcriterion concluded the weighting process. These results are presented below.

In the second workshop, the technical group with researchers and experts from INIA established project priorities. As ratings were needed for H2 and H3, the participants developed scales of intensity and evaluated the research projects against these scales (figures 5.4 and 5.5). In H1 the subcriterion "water" of the criterion "environmental" was eliminated because it lacked discrimination potential.

A need to revise H2 emerged when participants noted that the chances of research success were not satisfactorily assessed because the element "quality of the proposal" did not belong at the subcriteria level. The researchers agreed that a well-developed and well-formulated project proposal is crucial for achieving the expected research results. The element was consequently moved to the criteria level, which meant that the weights of H2 had to be redone. A second weakness was the absence of subcriteria in H2, which meant the participants could make only very general evaluations. After the introduction of subcriteria, the outcome made much more sense.

The final priorities were obtained by combining the outcomes of the individual hierarchies. To test the stability of the ranking, several scenarios were generated using the criterion weights of individual members of the strategic group. Another scenario distinguished between different types of research. These scenarios are discussed below.

The final step of the procedure took place in a meeting to which all stakeholders were invited. After the exercise and the results were presented, the approach was discussed. The discussion mainly centered around the future application in the NBP.

# Setting Research Priorities for the Chilean Biotechnology Program 59

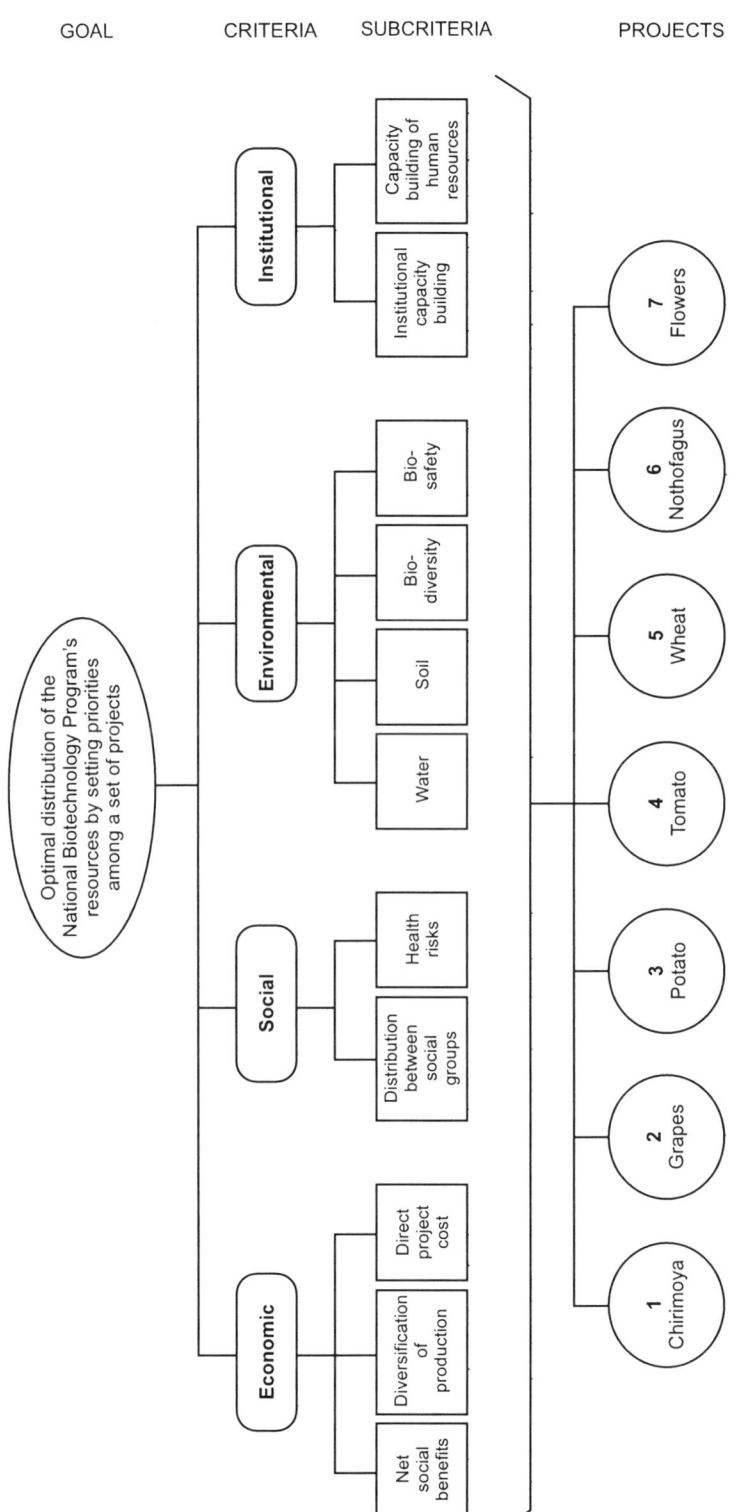

**Figure 5.3.** Hierarchy 1: Potential impacts (H1)

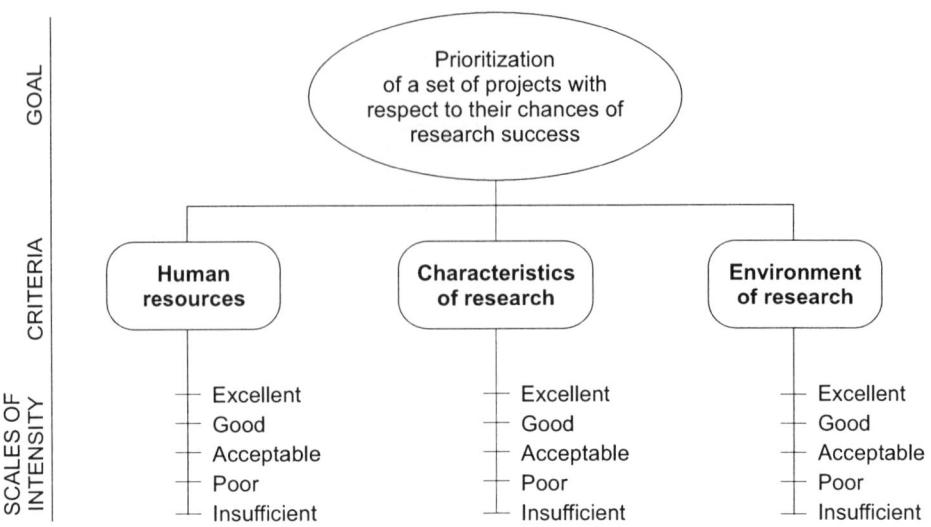

**Figure 5.4.** Hierarchy 2: Success of research (H2)

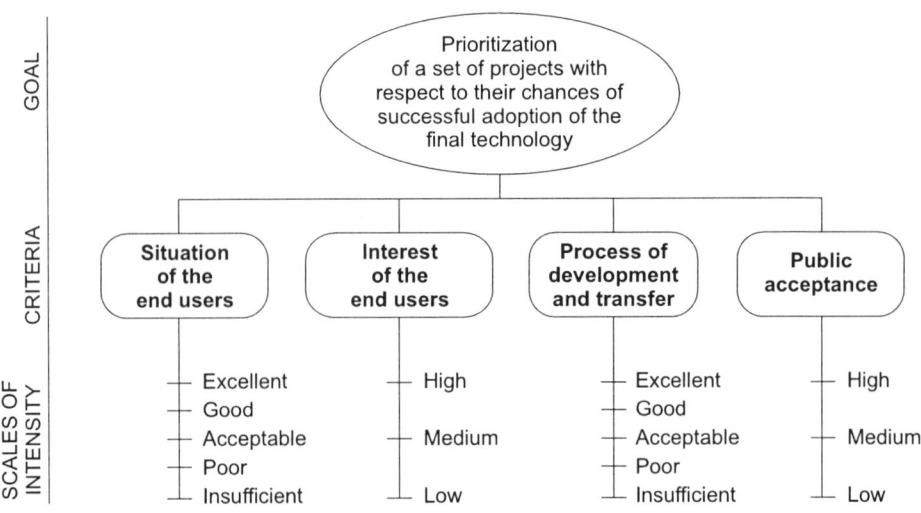

**Figure 5.5.** Hierarchy 3: Success of adoption (H2)

## Weights and priorities

Figure 5.6 displays the criteria and subcriteria weights of H1 ("potential impacts"). The criteria "economic" and "environmental" were given equal importance. The comparatively low weight of the criterion "institutional" is surprising given the strong emphasis on human and institutional capacity building of the NBP. The low weight of the subcriterion "direct project costs" is due to the fact that it only refers to the money solicited from the funding agency.

The criteria weights of H2 ("success of research") are presented in figure 5.7. The criterion "quality of the proposal" received by far the highest importance. The low significance of the criterion "characteristics of research" is based on the upgrading of "quality of the proposal" (meaning, "expected ability to do quality research"), which originally was part of this criterion. The negligible weight of "environment of research," seems rather strange because it includes the elements "infrastructure," "collaboration

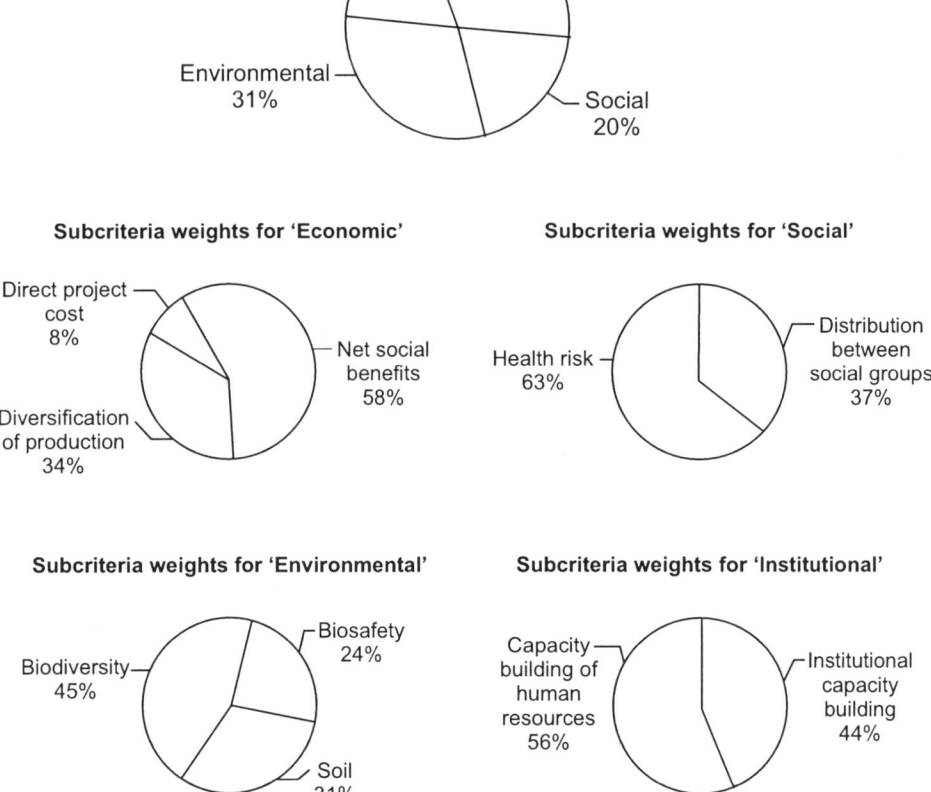

**Figure 5.6.** Criteria and subcriteria weights for the hierarchy "potential impacts" (H1)

between researchers," and "project management." One explanation may be that these elements are implicitly included in the proposal.

Finally, figure 5.8 gives the criteria weights for H3 ("success of adoption"). The criterion "interest of the end users" is the most important factor influencing the adoption process. Rather surprising is the relatively low importance of "public acceptance." One reason for this may be that scientists weighted these criteria. Another reason may be that public acceptance does not seem to be a major issue in Chile.

The values presented above are averages of the widely different weights provided by the members of the strategic group. For each of the four criteria of H1, at least one individual gave it more than half of the total weight.

The final priority of the seven projects is shown in figure 5.9. The priorities resulted from a selective multiplication of the outcomes of H1 with those of H2 and H3. "Grapes" is clearly the most preferred research project, and the set of proposals is divided into two groups. Next are "chirimoya," "tomato," and "potato," which are very close together in the ranking. "Nothofagus," "wheat," and "flowers" are at the bottom of the classification.

To test the stability of the ranking several scenarios were formulated based on the individual weighting of the criteria in H1. The average scenario and six other scenarios were tested that were based on the more extreme criteria weights obtained from individuals in the strategic group. "Grapes" remained at the top in six of the seven scenarios. In five of the seven scenarios the four top priority projects were the same. The conclusion was that the ranking was relatively insensitive to the criteria weights.

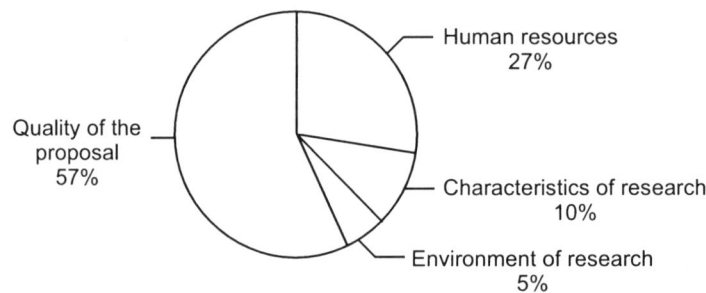

**Figure 5.7.** Criteria weights for the hierarchy "success of research" (H2)

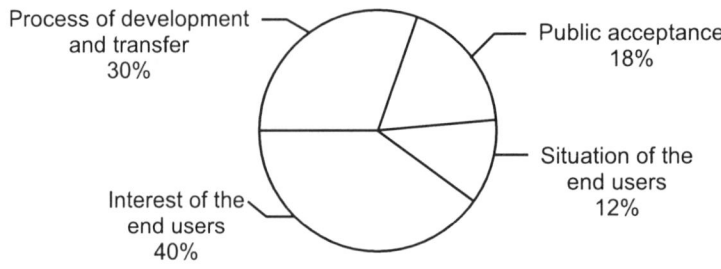

**Figure 5.8.** Criteria weights for the hierarchy "success of adoption" (H3)

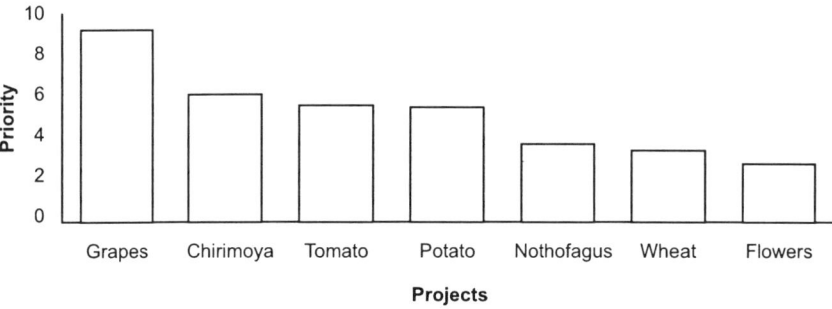

**Figure 5.9.** Ranking of the biotechnology projects

## Conclusions

The objectives of the Chilean case study—to evaluate the AHP process as a tool to support group decision making in biotechnology research for Chile's NBP—have been achieved. The excellent institutional support by INIA and the motivated participation of the individuals involved were key to the successful completion of the pilot application. We will take a brief look at how the seven management issues, mentioned in the introduction, were handled.

1. *Uncertainty.* The uncertainty caused by the forward-looking nature of ex ante evaluation was addressed by including two separate hierarchies, which allowed a detailed analysis of the chances of the projects to be successful both in terms of research and adoption by the end users.
2. *Lack of data.* The data constraint for evaluating modern biotechnology research was solved by a group decision-making approach that tapped the expertise, intuition, and experience of knowledgeable individuals. The pairwise comparisons facilitated the elicitation of judgments on the weights of the criteria and subcriteria. Group sessions generated comprehensive information for evaluating, eliminating individual biases, promoting ownership of the decision, and providing researchers with insights into the potential impact of their projects.
3. *Public acceptance.* The public perception of new biotechnology products was assessed by the subcriteria "transgenic products" and "chemical residues" under the criterion "public acceptance." Experts considered that chemical residues are more relevant for public acceptance than are transgenic products. The question is whether researchers who are involved in biotechnology research are qualified to make these assessments. It may be useful to involve other people without a bias in favor of biotechnologies.
4. *Investments.* High start-up costs for biotechnology research were not relevant in Chile because the facilities already existed. If initial investments are required, then they could be incorporated as a subcriterion under the criterion "economic."
5. *Link with other research.* The degree of integration of biotechnology into the national research system was not easy to evaluate because all projects were INIA projects. The issue of integration was implicitly treated in the subcriterion "technological challenge"— more complex biotechnologies imply larger potential difficulties for interacting with traditional research fields such as plant breeding.

6. *Legal obligations*. The importance of a legal framework for biotechnology research was captured by the subcriterion "regulation." Its low weight reflects that the legal system in Chile does not impede biotechnology developments. Biosafety legislation has only just been established.
7. *Institutional development*. NBP also focuses on capacity building. Through the criterion "institutional," the concerns for human resource development, institutional linkages, and infrastructure have been taken into account. The relevance of the criterion was confirmed by its weight of almost 20%.

## Synthesis and recommendations

Future applications of AHP should address the following issues concerning participation and costs:

The participation of both a strategic and a technical group was considered very useful. The widely divergent opinions in the strategic group revealed a need for a clear institutional policy on science and technology in general and biotechnology in particular. INIA is planning a workshop on how to address this need. AHP helped the participants determine what makes a project valuable. The participation of various stakeholder groups strengthened the ownership of the outcome of the decision. The approach provided a clear rationale for the choice. This transparency improved the communication and facilitated the acceptance of the results.

End users of the research results should be more directly involved in the process. They are important stakeholders of such decision processes and they can contribute valuable information on potential project impacts and the rate of adoption. This concerns not only farmers but also representatives from the private sector and consumer groups, as well as researchers such as plant breeders who apply the resulting technologies. It would be useful to collaborate with more specific experts in defining the social criteria, determining the respective indicators, and assessing the project's impact on these criteria. For evaluating the economic impact, more elaborate forecasting methods (not just extrapolations of the actual situation) and the use of scenarios may be justified in the light of fast economic changes and progressing globalization.

AHP can be applied as a logical exercise by an analyst, or carried out in a more extensive participatory team exercise. The costs can be kept within reasonable limits. AHP also helps keep costs low because (1) the number of criteria and subcriteria is variable and (2) criteria assessments can be based on (cheap) subjective information or on (more expensive) collected information. AHP is frequently used because managers can apply it as a quick, individual brain exercise or a tool to improve collective thinking, reasoning, and decision making. Playing with these factors has of course implications in terms of the quality of reasoning, ownership, and empirical grounding. The precise detail and participation of any specific exercise depends on the need to justify the decisions and to find a ground for implementing them. AHP is certainly not more costly than other priority setting methodologies, for which ISNAR (1998) has suggested 1.7% of the research resources to be spent as a reasonable up-front investment to improve decision making.

A key issue is to limit the time span of the participatory process. Efficiency gains can be expected from various modifications. First, the extended use of the ratings mode to elicit judgments (as opposed to the pairwise comparisons) will save considerable time. Second, a sharper focus on the most relevant decision criteria will result in a more specific data-gathering process and, at the same time, shorten the time needed for the group sessions. Third, fine-tuning the sequence of the individual steps of the procedure carries a substantial potential for a more efficient decision-making process. However, further research is required to analyze the trade-off between time saving and the quality of the decision outcome. ISNAR and ETH have initiated a follow-up project to address these issues.

## References

Dyer, R.F. and E.H. Forman. 1992. Group Decision Support with the Analytic Hierarchy Process. *Decision Support Systems* 8: 199-124.

Gil, L., C. Irarrázabal, M. Krauskopf, A Hargreaves y C. Muñoz. 1996. La Agricultura Chilena, Prioridades en la Investigición Biotecnológica en el Sector Agropecuario y Forestal. Paper prepared for *Turning Priorities into Feasible Programs*. Policy Seminar on Agricultural Biotechnology for Latin America and the Caribbean. Lima, Perú, 6-10 October 1996.

Golden, B.L., E.A. Wasil and D.E. Levy. 1989. Applications of the Analytic Hierarchy Process: A Categorized, Annotated Bibliography. *In* The Analytic Hierarchy Process: Applications and Studies, edited by B.L. Golden, E.A. Wasil and P.T. Harker. New York: Springer Verlag.

ISNAR. 1998. http://www.cgiar.org/isnar/fora/priority/PrTime.htm

Janssen, W. and A. Kissi. 1997. Planning and Priority Setting for Regional Research: A Practical Approach to Combine Natural Resource Management and Productivity Concerns. Research Management Guidelines No. 4. The Hague: International Service for National Agricultural Research.

Liberatore, M.J. 1989. A Decision Approach for R&D Project Selection. *In* The Analytic Hierarchy Process: Applications and Studies, edited by B.L. Golden, E.A. Wasil and P.T. Harker. New York: Springer Verlag.

Lockett, G., B. Hetherington, P. Yallup, M. Stratford and B. Cox. 1986. Modeling a Research Portfolio Using AHP: A Group Decision Process. *R & D Management* 16(2): 151-60.

Muñoz, C. 1998. Planificación de la Biotecnología Agrícola: Experiencia de Chile. *In Transformación de las prioridades en programas viables*. Actas del seminario de política biotecnológica agrígola para América Latina, edited by J. Komen, C. Falconi and H. Hernández. Perú, 6-10 October 1996. The Hague / Mexico, D.F.: Intermediary Biotechnology Service / CamBioTec.

Muñoz, C. 1997. Chile's experiences in planning agricultural biotechnology. *Biotechnology and Development Monitor* 31(June): 12-14.

Saaty, T.L. 1977. A Scaling Method for Priorities in Hierarchical Structures. *Journal of Mathematical Psychology* 15(3): 234-81.

Saaty, T.L. 1980. The Analytic Hierarchy Process: Planning, Priority Setting, Resource Allocation. New York: McGraw-Hill.

Saaty, T.L and L.G. Vargas. 1991. Prediction, Projection and Forecasting: Applications of the Analytic Hierarchy Process in Economics, Finance, Politics, Games and Sports. Boston: Kluwer Academic Publishers.

Shumway, C.R. 1981. Subjectivity in *Ex ante* Research Evaluation. *American Journal of Agricultural Economics* 63(1): 169-73.

Villalobos, V. 1995. La Biotecnología Vegetal en Chile. Análisis de sus Oportunidades y Limitaciones. Oficina Regional para América Latina y el Caribe. Santiago, Chile: FAO Regional Office for Latin America and the Caribbean.

Vio, M. (ed). 1995. Antecedentes y Directrices para el Programa Nacional de Biotecnología Agropecuaria y Forestal. Documento de Trabajo. Santiago, Chile: Consejo para la Innovación Agraria.

Zahedi, F. 1986. The Analytic Hierarchy Process—A Survey of the Method and Its Applications. *Interfaces* 16(4): 96-108.

# 6 Managing Biotechnology in AARD, Indonesia: Priorities, Funding, and Implementation

*Sugiono Moeljopawiro*

## Abstract

*When the Government of Indonesia recognized the potential of biotechnology to contribute to agricultural development, it faced major constraints in developing biotechnology research in the country. These included (1) inadequate infrastructure to support graduate research and education, (2) inadequate laboratory infrastructure for applied agricultural biotechnology, (3) new demands on human resource development, and (4) poor skills in managing research that created an environment for innovative research addressing national development goals. The Agency for Agricultural Research and Development (AARD) took specific steps to address these constraints. The process began in conjunction with a national priority setting exercise to develop an agricultural biotechnology strategy in Indonesia. This chapter highlights key managerial and policy steps addressing these constraints, helping to gradually evolve a strategy for biotechnology.*

## Setting the stage for biotechnology investments

In Indonesia's second long-term development plan (1994–2019), food security emerged as one of the most important issues in the national agricultural development framework. The rise in the consumption of rice is largely determined by the growth in population. With an expected population growth rate of 1.4% per year between 2000 and 2010, rice consumption is expected to grow from between 48.5 and 50.0 million tons in the year 2000 to between 50.0 and 57.5 million tons in the year 2010 (AARD 1996). The question is whether this growth can be met by expanding arable land—the total available land decreased from 66 million hectares in 1986 to 56 million hectares in 1994, despite the growing demand for agricultural products (see table 6.1).

Maintaining food self-sufficiency and improving human nutrition are therefore challenges requiring special attention. There are many physical, biological, and socioeconomic constraints to increasing agricultural productivity, including the current high population growth rate (1.9%), reduction of fertile agricultural land, outbreaks of insect pests and diseases, declining productivity, and natural disasters such as floods and drought. Although productivity in some parts of Indonesia has benefited from the use of improved technologies, low yields still persist in certain areas due to, for example, Indonesia's wide environmental diversity. This means that generic solutions for improving

© CAB *International*. 1999. *Managing Agricultural Biotechnology—Addressing Research Program Needs and Policy Implications* (ed. J.I. Cohen)

**Table 6.1.** Land Use for Agriculture in 1986 and 1994 (in '000 ha).

| Type of use | 1986 | 1994 |
|---|---|---|
| Agricultural estates | 8,910 | 13,046 |
| Dryland/gardens for crop cultivation | 12,235 | 11,245 |
| Woodland | 19,688 | 9,506 |
| Paddy fields | 7,774 | 8,439 |
| Temporarily unutilized land | 9,508 | 6,921 |
| House compounds & surroundings | 4,836 | 5,006 |
| Meadows | 2,922 | 1,893 |
| Dykes | 211 | 407 |
| Freshwater ponds | 121 | 199 |
| **Total land area** | **66,205** | **56,662** |

*Source:* CBS 1987 and 1996.

varieties and management practices are of little use and that location-specific approaches are needed that fit the widely different agricultural production systems.

Wardani and Budianto (1995) identified three major constraints to research productivity in Indonesia in both economic terms and in terms of science and technology: (1) human resources, especially research scientists, (2) research facilities, particularly well-equipped laboratories, and (3) research management that provides an environment conducive to innovative research to support national development goals. Another constraint may be information technology, which helps researchers access essential information resources (e.g., genomic databases) and helps them communicate and collaborate with advanced research programs throughout the world.

## Setting priorities for biotechnology development

In light of the potential advances that biotechnology offers, the Government of Indonesia (Ministries of Agriculture and Research and Technology) identified biotechnology research as the main tool to support national development, especially in the industrial, agricultural, and health sectors. In 1988, an assessment of all institutions engaged in biotechnology showed that the institutions were in very different stages of development. Three were selected to be the centers in a biotechnology research network: (1) University of Indonesia, Jakarta, for medical biotechnology, (2) Central Research Institute for Food Crops (CRIFC) of the Ministry of Agriculture for agricultural biotechnology (in 1993, the Center for Research and Development of Biotechnology (CRDB) of the Indonesian Institute of Sciences was assigned as the second center for agricultural biotechnology), and (3) Agency of Assessment and Application of Technology, for industrial biotechnology.

In 1989, when CRIFC's biotechnology division was established, Indonesia was recovering from economic turmoil caused by a crisis in oil prices. This meant that even though the government had designated biotechnology as a priority area, the economic situation was unfavorable to pursuing that goal. However, options for developing biotechnology continued. Attempts were made to address key questions concerning the use of agricultural biotechnology. These include the following:

- What major problems need to be solved in the country to increase agricultural productivity?
- What new products or processes are needed to solve these problems?
- Do these products and processes exist, or are they being developed elsewhere?
- If the required products or processes exist elsewhere, can they be transferred?
- If the required products and processes do not exist, do they need to be developed specifically in the country that needs them?
- If the required products and processes need to be developed, what is the most efficient way of developing them, and where? (Persley 1990).

In 1991, CRIFC, as the focal point of an agricultural biotechnology network, held a national biotechnology meeting to informally set priorities and determine problems to be solved by biotechnology. The meeting gathered information on Indonesia's biotechnology research capability and further shaped the agricultural biotechnology network. Working groups consisting of research directors and scientists engaged in biotechnology as well as conventional agricultural research were set up to (1) prioritize research on agricultural commodities on the basis of their importance relative to food security and national income, (2) identify problems associated with the production of each commodity, and (3) identify biotechnological approaches needed to overcome the problems.

Priority setting can be done formally, using the methods described by Evenson (1997), Widawsky and O'Toole (1990), and chapter 4 of this volume, or informally through a consultative process. In this case, the method used was a combination of top-down and bottom-up approaches: research directors provided information on the most important commodities, while scientists working with conventional methodologies provided information on problems encountered for each commodity. Scientists engaged in biotechnology research suggested how biotechnology might be used to overcome the problems. An informal priority setting exercise was then conducted, with table 6.2 showing the results. Asterisks in the table indicate areas of research that have now been implemented.

Setting priorities alone does not guarantee that a biotechnology program will be developed. But one of the lessons learned from the priority setting exercise was that biotechnology should focus on the products and processes needed to solve agricultural problems (AARD 1994). The priorities provided a clear basis for a strategy that identified projects. This included identifying appropriate funding and collaborators, developing human resources, and providing facilities and equipment. More recently, the market for agricultural biotechnology research products (agricultural producers, buyers, and consumers) has become an important strategic aspect to consider. A third part of the strategy was forming the strongest possible linkages between public-sector biotechnology research and the private sector as well as universities. After the priorities had been set and other elements of the strategy were taken into acccount, funding realities and options were considered.

**Table 6.2.** Priority Areas for Agricultural Biotechnology in Indonesia

| Problems | Biotechnology Solution | Expected Output |
|---|---|---|
| | **FOOD CROPS** | |
| | **1. Rice** | |
| 1. Productivity | • Cell and tissue culture, embryo rescue* and protoplast fusion | • High productivity and quality |
| 2. Insect (Stem borer) | • Recombinant DNA (use of Bt gene)* | • Resistant variety |
| 3. BLB and Blast | • RFLP markers* | • Assistance to breeding programs |
| 4. Tungro virus | • Recombinant DNA (use of viral coat protein)<br>• Monoclonal antibody | • Resistant variety<br>• Diagnostic kit |
| 5. Environmental stress | • Cell and tissue culture | • Tolerant variety |
| | **2. Soybean** | |
| 1. Productivity | • Rhizobium and mycorrhizae* | • High productivity |
| 2. Insects (pod borer and sucker) | • Recombinant DNA (use of Bt gene) | • Resistant variety |
| 3. Virus | • Recombinant DNA (use of viral coat protein)<br>• Monoclonal antibody | • Resistant variety<br>• Diagnostic kit |
| 4. Environmental stress | • Cell and tissue culture | • Tolerant variety |
| | **HORTICULTURAL CROPS** | |
| | **1. Garlic** | |
| 1. Virus | • Tissue culture for virus eradication<br>• Protoplast fusion<br>• Monoclonal antibody* | • Clean material<br>• Resistant variety<br>• Diagnostic kit |
| 2. Yield improvement | • Protoplast fusion | • Superior seeds |
| | **2. Pepper** | |
| 1. Virus (CMV) | • Control of viral genome by competitive RNA | • Vaccine application |
| 2. Lack of resistance traits for CMV | • Recombinant DNA | • Resistant variety |
| | **3. Potato** | |
| 1. Planting material | • Cell and tissue culture* | • Rapid multiplication |
| | **4. Citrus** | |
| 1. Virus (CVPD, tristeza, and others) | • Shoot tip grafting<br>• Monoclonal antibodies | • Diagnostics kit |
| | **5. Banana** | |
| 1. Stock seed | • Cell and tissue culture* | • Rapid multiplication |
| | **6. Mango** | |
| 1. Stock seed | • Cell and tissue culture | • Rapid multiplication |
| | **7. Ornamental Crop (orchid, crysant, carnation, palmae, anaceae)** | |
| 1. Seed material | • Cell and tissue culture* | • Rapid multiplication |
| 2. Virus | • Monoclonal antibody<br>• Protoplast fusion | • Diagnostic kit<br>• Resistant variety |
| 3. Fungi and bacteria | • Protoplast fusion | • Resistant variety |
| 4. Long-life flower | • Protoplast fusion | • Superior variety |
| 5. Yield improvement | • Protoplast fusion | • Superior variety |

*(continued on next page)*

*(table 6.2 continued)*

| Problems | Biotechnology Solution | Expected Output |
|---|---|---|
| | **ESTATE/INDUSTRIAL CROPS** | |
| | **1. Oil Palm** | |
| 1. Planting material not uniform | • In vitro cloning* | • Clonal planting material |
| 2. Oil quality low | • Protoplast fusion | • Better oil quality |
| 3. Parental lines highly heterozygous | • Microspore culture* | • Isogenic lines |
| | **2. Cocoa** | |
| 1. Planting material not uniform | • Cell and tissue culture* | • Uniform planting material |
| 2. Susceptibility to VSD | • In vitro of resistant cell lines | • Clones resistant to VSD |
| 3. Parental lines highly heterozygous | • Microspore culture | • Isogenic lines |
| | **3. Rubber** | |
| 1. Scion-root-stock incompatibility | • Somatic embryogenesis | • Complete plant |
| 2. Susceptibility to fungal disease | • In vitro selection of resistant cell lines* | • Clones resistant to disease |
| | **4. Coconut** | |
| 1. Planting material | • Cell and tissue culture* | • Rapid multiplication |
| | **5. Clove** | |
| 1. Planting material | • Cell and tissue culture | • Rapid multiplication |

*Source:* Brotonegoro et al. 1992.
\* Research has been implemented.

## Funding for research

The planning cycle for public research in Indonesia (see table 6.3) is related to the government's annual budget cycle. This is also a bottom-up and top-down process. Researchers propose research projects, which are evaluated by a selection team from the Agency for Agricultural Research and Development (AARD) using AARD's research priorities. The funding of agricultural research is based on the following (top-down process): (1) the total amount available at the Ministry of Finance, (2) interdepartmental priority setting done by the government as to which projects (in the ministries) receive what level of priority, (3) the budget allocated to the Department of Agriculture and to AARD, and (4) the budget allocated to each research proposal. Since few people understand that biotechnology research requires high investments, the annual budget awarded to biotechnology research has been always inadequate.

Indonesia's Sixth Five Year Development Plan (1993–98) stated that the ultimate goals of agricultural biotechnology research are to achieve and maintain self-sufficiency in food production, develop agroindustry, increase the efficiency in using biotic and abiotic resources, and increase crop and animal production. To achieve these goals and to raise awareness that research and development in biotechnology require very substantial resources, the National Development Planning Agency (Bappenas) began a survey in 1991 of the human resources and facilities available for biotechnology research in Indonesia.

The government used the data obtained through the survey to establish the Integrated Superior Research grant program, which aimed to address inefficiencies in funding, which are particularly due to poor coordination and teamwork among scientists, where research

**Table 6.3.** Research Planning Cycle at AARD

| Institution | Activity | Period |
| --- | --- | --- |
| RIFC Biotechnology | Development of proposal | April – May |
| AARD | Evaluation of proposal | May – June |
| Ministry of Research & Technology | Evaluation of proposal | June – July |
| AARD | Prioritization of proposal | July – early Aug |
| AARD | Budgeting | mid Aug |
| Agricultural Department | Evaluation of budget | end of Aug – Sep |
| National Development Planning Agency (NDPA) | Negotiation of budget | Sep – Oct |
| AARD, NDPA, Directorate General of Finance | Development of operational project | Dec – early March |

activities overlapped between institutes. The purpose of the program was also to encourage teamwork and to involve two or more institutes in a research activity. The importance of such interdisciplinary teams and research partnerships is highlighted in other chapters of this book.

The program was established to fund research in agricultural biotechnology and production technology. Scientists submit proposals for research on priority issues. In coordination with the Ministry of Research and Technology and the National Research Council, an evaluation committee assesses the relation of each proposal (using the criteria in table 6.4) to Indonesia's national development program. It then decides which proposals are awarded a grant.

After an initial screening, the committee's coordinator submits the proposals for peer review by qualified research scientists with experience in Indonesia or in countries with similar characteristics. Proposals that pass the review process are further scrutinized for

**Table 6.4.** Selection criteria for research proposal evaluation

| Selection criteria | Weight |
| --- | --- |
| Technology development | |
| 1. Contribution to technology development for commercialization | 40 |
| 2. Appropriateness of the approach used | 30 |
| 3. Probability of reaching the objectives based on the available manpower, funds and research facilities | 20 |
| 4. Institutional improvement | 10 |
| Applied science | |
| 1. Meeting the scientific prerequisite, problem identification and its solution | 35 |
| 2. Leading to science & technology development & improvement | 30 |
| 3. It has competitive advantage in solving development problem | 25 |
| 4. Institutional improvement | 10 |
| Basic science | |
| 1. Meeting the scientific prerequisite, problem identification, and its solution | 45 |
| 2. Leading to science and technology development and improvement | 30 |
| 3. It has a competitive advantage in solving development problem | 15 |
| 4. Institutional improvement | 10 |

their scientific merit and appropriate scope, time frame, pertinence to the needs of the community served by the research grants program, and the budget's limits of funding. The evaluation committee monitors progress of each research project every year. Funding is terminated for unsuccessful projects.

In order to accelerate the commercialization of products coming out of this program, a new grant system called Cooperative Research was begun in 1995. In this program, public-sector and private-sector institutions must jointly develop and submit research proposals. This is expected to speed up the production process from the laboratory level to the commercial level and begins to address the more recent needs of market-oriented research.

## Implementation, collaboration, and institutional restructuring

In the early 1990s, the only research facility with some basic equipment for biotechnology research was the Pioneering Research Laboratory for Palawija Crops. Built with assistance from the Japanese government and opened in 1988, this facility was established to do research on field crops (crops other than rice, such as soybean, maize, sorghum, peanut, and mungbean) and not specifically biotechnology research. Then CRIFC director Dr Ibrahim Manwan had it converted into a biotechnology laboratory, which was later selected as center of excellence of the agricultural biotechnology research network by the then minister for research and technology in 1990.

Calls for funds to upgrade the institute to a biotechnology center were honored by the Rice Biotechnology Network of the Rockefeller Foundation, the Applied Agricultural Research Project (AARP—funded by the United States Agency for International Development [USAID]), and the World Bank-funded Agricultural Research Management project.

However, more advanced technologies require more careful and efficient management: biotechnology research requires specialized skills and equipment, and recruitment and training must be systematically planned. To achieve this, opportunities were explored to obtain funding from international donor agencies, in addition to the limited funding received from the government.

The human resources needed to do biotechnology research were trained at CRIFC in a course from the Tissue Culture for Crop Project. This provided hands-on training exercise in tissue culture techniques for scientists and technicians from CRIFC, the Central Research Institute for Horticulture, and the Central Research Institute for Industrial Crops. The project was funded under the AARP project. By the end of the training project, some of the unspent AARP funds were relocated to "buy-in" on the Agricultural Biotechnology for Sustainable Productivity (ABSP) project, managed at Michigan State University. Table 6.5 lists the main training events, funding agencies, and the institutions conducting the training.

This kind of collaboration may not be the most appropriate way of addressing urgent national problems, but it did provide access to a training program scientist and advanced research institutes. In this way, human resources were strengthened, and national scientists could use available technology as models for more applied research in the future.

**Table 6.5.** Training Program for Human Resource Development

| Funding source | Training program | Institution |
| --- | --- | --- |
| Applied Agricultural Research Project | • Individual training in tissue culture<br>• Hand-on training in tissue culture<br>• Individual research project | Colorado State Univ.<br>CRIFC<br>CRIFC |
| Rockefeller Foundation | • Transformation<br>• Biotechnology career fellowship<br>• Transformation (postdoctoral)<br>• R&D management<br>• Degree training (MS and PhD) | Purdue University<br>Cornell University<br>Scripps Institute<br>CSIRO<br>Bogor Agric. Univ., Purdue, Cornell, and UPLB |
| Agricultural Research Management Project | • Group training in tissue culture<br>• Technician training (tissue culture) | CRIFC<br>CRIFC |
| Asian Rice Biotechnology Network (ARBN) | • Rice biotechnology<br>• Specialty training | IRRI<br>IRRI |
| Agricultural Biotechnology for Sustainable Productivity (ABSP) | • Maize transformation<br>• Potato transformation (postdoctoral)<br>• Sweet potato transformation (postdoctoral)<br>• Biosafety internships<br>• IPR internships | ICI Seeds, Iowa<br>MSU<br>MSU, Monsanto<br>MSU<br>MSU |
| Australian Centre for International Agr. Research | • Peanut transformation<br>• Pseudomonas | Prime Industry Dept. |
| ISNAR | • IBS regional seminar on planning, priorities, policies for agricultural biotechnology in Asia | ISNAR |
| Swiss Development Coop.; Official Development Assistance (Japan) | • New technologies for agricultural research | ISNAR |

## RIFCB—Identifying and addressing management challenges

In 1995, in accordance with the national development plan, biotechnology research was strengthened when AARD was reorganized. This resulted in a merger between the Biotechnology Division of CRIFC and the Bogor Research Institute for Food Crops to become the Research Institute for Food Crops Biotechnology (RIFCB). RIFCB's mandate is to (1) conduct biotechnological research for plant microbes, insect pests and diseases, (2) conduct pioneering research on food crops, (3) explore, conserve, characterize, and evaluate germplasm. RIFCB executes these functions in five divisions:

The **Genetic Resources Division** designs and conducts exploration, and it is responsible for conserving, characterizing, evaluating, and managing the database of genetic resources. It also provide information and gene sources for genetic improvement through either conventional methods or biotechnology.

The **Molecular Biology Division** designs and conducts research on molecular markers of economically important traits. It identifies, maps, and isolates genes, and it studies the structures, functions, and products of genes. It also develops gene construct and efficient transformation techniques for genetic engineering.

The **Growth and Reproduction Division** designs and conducts research on in vitro propagation through somatic embryogenesis, varietal improvement through in vitro culture, secondary metabolite production, and in vitro germplasm preservation.

The **Protein Engineering and Immunology Division** designs and conducts research on purification, characterization, and sequences of natural protein. It identifies insect pest biotypes, races/strains of diseases and microbes, and biocontrol agents. It also develops early diagnostic kits for biotic and abiotic stresses using molecular and immunological techniques. It engineers protein and antibody.

The **Microbiology and Processing Technology Division** identifies, selects, and isolates microbes for biofertilizer, bioconversion, and bioprocess/bioenzymatic. It improves the effectiveness of microbes through selection using recombinant DNA technology. And it improves microbial products for bioconversion, bioprocessing, production techniques and pilot scale application.

To address some of the weaknesses of the former components of RIFCB, management identified the following main problem areas:

1. managerial skills: needed to be strengthened
2. unskilled senior scientists: needed to be trained
3. coordination and teamwork among scientists: needed to be improved

Today RIFCB has a better infrastructure and, thanks to collaborative research and overseas training, better-trained human resources. It currently undertakes 14 research projects (see box 6.1).

---

**Box 6.1.** Current RIFCB Research Projects

1. genetic improvement of rice through the use of molecular marker, anther culture and embryo rescue
2. development of Bt microbial pesticide
3. molecular marker for drought tolerance in rice
4. genetic engineering of Rhizobium for Rhizo-plus development
5. study on Azospirillum and bacteria for biofertilizer development
6. increasing genetic variability of vanilla, pepper, and ginger through in vitro culture
7. in vitro propagation of medicinal and ornamental plants
8. DNA fingerprinting as an identification tool for brown planthopper biotypes
9. production of antagonistic microbes against rust fungi, and NVP biopesticide development
10. conservation and utilization of food crops gene bank
11. in vitro conservation of root crops germplasm
12. maize transformation for resistance to Asian corn borer
13. improvement of rice bran quality by reducing phytic acid and lypase activities through enzymatic processes
14. development of transgenic rice resistant to stemborer

Research projects were assigned to divisions in a top-down and bottom-up process. Researchers participated in allocating the research projects. Projects 1, 10, and 11 were assigned to the Genetic Resources Division; projects 3, 12, and 14 to the Molecular Biology Division; projects 6 and 7 to the Growth and Reproduction Division; projects 2, 8, and 9 to the Protein Engineering and Immunology Division; projects 4, 5, and 13 to the Microbiology and Processing Technology Division.

Agricultural biotechnology raises issues of intellectual property and biosafety that are specific to agriculture. This argues for a capacity to deal with them internally to AARD. Some more general management problems have emerged for AARD and RIFCB in particular: intellectual property protection. In Indonesia, a patent law protects new inventions, including biotechnology. This has been administered by the Directorate General of Intellectual Property Rights, Ministry of Justice. In the case of plants, however, the inventor may apply for plant variety protection with the Ministry of Agriculture. A recently established intellectual property and transfer technology office at AARD will serve as the intermediary agency to deal with business development and IPR-related technology transfer.

Indonesia established a Biosafety Commission for agricultural products developed through genetic engineering. A 1997 ministerial decree listed the provisions regarding biosafety of genetically engineered products from agricultural biotechnology. With this important regulation, the possible negative impacts of biotechnology on the environment may be minimized, and collaborative research could be carried out. In addition, a containment facility was built to carry out risk assessment and to do further risk analysis and risk management.

## Conclusion

A relatively new science in Indonesia, biotechnology can help address problems in agricultural production that conventional research cannot solve. Four activities are key to the successful application of agricultural biotechnologies in Indonesia. These are (1) the identification of priority projects, (2) the development of human resources, (3) the provision of facilities and equipment, and (4) the identification of appropriate collaborators. Identifying the problems that biotechnology is likely to solve was the first step taken in defining a strategy. National priorities should be identified and addressed through technologies identified at the research level that address the problems. After setting priorities, activities are identified as well as the resources (human, financial, and physical) needed and available to carry out those activities. In addition, a legal and institutional framework has to be established to provide an environment that is favorable to innovative research. In order to implement a strategy, it is also crucial to establish linkages with the various players of the international biotechnology system, not only to make scientific contacts but also to contribute to developing human resources, provision of facilities, and overcoming financial constraints.

## References

AARD. 1992. This is AARD. Jakarta: Agency for Agricultural Research and Development.

AARD. 1994. Strategic Plan Agency for Agricultural Research and Development 1995-2005. Jakarta: Agency for Agricultural Research and Development.

AARD. 1996. Agency for Agricultural Research and Development. Jakarta: Agency for Agricultural Research and Development.

Brotonegoro, S., J. Dharma, L. Gunarto, M. Kosim Kardin. 1992. Agricultural Biotechnology. Proceedings of a workshop on agricultural biotechnology. Bogor, Indonesia, May 21-24, 1991. Indonesia: Central Research Institute for Food Crops.

CBS. 1987. Statistical yearbook of Indonesia 1986.

CBS. 1996. Statistical yearbook of Indonesia 1995.

Evenson, R. E. 1997. Priority-Setting Methods. *In* Rice Research in Asia: Progress and Priorities, edited by R.E. Evenson, R.W. Herdt and M. Hossain. Wallingford: CAB International.

Kasryno, F. 1995. Current status of agricultural biotechnology research in Indonesia. Paper presented at the Second Conference of Biotechnology, 13-15 June, 1995, Jakarta, Indonesia.

Persley, G. J. 1990. Beyond Mendel's Garden: Biotechnology in the Service of World Agriculture. Biotechnology in Agriculture No. 1. Wallingford: CAB International.

Rogers, T. F. 1970. Small Grant Projects of the Regional Research Program: Final Report. New York: Bureau of Applied Social Research, Columbia University.

Wardani, A. and D. Budianto. 1995. Priority setting and program development in agricultural biotechnology: A perspective for Indonesia. Paper presented at the Second Conference on Agricultural Biotechnology, 13-15 June, 1995, Jakarta, Indonesia.

Widawsky, D.A. and J.C. O'Toole. 1990. Prioritizing the Rice Biotechnology Research Agenda for Eastern India. New York: The Rockefeller Foundation.

# SECTION III

# Maximizing Benefits from Resources

A national or institutional research agenda that includes biotechnology requires additional human, financial, technical, and biological resources. Mobilizing resources and maximizing access to, use of, and benefits from them entails many challenges for managers to consider. This section pays particular attention to the following:

- developing the wide range of skills required for research, product development and approval
- creating, using, and managing multidisciplinary teams
- dealing with complexities of legal and regulatory frameworks for biotechnology and biodiversity
- applying advanced applications of biotechnology to broader arrays of bioresources
- maximizing opportunities from and addressing problems in international collaborative research and linkages with commercial organizations.

The first chapter of this section ("Issues in Human Resource Management and Development") takes a preliminary look at strategic issues on developing human resources, focusing on building biotechnology capacity and competency. The chapter emphasizes the need for a comprehensive human resource development strategy that includes biotechnology and reflects the skill mixture that is needed to do multidisciplinary research. It examines key management issues for deriving targets for capacity building directly from research and development program needs, ensuring a portfolio of skills for multidisciplinary teams, and providing training for mid-career professionals. The chapter also provides examples of capacity building opportunities available from international training programs.

Chapters 8 ("Managing Bioprospecting and Biotechnology for Conservation and Sustainable Use of Biological Diversity") and 9 ("Managing Genetic Resources and Biotechnology at IRRI's Rice Genebank") explore management challenges arising from the application of biotechnology to new sources of biological and chemical diversity. Sustained advances from new technologies must build on a strong foundation of conventional research and conservation. When applying biotechnology to biodiversity, it must complement conservation needs and be implemented in the context of the sustainable use of biological diversity, including that contained in ex situ genebanks.

With regard to bioprospecting and bioresources, it is necessary to define and implement a bioprospecting framework, create multidisciplinary teams, and provide mechanisms that distribute the benefits obtained from bioproducts. Chapter 8 discusses bioprospecting and biotechnology from an institutional perspective, with specific lessons for agricultural biotechnology. It cautions that bioprospecting can only complement ongoing research and development. It is not a long-term solution for the financial needs of conservation.

In chapter 9, other factors are taken into account from the perspective of a genebank manager. These include the added value of biotechnology over other approaches, the cost of investment, the trade-off in not using biotechnology and the way in which resources were allocated over all genebank operations. Investments in biotechnology were made after basic conservation operations had been significantly upgraded.

In chapter 10 ("International Collaboration in Agricultural Biotechnology") opportunities for international collaboration are explored, including topics that managers need to pay specific attention to. International collaboration contributes directly to agricultural and social development objectives and provides hands-on experience for national research managers through collaborative research, training programs, and international technology transfer. Deriving greater benefit from this collaboration requires that national scientists and decision makers be involved early on in project planning. Projects should provide support for policy and management aspects of biotechnology as well as technical research. Finally, support for developing local research capacity should be increased to ensure impact of tangible products resulting from collaboration.

The next chapters provide experiences gained from the design and implementation of international initiatives in biotechnology. Chapter 11 ("Public- and Private-Sector Biotechnology Research and the Role of International Collaboration") focuses on how international development agencies can facilitate the application of commercial research addressing the needs of developing counties. It analyzes steps taken by a development agency to expedite private- and public-sector partnerships in a new biotechnology initiative. Commercial as well as national and university research entities from both developing and developed countries were involved equally and from the beginning in project design and priority setting.

Another case of international collaboration is described in chapter 12 ("Indo-Swiss Collaboration in Biotechnology: Lessons Learned and Future Strategies"). The chapter discusses challenges associated with managing bilateral research and capacity development. It describes the lessons learned during the project and how planning for a next phase is based on an evaluation of the first phase. Advantages are expected to be gained from using the concept of an "integrated value chain" and the main organizational changes which occur as a result of its use. The new organizational arrangements will facilitate greater sharing of administrative and managerial tasks, define clearer rules for partnerships and create a verifiable scientific monitoring system.

## Recommendations

Using resources efficiently and maximizing benefits from resources is the primary focus of this section, with the six chapters presenting individual management experiences based on personal and institutional experiences. Overall recommendations, as taken from the lessons learned and strategies presented, are summarized here:

- Initiate a comprehensive strategy for human resource development, taking into account needs for specialized training for research managers, building multidisciplinary research teams and providing for a portfolio of skills for biotechnology.

- Undertake an audit of existing resources and future needs as the foundation for the above-mentioned strategy, with particular focus on providing complementarities needed for product development.
- Ensure that a comprehensive understanding of bioprospecting and biodiversity frameworks exists among scientists and managers, including expectations and mechanisms for achieving benefits from biological and genetic resources.
- Develop applications of biotechnology, as regards ex situ genetic resource collections, that enhance existing genebank procedures and are applied only after all elements of a strong conservation program are in place.
- Ensure early participation of potential partners in design and implementation of international initiatives for biotechnology, with attention given to ensure integration of policy and managerial dimensions with collaborative research.
- Provide opportunities for private-sector participation in international initiatives to facilitate collaborative partnerships.
- Enhance bilateral collaboration by adopting an explicit product-oriented focus, for example, by adopting the "integrated value chain" concept.

# 7   Issues in Human Resource Management and Development

*Bruce W. Holloway*

## Abstract

*Biotechnology enterprises—whether they involve research organizations, government departments, the private sector, or a combination of all of these—require a wide range of skills to take novel scientific data and convert it to useful products. This chapter presents preliminary data and initial suggestions as to how human resource management and development can help build capacity for biotechnology. Human resource management requires specific planning and allocation of resources in ways that will be new and difficult, including an assessment of available staff skills, their utility for agricultural biotechnology, and accounting for the transition of national scientists to responsible biotechnology managers.*

## Introduction

Developments in genetics, microbiology, immunology, biochemistry, and reproductive biology made over the last 30 years have changed the nature of agricultural research. These changes are still occurring as biological knowledge is used increasingly to modify living organisms. Developed countries with an established research, teaching, and industrial infrastructure have embraced these scientific developments to improve their agricultural knowledge. As a result, agricultural productivity is benefiting from these prior investments.

One of the essential features of the growth of modern biotechnology in developed countries has been the existence of a general scientific infrastructure, which has provided the stock of skilled personnel required by research institutes, universities, government agencies, and industry as they became part of this biological revolution. Some countries, particularly the USA, gained experience in organizing the necessary human resources for new technologies through their work in high-technology areas such as atomic energy and aerospace.

In Europe, preparatory meetings to address skills shortages in biotechnology led to the formation of Biotechnology in Europe, Manpower, Education & Training (BEMET). The preparatory meeting identified four factors limiting the competitive development of European biotechnology (Walker 1997). There was particular concern that current European infrastructures for science were poorly developed in comparison with the USA and Japan. Particularly limiting factors included (1) decline in the quality of students at

© CAB *International*. 1999. *Managing Agricultural Biotechnology—Addressing Research Program Needs and Policy Implications* (ed. J.I. Cohen)

undergraduate level, (2) decrease in number of graduates proceeding to postdoctoral training, (3) migration from southern to northern Europe for training, and (4) variation among European countries in completing PhD degrees, inhibiting mobility of labor across Europe. BEMET facilitated a pan-European network for human resources and training issues in biotechnology. It had two initial objectives: to identify shortages in biotechnology skills in special areas and disciplines across the European Union and to investigate existing training provisions by creating an up-to-date inventory of biotechnology courses across Europe.

Few developing countries, however, have benefited from such national efforts to increase human and infrastructure resources. As regards biotechnology, developing countries are challenged as to how best to build capacity that facilitates their participation in these global developments in agriculture, allowing them to take advantage of knowledge and products from other countries.

The level of human resource development is therefore becoming an increasingly important indicator of economic success or failure. Countries not able to take advantage of new research opportunities will find that reliance on traditional agricultural practices alone limit their ability to provide sufficient food to maintain their growing population. A key factor to ensuring success of agricultural biotechnology is the management of human resources that increases the effectiveness of the acquisition of new knowledge and ensures that it is used in a focused and timely manner.

Human resource management depends on effective planning. Brush (1993) identified three inhibitors to effective human resource planning:

- weak capacity of a national agricultural research system (NARS) to commit staff to planning activities
- lack of information needed for planning
- planners losing sight of how their plans are to be used, with the result that not all plans are implemented, with loss of confidence in the planners and the process

Effective planning must take into account features particular to biotechnology. Of particular significance is that the development of products from biotechnology requires scientists and others with skills and expertise in quite different subjects. All of those involved in biotechnology must be recruited to perform not only at their individual best but also collectively in a team environment. Different members of the team will need different compensation and rewards. For example, publications, professional recognition, and patents are important to scientists, but adequate mechanisms for financial rewards must also be ensured and made part of a comprehensive human resource development (HRD) package for biotechnology research.

Within this context, this chapter explores the main issues involved in human resource management and development, and the main steps to develop effective HRD plans. It emphasizes the notion of building teams that cover the range of skills needed for both research and development, as this is a critical factor of success in biotechnology.

## Human resource management and development in biotechnology

One essential strategic aim of agricultural biotechnology is to develop research results into products that increase productivity, reduce losses due to pests or disease, or increase the range of food or fiber products. Another aim is to enhance the efficiency of research, saving time and money in achieving successful results. In developing countries, there are very few companies dedicated to producing biotechnology products as described for the first strategic aim. This is a major impediment to achieving new product development. Even though governments are capable of carrying out this role, there is little evidence from either developed or developing countries that this is done efficiently or indeed done at all. To do so, government agencies would require the same range of skills as that needed for the private sector, including a distribution and marketing system.

This situation leads to two problems: one arising from the lack of a "skill conduit[1]" in the public sector and the second from a limited base of participation in such research by the private sector. In general, an effective environment for human resource management and development is lacking: individuals with career aspirations in biotechnology are often frustrated due to the lack of a critical mass of scientists; very few institutions are able to apply biotechnology; there are few companies to develop products; and there are very limited marketing avenues available for products. One effect feeds on the other, with detrimental outcomes on the supply of the necessary human capacity.

Another problem facing all institutions involved in biotechnology, be they located in developed or developing countries, is the rate of change in the knowledge base and the increasing range of technical procedures. To keep up with these developments necessitates ongoing learning opportunities at various levels. Providing such opportunities is a challenge to all human resource management and development strategies, with the need for international linkages being critical.

Finally, there is the need to manage the transition from scientist to manager. Biotechnology involves managing the long time spans required to develop biotechnology products, the greater financial resources needed for development compared to conventional research, and the intellectual property advantages of the developed countries. All of these factors require that the portfolios of projects in agricultural biotechnology in developing countries must be tailored to very specific and achievable goals. Understanding these requirements means that scientists will also undergo a transformation to become research managers and leaders, building on skills they use when facing technical problems, by solving them in an analytical manner (Vitiello et al. 1998).

Given these challenges to developing and maintaining a conduit of skilled professionals in national agricultural research organizations, the following strategic planning tasks are suggested in order to achieve the objective of building human resource capacity for biotechnology within public agricultural research organizations.

## Capacity building

There are five key tasks in human resource management that are relevant to the challenges imposed by capacity building in biotechnology research: strategic planning, assessing needs, analyzing staff capacity, building a plan, and building research teams.

---

[1] As used in this chapter, a conduit of skills refers to the range and sequence of skills needed for scientific and technical research, product development and testing, and assurance of quality. These skills are needed in both the public and private sectors, each emphasizing a different combination.

## Strategic planning

The purpose of a strategy is to define the direction in which an organization, small or large, intends to move and a framework for reaching its objectives. Strategic planning for HRD emphasizes the fact that the ultimate source of innovation is people. This is especially important given that biotechnology projects of any value are competitive; a rival group can copy the technology, but it is much more difficult to copy the people who produced the technology.

Once an organization has formulated a strategic plan for research, it needs to plan its human resources in the context of the overall corporate plan or organizational mission (Abe and Marcotte 1990). The human resources plan addresses all programs and activities related to human resource management, including staffing, recruitment and selection, performance assessment, career development, and compensation and reward systems.

One potential outcome of a strategic planning exercise is that it justifies resources specifically for agricultural biotechnology. This may well mean deprivation in other areas. The allocation of resources begins by identifying a conduit of the skills needed in relation to the overall objectives of the institution. Once this has been done, three basic components are necessary for strategic planning for human resources:

1. identifying **all skills necessary** for success on a continuing basis, and knowing when they will be needed in the life of a project and how they relate to institutional objectives
2. a process for **auditing** the skills already present and what is needed to change them to reach the strategic objectives
3. merging **total resource management** of people, equipment, consumables, and financial provision into comprehensive institutional plans

## Assessing needs

Assessing needs is critical when beginning a strategic planning exercise to build human resource capacity for biotechnology. A preliminary assessment of management needs was one outcome of the first agricultural biotechnology policy seminar organized by the International Service for National Agricultural Research (ISNAR) in 1994. This seminar was arranged to identify and sensitize managers to the needs for the development of agricultural biotechnology in Southeast Asia (Komen et al. 1995). Participants also identified specific HRD gaps for the countries (see table 7.1) and the possible actions that should be taken to remedy them. While country and institutional needs differ in their approach to developing a skill conduit, there are common steps that together form a strategic response to the HRD needs identified. These steps are presented below.

## Analyzing staff capacity

As part of ISNAR's research on biotechnology indicators (see chapter 3 of this volume), information has been obtained about the degree levels of agricultural biotechnology researchers and the ratios of researchers to technical support staff and to managers (table 7.2). During the period covered in the various country studies, the number of researchers more than doubled in Kenya and Indonesia, with those holding a PhD degree increasing by almost three times. In Mexico and Zimbabwe, the numbers of research scientists quadrupled, with a fivefold increase in PhD holders. The growth in numbers of researchers is probably explained by the significant increase in the number of postgraduate programs in disciplines related to biotechnology:

- In 1985, **Indonesia** established the Inter-University Center for Agricultural Biotechnology at the Bogor Agricultural University, the main center for developing human resources. It also established specialized biotechnology research centers that required more scientists trained in biotechnology. In addition, the government launched a special grant program to encourage scientists to become involved in biotechnology research.

**Table 7.1.** Human Resource Aspects of Biotechnology Management in Asia

| Country | Identified gaps | Possible actions |
|---|---|---|
| Indonesia | • Insufficient human resources in particular areas (molecular biology, biochemistry, genetics) scattered, commodity oriented | • Assessment of human resources<br>• Training and exchange programs<br>• Networking |
| Malaysia | • Lack of project management skills | • Training of project managers |
| Philippines | • Project management skills for researchers and administrators | • Exposure to advanced laboratories |
| Singapore | • Need for more scientists in agro-related areas (upstream and downstream)<br>• Project management | • Enhance attractiveness of agro-industrial sector<br>• Project management training |
| Thailand | • Human resource development<br>• Personnel in policy, planning and management | • International cooperation<br>• Develop attractive career paths<br>• Management training program |
| Vietnam | • Duplication of efforts at different levels (research, development, and production)<br>• Lack of coordination across ministries<br>• Regional and international cooperation in biotechnology still limited | • Organization of annual national workshops on different subjects of biotechnology<br>• Interregional workshop on management of biotechnology programs<br>• Participation in regional and international workshops and training activities |

**Table 7.2.** Agricultural Biotechnology Research Personnel in Four Developing Countries

| | Indonesia | | Kenya | | Mexico | | Zimbabwe | |
|---|---|---|---|---|---|---|---|---|
| Researchers[1] | 1989 | 1997 | 1989 | 1996 | 1985 | 1997 | 1989 | 1998 |
| PhD | 50 | 102 | 14 | 41 | 14 | 127 | 5 | 27 |
| MSc | 28 | 93 | 12 | 15 | 12 | 49 | 5 | 31 |
| BSc | 47 | 154 | 6 | 9 | 25 | 62 | 9 | 23 |
| **Total** | 125 | 349 | 31 | 64 | 51 | 238 | 19 | 81 |
| Researcher-technical support staff ratio | 1.3 | 1.4 | 2.0 | 1.4 | 3.1 | 2.1 | 1.1 | 2.1 |
| Researcher-manager ratio | 4.4 | 4.5 | 2.2 | 3.3 | 1.6 | 2.2 | 2.2 | 7.1 |

*Source*: Falconi 1999.
[1] Researchers on leave are not included.

- Mexico conducted eight PhD and 12 Master's biotechnology programs in 1997. In addition, the Mexican government began programs that increased the number of scientists in major technological areas and provided incentives to promote scientists' career paths.
- The University of **Zimbabwe**, with donor support from Sweden and the Netherlands, founded the Master's Program in Biotechnology, the first and most relevant initiative to promote human development in agricultural biotechnology. Zimbabwe also created a Biotechnology Research Institute.

In the four countries, the ratio of technical support staff to researchers is on average one technician for every two researchers. Most of the research organizations show a low ratio of technical support staff available for research, which could affect the potential development of research outputs. A second ratio concerns the number of researchers per manager. On average there is one manager for every two researchers in Mexico and Kenya. Indonesia and Zimbabwe have almost five researchers per manager.

The policy seminars confirm that governments are keen to identify needs regarding development of national capacity for biotechnology, providing a conduit of skills. The indicator studies highlight steps that have been taken to develop staff capacity as one way of addressing these needs. In many cases, collaboration with international biotechnology programs has helped achieve this development, as will be shown later. However, comprehensive HRD for biotechnology requires training beyond academic degrees, building capacity in areas such as biosafety, regulatory mechanisms, commercial development, and management of intellectual property. Addressing these needs must also be anticipated, complementing training received for research. Specific steps have been identified to ensure greater strategic use and development of this growing capacity, as discussed below.

### *Building a plan*

Planning for HRD should aim at using human resources more effectively and efficiently and producing more satisfied and better experienced personnel at all levels. Through planning, clear objectives are set and human resources are analyzed in relation to other resources. Questions to stimulate such planning include the following:

- Can a detailed plan be produced covering the major skills and disciplines?
- Have human resource needs been assessed for different stages of the program or project?
- Is there adequate financial planning to ensure the completion of the program or project?
- What skills are lacking and how should those skills be obtained?
- Should linkages with a commercial organization be initiated?
- Should any new collaboration be initiated to gain access to equipment or techniques?

There are more detailed questions that supervisors can ask, depending on the research area, the type of organization, and other circumstances. Perhaps the most important of the above questions relates to collaborations. There are now increasing pressures to initiate both national and international collaborations, simply because any one organization does not have all the human and other resources to adequately address all the needs of a project.

Modern electronic communication has made such collaboration much more feasible, even at a distance.

### Building research teams

Finally, it is important to recognize that a traditional academic research approach may not obtain the desired outcomes in biotechnology research. While the role of the individual researcher following ideas in isolation has been a productive and dominant aspect of research over the last 50 years, no single individual has the range of knowledge or skills to bring a biotechnology project to completion. This can only be achieved by teamwork at all stages from the initial basic research through the developmental stages to the release of the final product to the agricultural community. Human resource management and development must therefore take into account early on how to build effective research teams.

## Management tasks

In terms of the central concern of this chapter—how to manage and develop human resources to build capacity for biotechnology in agricultural research organizations—three management tasks can be identified:

1. **Derive HRD targets from R&D program needs.** This reflects the need to enhance R&D planning for all resources, as human resources are only one of a number of limiting factors that need to be factored in to improve the likelihood of success for biotechnology programs and projects.
2. **Ensure a portfolio of skills in biotechnology R&D.** This is critical for reaching desired goals, but some of the skills may be difficult to obtain. Decisions to assign financial and other research resources to a particular project are often not related to the quality or quantity of human skills required. They often reflect the skills available at a particular time for a research group or institution. Setting priorities or competing for research grants where the experiences and professional qualifications of such teams are taken into account can reinforce this team approach. Measures such as quality of research, experience of the team leaders, and critical mass are factors that can and should be evaluated (see chapter 5 of this volume). These approaches will serve to strengthen the desirability of such teams and help ensure their use.
3. **Focus on the most important skills for the research team.** A lack of certain critical skills will have a disproportionately strong effect on the outcome of the organization's R&D. There are two areas of particular importance. First is the need for a *resource manager*, preferably with scientific, management, and financial training and experience. This individual would coordinate resource activities of the group, saving the time of research scientists and providing invaluable information for senior decision makers. By recognizing where resources are best used, even on a day-to-day basis, potential major impediments can be identified and resolved.

    A second important skill area is the *project manager*. While this requires special training, there are now inexpensive computer programs that make the task of the project manager easier. The main objective of this manager is to take research findings through to a product or process that impacts on agricultural production in the field. An essential task of a project manager is to ensure that the value and ownership of the research is retained by its inventors through proper handling of the intellectual property. The ability of such an individual to act as an interface between researchers, industry personnel, financial advisors, and government departments is essential.

## Addressing HRD constraints: Sri Lanka's Coconut Research Institute

Developing a critical number of researchers representing a multidisciplinary approach for biotechnology research is a major constraint in R&D centers. One example of a national institute with this problem is the Coconut Research Institute in Sri Lanka. The following limitations for HRD were identified for the institute (Jayasekara and Fernando 1998):

### Training on crop-specific biotechnology

Training of researchers as well as technical staff in advanced techniques is of utmost importance. Many R&D centers and universities in Sri Lanka are skilled in basic tissue culture methodologies required for perennial tree species. However, most R&D in developed countries focuses on annual and semi-annual crops. While tropical perennials are important locally, the lack of international resources and expertise in this area hampers specific training on coconut-related work in advanced laboratories. Trainees are often compelled to work on any available plant without crop-specific training. Crop-specific problems are therefore handled subsequent to postgraduate training, requiring long periods to conduct research and additional resources locally.

### Resource limitations for research

Applications of molecular genetics to crop improvement advance rapidly, with certain techniques becoming outdated rapidly. After receiving training in molecular biology at advanced institutions, researchers find that such applications are often unrealistic under local circumstances. National funding for research is far below investments required for setting up laboratories with modern equipment. The recurrent cost for consumables is too high for the institute without international funding support. Thus, resources are lacking to implement techniques learned abroad, leaving local scientists unable to keep abreast of biotechnology's rapid developments.

### Sustaining capacity

Recognizing the importance of integrating biotechnological methods in coconut breeding, the Coconut Research Institute began to develop its research and technical staff. This would result in a critical mass of experts that would help attain a sustainable biotechnology program. With respect to postgraduate training, negotiations with postgraduate training institutions enabled the trainees to work on coconut while using routine supplies of material for their experiments from Sri Lanka.

To keep abreast of new developments in molecular biology, a well-developed library with facilities to get connected to the Internet is now available to the researchers, and a molecular biology laboratory is under development with assistance from the Coconut Cess Fund and the Asian Development Bank. The Coconut Genetic Resource Network (COGENT) of the International Plant Genetic Resources Institute assisted the institute by providing short-term training to researchers to gain hands-on experience in molecular biology, tissue culture, and embryo-culture technologies. Further financial assistance from COGENT is available for researchers to conduct coordinated research programs and provide access to the information from neighboring counterparts.

## International activities supporting HRD for biotechnology

It takes considerable time to acquire all or most of the skills necessary in a portfolio for biotechnology. A number of activities can be initiated to create momentum while the necessary human resource structure is being built:

### International networking

Once success has been achieved at a national level, cooperating with international networks and contributing skills to their projects is easier. It may be possible to exchange personnel, apply for joint funding from international granting agencies, or gain access to unpublished databases (see chapter 10 of this volume). International collaboration is facilitated through electronic communication, with linkages between research teams and commercial organizations providing one stimulus to improve skill levels. Decision makers and senior scientists must be proactive in establishing networks as a critical element of their HRD strategy.

### Midcareer training

One of the major resources of most developing countries is the strength of their agricultural scientists. Many of these midcareer or more advanced professionals have had overseas training, and most developing-country universities have high-quality agricultural training. However, many of these individuals, both in the universities and the national agricultural services, have missed out on training in modern aspects of biology, in particular recombinant DNA technology. This affects the effectiveness of agricultural research in at least two ways: (1) failure to use restriction fragment length polymorphic markers in plant breeding programs will result in delays in getting more productive or disease-resistant cultivars into the field, (2) younger individuals benefiting from recent training in these techniques pose new challenges when they join a research team. It is difficult for senior individuals who lack such training to incorporate the new approaches into research, which means the skills of the younger team members are not used effectively.

Allocating resources for the shorter-term training of midcareer agricultural scientists in new areas of biology gives these scientists a deeper appreciation of their potential and helps them work with and encourage younger team members. Also, such training creates a better environment for establishing research linkages with other laboratories that are more experienced in the recombinant DNA procedures but may be less experienced in the more traditional plant breeding or phytopathological areas. Short courses do not produce professional molecular biologists, but they do raise the perception of the participants and serve as a platform by which additional experience can be gained. The research institutes or universities where these midcareer scientists are employed will have to contribute to the training process. They can, for example, lighten the teaching or administrative duties of these individuals so that they can devote sufficient time to expanding their own skill base.

Two examples of training opportunities are provided below, indicating how international collaboration contributes to building a conduit of skilled professionals.

### The Crawford Fund Master Class program

The Master Classes in Biotechnology supported by the Crawford Fund[2] is one scheme that focuses on upgrading midcareer scientists' skills in biotechnology. The purpose of the training is to provide participants with a better understanding and appreciation of recent developments in biotechnology, so that they can encourage and support the use of these new technologies in their own institutions.

The geographic focus for the Master Classes has been the developing countries in East and Southeast Asia. Sixteen Master Classes have been held since the program began in 1992—11 in Australia and the rest in Malaysia, Thailand, Taiwan, and the Philippines. A total of nearly 250 midcareer agricultural scientists and administrators have been trained in topics such as microbial and plant molecular genetics, beef cattle reproductive technology, biodiversity, and diagnosis of tropical plant disease.

Determining impact of such classes and other HRD options is critical both for those running such exercises and for the participants. A recently released review of the Master Class program noted the following (McWilliam et al. 1997):

1. The Classes achieved the major objective of helping midcareer professionals in developing countries to gain access to modern approaches in biotechnology.
2. The response of the clients was very positive, with demand for more frequent classes covering a wider range of aspects of agricultural biotechnology.
3. The strength of the program lies in its innovative approach to training, the importance of the target group, and the high quality of those providing the material.
4. Although the Master Class program is relatively small, its potential catalytic effect through targeting key midcareer professionals in the developing countries is quite considerable
5. The most urgent task is to strengthen the follow-up activities to ensure that the information and skills developed as a result of the classes are implemented in the participants' home countries.

### Asian Rice Biotechnology Network

Another example of international collaboration in HRD comes from the Asian Rice Biotechnology Network (ARBN). The main activities of ARBN are situated at the International Rice Research Institute (IRRI), but it also has activities in the member countries of the network, including Bangladesh, China, India, Indonesia, Pakistan, Philippines, Sri Lanka, Thailand, and Vietnam. Comprising 20 institutes, the network has provided a focus for capacity building in three key areas: institutional development, scientific skills, and project management. ARBN has established a network among the participating institutions and an administrative framework in which to initiate activities. It has also contributed to developing the infrastructure for agricultural biotechnology at a number of the participating national agricultural research institutes.

---

[2] The Crawford Fund is an organization associated with the Australian Academy of Technological Sciences and Engineering.

Scientific capacity has been built through low-cost DNA marker technology for rice, establishing transformation protocols for popular Asian rice, and facilitating technology transfer. The network established a Training and Shuttle Research Laboratory at IRRI to help with advanced steps of a particular biotechnology and conducted collaborative research on protection of rice from blast, bacterial blight, gall midge, and stem borers. For this work, ARBN has developed a specific focus on team training. The members of the teams include senior PhD scientists with breeding and pathology experience but little molecular experience, and more junior MSc and PhD candidates. Technicians are generally trained during in-country workshops and backstopping exercises.

Lessons learned from IRRI and ARBN experiences regarding HRD strategies and biotech include:

1. Develop collaborative research activities using strengths of both IRRI (in breeding, molecular biology, phenotypic analysis, biotechnology, bioinformatics, and research management) and its NARS partners (in breeding and phenotyping, e.g., pathology and entomology).
2. Create teams whose membership is as constant as possible.
3. Obtain agreement from the government for providing opportunities for regulatory training.
4. Develop national and international mini-networks for creating critical mass in key traits (e.g., pest and disease resistance).
5. Ecourage sustainability through development of skills in priority setting, proposal writing, and report writing (Bennett 1999).

## Conclusion

Agricultural biotechnology in both developed and developing countries benefits from strategic plans for managing and developing human capacity. Often, such planning for biotechnology is treated as a distinct activity or handled in an opportunistic manner, and it is not associated with human resource development for agricultural research. However, combining strategic needs for biotechnology capacity with broader capacity-building programs can mobilize talent across the broader skill base.

International organizations can play an important role in such strategies by stimulating countries to develop their own individual approaches that address capacity-building needs. This will include gathering information on the factors limiting human resource management in agricultural biotechnology, leading to efforts to integrate human resource mangement with the management of all resources necessary for agricultural biotechnology. International collaboration in biotechnology can play a vital role in this regard and should become a paramount objective for public agricultural research organizations, as well as interactions with commercial organizations if agricultural biotechnology is to achieve its potential to raise international agricultural production.

## References

Abe, L. and P. Marcotte. 1990. Strategic Planning for Human Resources in National Agricultural Research Systems. ISNAR Training Series No. 1. The Hague: International Service for National Agricultural Research.

Bennett, J. 1999. Personal communication.

Brush, E.G. 1993. Human Resource Planning. *In* Monitoring and Evaluating Agricultural Research, A Sourcebook, edited by D. Horton, P. Ballantyne, W. Peterson, B. Uribe, D. Gapasin and K. Sheridan. Wallingford: CAB International.

Falconi, C. 1999. What is agricultural biotechnology research capacity in developing countries? ISNAR Discussion Paper No. 99-10. The Hague: International Service for National Agricultural Research.

França, Z. P. 1994. Irrigation Management Training for Institutional Development: A Case Study from Malaysia. Colombo, Sri Lanka: International Irrigation Management Institute.

Jayasekara, C. and W.M.U. Fernando. 1998. Human resource development for biotechnology in the Coconut Research Institute of Sri Lanka. Paper presented at the course "Managing Biotechnology in a Time of Transition". Haikou, China, 2-13 November, 1998.

Komen, J., J.I. Cohen and S.K. Lee (eds). 1995. Turning Priorities into Feasible Programs: Proceedings of a Regional Seminar on Planning, Priorities and Policies for Agricultural Biotechnology in South East Asia. Singapore, 25–29 September 1994. The Hague / Singapore: Intermediary Biotechnology Service / Nanyang Technological University.

McWilliam, J.R., R. Sdoodee and L. Marlow. 1997. Evaluation of the Crawford Fund Master Classes in Biotechnology and a tracer study of participants, their sponsoring institutions and course providers. ACIAR Economic Evaluation Unit Working Paper Series No. 27. Canberra: Australian Centre for International Agricultural Research.

Qaim, M. and C. Falconi. 1998. Agricultural Biotechnology Research Indicators: Mexico. Discussion Paper No. 98-20. The Hague: International Service for National Agricultural Research.

Vitiello, E., R. Ciliberti and T. Ruzzon. 1998. Transformation: scientist to manager. *Journal of Commercial Biotechnology* 5 (2): 123-129.

Walker, V. 1997. Characteristics of the biotechnology industry. *In* Biotechnology Self Study. Athens, Greece: Agro-Biotech Project Training Centre.

# 8 Managing Bioprospecting and Biotechnology for Conservation and Sustainable Use of Biological Diversity

*Ana Sittenfeld and Annie Lovejoy*

## Abstract

*The biological and chemical diversity of nature represents a source of new medicines and products for agriculture and industry. Today, countries and institutions use the guidelines offered by the Convention on Biological Diversity and recent advances in biotechnology to implement bioprospecting programs that benefit the conservation and sustainable use of biological resources. Bioprospecting is a complex and long-term undertaking that requires careful management of bioprospecting frameworks, favorable macropolicies, biodiversity inventories, information management systems, technology access, and business development. It requires the creation of interdisciplinary and multidisciplinary teams of scientists, lawyers, conservation managers, and business developers. Benefits also need to be distributed in order to build biotechnology capacity and improve biological resource management and conservation. At the same time, bioprospecting is an instrument that, supported by science and technology, can improve national capacities, support economic growth, and generate financial resource for conservation activities.*

## Introduction

The search for useful products derived from biological and genetic resources coupled with innovative ways to link benefits with conservation of biodiversity and economic development is attracting attention worldwide. Using the guidelines of the Convention on Biological Diversity (UNEP 1992), modern bioprospecting[1] involves the systematic search for valuable biochemical and genetic entities from nature. Today many countries and institutions are implementing bioprospecting agreements with private and public sectors, based on the opportunities and obligations offered by the Convention on Biological Diversity and the new developments in biotechnology and molecular biology, which are

---

[1] Bioprospecting describes the systematic search for and development of new sources of chemical compounds, genes, micro- and macroorganisms, and other valuable products from nature. Bioprospecting incorporates two fundamental goals, (1) the sustainable use through biotechnology of biological resources and their conservation, and (2) the scientific and socioeconomic development of source countries and local communities (Sittenfeld 1996).

© CAB *International.* 1999. *Managing Agricultural Biotechnology—Addressing Research Program Needs and Policy Implications* (ed. J.I. Cohen)

rapidly generating new tools and bioproducts. Bioprospecting collaborations are occurring in both developing and developed countries (Sittenfeld 1996; Varley and Scott 1998).

The term biodiversity appeared in recent years to describe the rich variety of plants, animals, and microorganisms in a region. The term encompasses species diversity, genetic diversity, and ecosystem diversity (UNEP 1992). For thousands of years, biodiversity has been the source of useful compounds and materials for food, energy, shelter, medicines, and environmental services. The overall economic value of biodiversity is not known. However, a recent attempt estimated the figures received from biodiversity ecosystem services[2] at US $2.9 trillion for the entire world. From those estimates, $500 million represents charges for ecotourism, $200 million for pollination, $90 million for nitrogen fixation, and $135 million for $CO_2$ sequestration worldwide (Gordon 1998).

The pharmaceutical industry has benefited from biodiversity through drugs developed from natural compounds, while the agricultural industry improves crops by breeding them with wild relatives, usually with little returns to nature conservation and to the providers of the raw materials (Reid et al. 1993). Today, about half of the best-selling pharmaceuticals are natural or related to natural products (Demain 1998). The combined market worldwide for pharmaceuticals, agrochemicals, and seeds is over $400 billion annually, and genetic resources provide the starting material for a portion of this market (Putterman 1994; Ten Kate 1995; Thayer 1998a; James 1997).

Agricultural biodiversity provides genetic resources for domestication, an activity used largely by native populations and plant breeders, based on intraspecific genetic variability of cultivated species and agricultural practices (Ferreira, Sampaio, and de Miranda Santos 1996). The introduction of molecular markers to characterize genetic variance (Caetano-Anollés 1996), together with the possibility of introducing genetic material from other species and genera, to improve crop yields and concentration of nutrients, as well as resistance to disease or environmental conditions, are increasing the potential value of biodiversity for agriculture (Feinsilver and Chapela 1996). It is then expected that from the manipulation of DNA in agricultural research, the world will obtain most of its food, fuel, fiber, chemicals, feedstock, and even some pharmaceuticals from genetically modified plants (Abelson 1998).

Other avenues in the biotechnology, biodiversity, and agricultural continuum include the possibility of cultivating wild or transgenic plants for the production of active compounds for use as drugs or agrochemicals. This is especially true when it is not possible to fully synthesize the compound in the laboratory, or when the cost-benefit analysis favors the domestication process. The introduction of novel biological control agents is also offering new possibilities, as well as prospecting for new "exotic" sources of fruits, cereals, flowers, ornamentals, construction materials, and fine chemicals.

Collecting bioresources, extracting and testing their constituents (either chemicals or genes) for biological activity, and further developing a product is a long and expensive process, with high opportunities for failure as well as success (Reid et al. 1993). The time and cost of developing new medicines is up to 15 years and more than $360 million per product (Thayer 1998a). On the other hand, bioprospecting requires careful design and strategic planning in order to maximize nondestructive uses. At the same time it encourages the investment of benefits for the acquisition of knowledge and the improvement of

---

[2] These ecosystem services refer to services given by the ecosystem as a whole. These include, but are not limited to, $CO_2$ fixation, $O_2$ production, protection of watersheds, insect pollination of fruits and flowers, and biological control.

biodiversity conservation and management. These factors present significant and immediate challenges to countries and institutions implementing bioprospecting programs.

## Selected management challenges

Using the experiences of the National Biodiversity Institute (INBio) in Costa Rica, this chapter will concentrate on the management aspects to ensure the conservation of the biological and genetic resources in-country and the generation of new knowledge. The lessons learned with regard to management will focus on the following aspects:

- defining and implementing a bioprospecting framework, meaning favorable macropolicies, biodiversity inventories, information management systems, technology access, and business development
- creating interdisciplinary and multidisciplinary teams of scientists, lawyers, conservation managers, and business developers
- distributing the benefits obtained from bioproducts into building biotechnology capacity and improving biological resource management

The experiences summarized in this paper provide inputs based on a learning-by-doing process. Most of INBio's bioprospecting activities are concentrated on developing new pharmaceutical products. However, the basic issues and strategies can be applied to the agricultural sector as well, with modifications to suit each country's different needs and circumstances.

### INBio: An institutional view

Established in 1989, INBio is a private, nonprofit research institute. Its mission is to raise awareness of the value of biodiversity and thereby promote its conservation and improve the quality of life of Costa Rican society. The institute generates knowledge about biodiversity and disseminates and promotes the sustainable use of biological and genetic resources. Approximately 25% of the Costa Rican territory is protected in a National System of Conservation Areas (SINAC) under the administration of the Ministry of the Environment and Energy (MINAE). SINAC was designed following the global biodiversity conservation strategy, based on the three interdependent and interrelated steps of saving, knowing, and using biodiversity (WRI 1992). The implementation of the strategy provided the guidelines for the establishment of INBio.

Several of INBio's programs (including its national biodiversity inventory, bioprospecting, information management, and information dissemination and conservation program) document Costa Rica's biodiversity, where it can be found, and how the country can find sustainable, nondamaging ways to use it and to conserve it (Tamayo et al. 1997). The collaborative agreement established between INBio and MINAE provides the framework for inventory and bioprospecting activities in collaboration with SINAC. Through specific access permits, INBio collects samples for its Inventory and Biodiversity Prospecting Divisions and shares intellectual and monetary benefits with MINAE.

Bioprospecting relies on specialists that collect "prospectable" specimens (sampling that will not promote biological and genetic erosion) and make sure that nondamaging resupply is possible. They also follow biological leads, contribute to the natural history of potentially useful organisms collected and add key information to databases.

Bioprospecting is carried out in collaboration with local and international research centers, universities, and the private sector. The set of criteria used by INBio to define its research agreements includes access, equity, transfer of technology, and training. Agreements stipulate that 10% of research budgets and 50% of any future royalties be awarded to MINAE for investment in conservation. The remainder of the research budget supports in-country capacity in biotechnology and value-added activities, also oriented to conservation and the sustainable use of biodiversity.

## Management challenge no. 1: Bioprospecting framework

### Defining and implementing a bioprospecting framework

Modern bioprospecting requires functional systems based on appropriate frameworks and the cooperation and involvement of governments, intermediary institutions, private enterprise, the academic sector, and local communities and organizations (Sittenfeld and Gámez 1993; Tamayo et al. 1997). It also requires adequate macropolicies at the national and international levels and an integrated set of activities, including inventories and information management, business development, and technology access, which all together form what we call a bioprospecting framework (Sittenfeld 1996).

### Macropolicies

In general, macropolicies represent the collection of regulations, policies, and strategies, which together provide the rules of the game for bioprospecting, regarding conditions for access, distribution of benefits, and conservation of biodiversity. The Convention on Biological Diversity represents macropolicy on a global scale. Recognition of sovereign rights of states over their biological and genetic resources and transfer of technology (particularly biotechnology) to developing countries in exchange for controlled access to their biodiversity are central issues addressed by the convention.

However, international conventions represent only guidelines and still leave the major responsibilities of designing adequate legislation and regulations to each individual country. The protected status of genetic resources, coupled with clear laws regulating land and biological resource ownership, intellectual property rights (IPR), and access to resources are an incentive for collaborative research activities. Clear procedures that simplify steps needed to obtain access permits will substantially reduce risks of those interested in conducting research into and development of natural resources. They will also promote in-country partner stability and attract private industry, academic, and scientific research counterparts (Ten Kate 1995; Tamayo et al. 1997).

Costa Rica's national policies on ownership, access to, and use of biological resources represent the point of departure for INBio's bioprospecting activities. These national policies are supported by a collection of laws, by-laws, and presidential decrees directed to regulate biological resource use. The key ones are the Wildlife Conservation Law approved by the Costa Rican Congress in 1992 and the more recently sanctioned Biodiversity Law (No.7788, from April 23 1998). The latter is one of the first of its kind worldwide and contains the regulations for access, management, conservation, research, bioprospecting, and other uses.

Macropolicies will also help governments determine how access to genetic resources, while ensuring benefit sharing, is accomplished. The need for new legislation and institutions as well as designated focal points should reflect a consensus among the various actors and stakeholders. These can include governmental agencies, industry, the scientific community, ex situ conservation facilities and in situ protected areas, indigenous and local communities or their representative organizations, and relevant nongovernmental organizations, as well as private individuals (Glowka 1998). Other policies also affect the dynamics of bioprospecting. For example, in Costa Rica significant investments in education have created a pool of highly competent and skilled technicians and scientists. These resources combined with the country's natural biodiversity provide a comparative advantage for Costa Rica.

## Biodiversity inventories and information management

Bioprospecting is the search for new chemicals or genes in living organisms. It involves the development and management of biological, ecological, taxonomic, geographic distribution, and related information on living species and ecosystems. Biodiversity inventories create catalogues of available resources and their location. They prevent damage to ecosystems, areas, species, and populations by indicating what resources are available and where they can be collected without damaging the environment (Raven and Wilson 1992). Microbial gene prospecting in order to search for new gene coding for proteins does not require a previous taxonomic knowledge of the biological resource. However, if information is available this facilitates the selection of potential collection sites and the sampling process. In general, a better knowledge of the biological resource contributes to bioprospecting by transforming the source-country collaborator into a more attractive, knowledgeable, and reliable business partner (Sittenfeld 1996; Tamayo et al. 1997; Ten Kate 1995).

## Business development

INBio builds on sound biodiversity knowledge, which helps define market needs, major actors, and national scientific and technological capacities (Sittenfeld 1996; Tamayo et al. 1997). The principal markets for bioprospecting, as highlighted in this chapter, are the pharmaceutical, agricultural, and biotechnological sectors. Important requirements for bioprospecting include knowledge of national and institutional strengths and weaknesses, market surveys, and evaluation of conservation needs.

A sense of what industrial partners require is part of a continuous exploration of markets and different business partners. Private enterprises look for reliability, correct taxonomic identification, quality control and access to resources not clouded by bureaucratic procedures and in line with national and international legislation. One of the early lessons from the private sector is the need to become good business associates, especially since isolating useful compounds and genes from nature is a process of many years and may depend on a stable and long-term relationship. Business development also tries to encourage local entrepreneurs to use biodiversity sustainably as part of source-country economic development. This requires knowledge of in-country markets as well as skills and economic goals that need to be harmonized with sustainable development (Tamayo et al. 1997).

## Opportunities for business development and biotechnology

Research that links biodiversity and biotechnology offers opportunities for biodiversity-rich countries to take advantage of the natural resources in a sustainable manner and opens new research and training fields that are compatible with local needs and values.

The pharmaceutical sector represents the larger market for nature-derived compounds. Global annual sales of pharmaceuticals are about $300 billion and the market is expected to grow 6% per year (Thayer 1998a) (see table 8.1). Modern biological discoveries in this sector have changed drastically in the last years with the use of new tools such as combinatorial chemistry, high through-put screening, structure-based drug design, combinatorial biology, and genomic sciences. Bioprospecting programs must adapt to the changing environment of drug research. As an example, the combination of genomics and biotechnology gives researchers a new tool where microbial genes and products represent an important source of tools and compounds (Short 1997).

Agricultural biotechnology, specifically the search for new genes for plant improvement, offers advantages to biodiversity-rich countries compared with pharmaceutical research. Infrastructure and capital equipment costs are higher for the pharmaceutical area than for agriculture research (Tamayo et al. 1997). The need for alternatives to production and protection of crops and livestock, and the increasing capacity in biotechnology (e.g., differential gene expression techniques and genetic engineering) offer new opportunities for bioprospecting. Biotechnology can facilitate the transfer of

**Table 8.1.** Estimates of Market Size as Derived from Biological and/or Genetic Resources

| Type of product | Revenue (in US$ billion) |
|---|---|
| Pharmaceuticals (worldwide, 1998, estimated growth of 6% per year through 2001)[1] | 300.0 |
| Nutraceuticals (dietary supplements, functional and medical foods) (USA) (estimated for 1998, varies according to definition)[2] | 86.0 |
| Seeds (worldwide) (1994), including commercial seed ($15b), farm saved seed, and seed from government institutions[3] | 45.0 |
| Agbiotech crop and food products (estimates for 2010)[4] | 40.0 |
| Pesticides (worldwide) (1996)[3] | 31.2 |
| Cosmetics and toiletries (USA) (1998)[5] | 25.6 |
| Animal health products (worldwide) (1995)[3] | 14.4 |
| Top selling herbs (Europe, Asia, North America) (1996) (average growth rate 12–15% per year)[6] | 12.9 |
| Soap and detergent ingredients (USA) (1996)[7] | 9.0 |
| Herbal remedies (Europe) (1998)[8] | 8.0 |
| Enzymes (1996)[9] | 2.5 |
| 2.5% natural cosmetics[10] | 0.6 |
| Detergent enzymes (USA) (1996)[7] | 0.16 |

[1] Thayer 1998a; [2] Brower 1998; [3] James 1997; [4] Thayer 1998b; [5] Kirschner 1997; [6] Wilkinson 1998; [7] Kirschner 1998; [8] Anton and Kuballa 1998; [9] Madigan and Marrs 1997; [10] Ten Kate 1995.

several traits from wild biodiversity into crops. However, as with traditional plant breeding, there is a need to select the precise traits that consumers would reward in the market (Carter 1996). Advances in biotechnology also provide choices of diversity beyond traditional use of ex situ collections in germplasm banks. Thus, it is important to incorporate in situ collections (in the form of wild biodiversity) into agricultural research. Together with this concept, the need for developing innovative systems to connect to agricultural practices, biodiversity conservation, and intelligent use of biological resources becomes apparent.

### Technology access

Successful technology transfer depends on complex and interdependent factors. Strong scientific capacity will attract research collaboration because it reduces investment risks. These collaborations must in turn provide additional technology, training, and information to build upon that base and expand capacity to target other areas, becoming a cycle of benefits for developing countries (Sittenfeld 1996). The key to this cycle's success lies in the ability to disperse benefits to diverse sectors of society, not merely channel them into one or two areas. For example, technology transfer should not be limited to bioprospecting activities; it should support the conservation sector in order to improve infrastructure and management techniques.

## Management challenge no. 2: Transdisciplinary and multidisciplinary teams

Modern bioprospecting depends on interactions of several actors to build cooperation among the international community, governments, institutions, private enterprise, and academia. Formation of interdisciplinary and multidisciplinary teams of scientists, lawyers, conservation managers, and business developers is one way of accomplishing this (Sittenfeld 1996; Tamayo et al. 1997). Incorporating local or indigenous communities may be appropriate for some bioprospecting efforts.

Governments of biodiversity-rich countries implement national legislation and international conventions and treaties, define IPR regimes, and develop incentives for economic development. Industry and private companies convert biodiversity into products and facilitate access to and transfer of technologies.

Given that "raw" biological samples have low market value (Reid et al. 1993), bioprospecting should seek to increase value by moving beyond simple resource collection and distribution services. Research contracts should concentrate on augmenting the value of biological resources by carrying out research in the source country. Additionally, involving national academics and researchers ensures that technologies transferred or accessed remain in the developing country. Increasing value is particularly important when negotiating royalty ranges. In general, royalties for raw samples and collecting information are very low, but adding information on activity, structure, and use of compounds and genes can increase sharing of profits up to 15% or even more, depending on the area of activity and market size of the product (Reid et al. 1993; Ten Kate 1995).

Joining skills of interdisciplinary and multidisciplinary teams of scientists, lawyers, conservation managers, and business developers is best done by an agency or coordinating institution (mediating organization) that is independent from agencies managing and protecting wildlands. A new breed of scientist, the "bioprospector" works closely with biologists, chemists, and molecular biologists. The bioprospector follows biological leads, contributes to the natural history of potentially useful organisms, adds key information to databases, and makes sure that biomaterials are collected using systems that will not cause damage. Conservation managers guide and facilitate the process for better resource use in connection with bioprospectors and other scientists. Information managers organize data and other resources into adequate formats. Business developers combine the information from biological inventories, market opportunities, national scientific and biotechnological capacities with goals of conservation and economic development.

Lawyers are a fundamental part of contract negotiation. Appropriate royalty rates, milestone payments, and IPR are major business issues discussed during biodiversity prospecting contract negotiation. These issues are generally affected by previous agreements in the biotechnology and pharmaceutical sectors. Bargaining power in these negotiations is enhanced by a clear knowledge of the user's industry, the resource market, the legal precedents, as well as the understanding of conservation needs and scientific capacity.

## Management challenge no. 3: Distributing benefits

The guidelines provided by the Convention on Biological Diversity and the research experiences with different commercial and academic entities allows INBio to follow basic rules such as the fair and equitable sharing of benefits, the implementation of collection methods with reduced impact on biodiversity, technology transfer, biotechnology capacity building and up front contribution to conservation activities. Examples of INBio agreements with academic and commercial entities are described elsewhere (Sittenfeld and Villers 1993 and 1994; Sittenfeld 1996; Mateo 1996).

Assessing the benefits accrued from bioprospecting is difficult given the inherent complexities of assigning value to the accumulated knowledge of biodiversity, the transfer of know-how and technology, and the enhanced capacity building. Up to this moment products obtained from samples processed by INBio have not reached the marketplace. In 1992–97 INBio has conducted bioprospecting agreements worth over $6 million: $3.5 for investments and research expenses at INBio and $2.5 million for MINAE, the conservation areas, and the universities. It is important to take into consideration that the latter figure is significant for a small country like Costa Rica, with a GDP of $9 billion for 1996 (Proyecto Estado de la Nación 1997). It is significant that MINAE has used its share to support the management and upkeep of Costa Rica's National Park at Coco Island, a unique site, and a clear-cut example of direct bioprospecting benefits flowing to conservation (Mateo 1996).

## Conclusions and management implications for the future

Bioprospecting is a complex and long-term undertaking. Supportive macropolicies tied to an integrated program for biological research, business development, and technology transfer can link biodiversity and biological resource users in modern bioprospecting programs. This new breed of research and development ultimately depends on implementation of adequate frameworks, good coordination, multisectoral collaboration, and transfer of benefits to conserve biodiversity and increase capacity in biotechnology.

The INBio experience in Costa Rica illustrates one particular framework for biodiversity prospecting. As a pilot project, it cannot be considered a model for all nations, but the experiences and results obtained provide applicable concepts for other countries, needs, and circumstances.

The path for future management will not be easy. INBio faces a number of challenges as it moves forward to catalog, investigate, use, and contribute to conserving Costa Rica's biodiversity. There is a clear need to establish a solid funding base for its various activities by means of a strong endowment fund. The presence of a small biotech private sector in the country, the lack of sufficient bioprospecting expertise in the area of agriculture, and the need to be competitive in an environment of globalization of economies and rapid advances of biotechnology represent important challenges for the institution to keep pace and make progress in bioprospecting.

Finally, the economic impact of bioprospecting should not be overestimated. Bioprospecting can only complement other activities designed to advance human development and therefore cannot solve conservation and development issues in and by itself. It is an instrument that, complemented by science and technology, can work together with other tools to improve national capacities, support economic growth and generate financial income to conservation activities.

## Acknowledgments

The authors gratefully acknowledge Nicolás Mateo, Manuel Ruiz, and Reynaldo E. de la Cruz for their advice, critical reading, and major improvements of the manuscript. This work is supported in part by grants No. 5U01 TW/ CA00312 from the National Institute of Health (Forgarty International Center) and No. 801-96-582 from Vicerrectoría de Investigación, Universidad de Costa Rica.

## References

Abelson, P.H. 1998. A Third Technological Revolution. *Science* 279: 2019.
Anton, R. and B. Kuballa. 1998. Status of Phytopharmaceuticals within the European Market. *In* Phytomedicines in Europe. Chemistry and Biological Activity, edited by L.D. Lawson and R. Bauer. ACS Symposium Series 691. Washington, D.C.: American Chemical Society.
Brower, V. 1998. Nutraceuticals: Poised for a healthy slice of the healthcare market. *Nature Biotechnology* 16: 728-731.
Caetano-Anollés, G. 1996. Scanning of nucleic acids by in vitro amplification: New developments and applications. *Nature Biotechnology* 14: 1668-1674.
Carter, M.H. 1996. Monsanto Company: Licensing 21st Century Technology. Harvard Business School Case No.596-034. Boston, MA: Harvard Business School.
Demain, A.L. 1998. Microbial natural products: alive and well in 1998. *Nature Biotechnology* 16: 3-4.
Feinsilver, J.M. and I. H. Chapela. 1996. Summary, Concluding Remarks, and Policy Options. *In* Biodiversity, Biotechnology and Sustainable Development in Health and Agriculture: emerging connections. PAHO Publication No.560. Washington, D.C.: Pan American Health Organization.
Ferreira, M.E., M.J. Amstalden Sampaio and M. de Miranda Santos. 1996. Biodiversity and Biotechnology: Approaches to Sustainable Agricultural Development. *In* Biodiversity, Biotechnology and Sustainable Development in Health and Agriculture: emerging connections. PAHO Publication No.560. Washington, D.C.: Pan American Health Organization.

WRI. 1992. Global Biodiversity Strategy: Guidelines for Action to Save, Study, and Use Earth's Biotic Wealth Sustainably and Equitably. Washington, D.C.: World Resources Institute.

Glowka, L. 1998. A Guide to Designing Legal Frameworks to Determine Access to Genetic Resources. Bonn: World Conservation Union.

Gordon, J. 1998. In brief. *Scientific American* 278: 22.

James, C. 1997. Progressing Public-Private Sector Partnerships in International Agricultural Research and Development. ISAAA Briefs No. 4. Ithaca, NY: International Service for the Acquisition of Agri-biotech Applications.

Kirschner, E.M. 1997. Boomer's quest for agelessness. *Chemical & Engineering News* 75(9): 19-25.

Kirschner, E.M. 1998. Soaps and Dertergents. *Chemical and Engineering News* 76 (4): 39-54.

Madigan, M.T. and B.I. Marrs.1997. Extremophiles. *Scientific American* 276: 82-87.

Mateo, N. 1996. Wild Biodiversity: The Last Frontier? The Case of Costa Rica. *In* The Globalization of Science: The Place of Agricultural Research, edited by C. Bonte-Friedheim and K. Sheridan. The Hague: International Service for National Agricultural Research.

Pace, N. 1997. A molecular view of microbial diversity and the biosphere. *Science* 276: 734-740.

Putterman, D. M. 1994. Trade and the biodiversity convention. *Nature* 371:553-4.

Raven, P. and E.O. Wilson. 1992. A Fifty Year Plan for Biodiversity Surveys. *Science* 258: 1099-1100.

Reid, W.V., S. Laird, C.A. Meyer, R. Gámez, A. Sittenfeld, D.H. Janzen, M.A. Gollin, and C. Juma. 1993. Biodiversity Prospecting: Using Genetic Resources for Sustainable Development. Washington, D.C.: World Resources Institute.

Short, J.M. 1997. Recombinant approaches for accessing biodiversity. *Nature Biotechnology* 15: 1322-1323.

Sittenfeld, A. and R. Gamez. 1993. Biodiversity Prospecting by INBio. *In* Reid et al.

Sittenfeld, A. and R. Villers. 1993. Exploring and preserving biodiversity in the tropics: the Costa Rican Case. *Current Opinion in Biotechnology* 4(3): 280-285.

Sittenfeld, A. and R. Villers. 1994. Costa Rica's INBio: Collaborative Biodiversity Research Agreements with the Pharmaceutical Industry. *In* Principles of Conservation Biology, edited by G.K. Meffe and C.R. Carroll. Sunderland, MA: Sinauer Associates.

Sittenfeld, A. 1996. Issues and strategies for bioprospecting. *Genetic Engineering and Biotechnology Monitor* 4: 1-12.

Tamayo, G., W.F. Nader and A. Sittenfeld. 1997. Biodiversity for the Bioindustries. *In* Biotechnology and Plant Genetic Resources: Conservation and use, edited by B.V. Ford-Lloyd, H.J. Newbury and J.A. Callow. Wallingford: CAB International.

Ten Kate, K. 1995. Biopiracy or green petroleum? Expectations and Best Practice in Bioprospecting. London: Overseas Development Administration.

Thayer, A.M. 1998a. Pharmaceuticals: Redesigning R&D. *Chemical & Engineering News* 76 (8): 25-37.

Thayer, A.M. 1998b. Living and loving life sciences. *Chemical & Engineering News* 76(47): 17-24.

UNEP. 1992. Convention on Biological Diversity. Nairobi: United Nations Environment Programme.

Varley, J.D. and P.T. Scott. 1998. Conservation of Microbial Diversity: A Yellowstone Priority. *ASM News* 64: 147-151.

Wilkinson, J.A. 1998. Data presented at IBC Fifth Annual Conference, Functional Food 1998. Copthorne Tara Hotel, London, 7-8 September, 1998.

# 9 Managing Genetic Resources and Biotechnology at IRRI's Rice Genebank

*Michael T. Jackson*

## Abstract

*Ex situ conservation in genebanks is a safe and cost-efficient method of preserving the genetic diversity of crops and their wild relatives, particularly for species whose seeds can tolerate desiccation and storage at low temperature. Biotechnology is particularly useful for in vitro culture, cryopreservation, and disease elimination in vegetatively propagated crops. The use of molecular markers to study genetic diversity, identify duplicate accessions, and increase utilization by more efficient screening of germplasm are recent developments. Management of biotechnology in a genebank depends on the value of biotechnology over other approaches, the cost of investment, the trade-offs in not using biotechnology, and the resource allocation prioritization over all genebank activities and operations. The principal applications of biotechnology in the International Rice Genebank of the International Rice Research Institute (IRRI) are in vitro culture of seedlings and the study of genetic diversity using a range of molecular markers. Investment in biotechnology was made only after conservation operations per se had been upgraded.*

## Introduction

For thousands of years, farmers worldwide have been cultivating many different crops. The combined effects of adaptation to different environments, the breakdown of reproductive isolation between domesticated species and their wild relatives, and selection by farmers over many generations led to a multiplicity of varieties, each with particular traits valued by the communities that developed them. These are the genetic resources of the agricultural crops that sustain the world's growing population, and the genetic building blocks for more productive crop varieties (Ford-Lloyd and Jackson 1986). They are the source of traits to transfer to commercial varieties through conventional breeding techniques or through genetic transformation.

In a broad sense, the genetic resources of a crop include not only the varieties developed by farmers in indigenous farming systems and maintained by them for generations (often referred to as traditional, landrace, or farmers' varieties) and the related wild species, but also modern commercial varieties, obsolete varieties, breeding lines, and genetic stocks. However, genebanks usually give priority to the conservation of the landrace varieties and wild species. Ex situ conservation is a safe and cost-effective method

© CAB *International*. 1999. *Managing Agricultural Biotechnology—Addressing Research Program Needs and Policy Implications* (ed. J.I. Cohen)

of preserving the genetic diversity of crops and their wild relatives, particularly for species whose "orthodox" seeds can tolerate desiccation and storage at low temperature. The long-term safety and integrity of genetic resources—seeds, living plants, cuttings, tissue cultures—are its primary goals.

An additional advantage of ex situ conservation in genebanks is the easy access to germplasm by breeders and researchers who wish to use these sources of genetic diversity in crop improvement programs, or to understand their reaction to biotic and abiotic stresses such as pests and diseases or drought, for example. Increasingly, the molecular basis of traits is being studied, which should facilitate their transfer to commercial varieties through genetic engineering.

Plant genetic resources are among the most vulnerable of all nonrenewable natural resources—once lost, they are lost forever. That is why, for several decades, there have been concerted international efforts to collect and conserve plant genetic resources for food and agriculture in genebanks worldwide. These efforts culminated in June 1996 during the Fourth International Technical Conference on Plant Genetic Resources in the adoption by 150 countries of a Global Plan of Action for the Conservation and Sustainable Utilization of Plant Genetic Resources for Food and Agriculture (FAO 1996). The framework of the plan is to ensure the long-term preservation of genetic resources at national, regional, and international levels, as well as the necessary actions to facilitate use of these valuable resources for the benefit of all humans. Biotechnology is recognized as an important component of implementing the global plan.

## Biotechnology and ex situ conservation

A genebank has several functions, including (1) collection and acquisition of germplasm, (2) the long-term conservation of germplasm, including multiplication and regeneration in whatever is the most convenient and accessible form (such as seeds, in vitro cultures, and living plants), (3) germplasm characterization and evaluation, (4) data management, (5) germplasm exchange, and (6) promotion of germplasm use to enhance crop productivity. There are many different applications of biotechnology that are useful in this respect, but the relative importance of different techniques depends on the particular characteristics of a specific crop and its wild relatives (Callow et al. 1997). For example, in vitro culture of explants is essential for plant species that produce so-called "recalcitrant" seeds that cannot be stored at low moisture content and low temperature. Long-term cryopreservation of vegetative propagules or culture in slow-growth culture media are biotechnology options that must be explored for species difficult to conserve as seeds. Furthermore, tissue culture methods are widely applied for elimination of systemic diseases such as viruses. Engelmann (1997) provides a comprehensive review of in vitro conservation methods.

Recent developments in the area of molecular biology hold the promise of more efficient management and study and exploitation of genetic resources in ways that could not be imagined only a few years ago. These include molecular technologies to assess and monitor biodiversity, facilitate critical decisions on what should be conserved, or increase utilization through more efficient screening of germplasm (Barlow and Tzotsos 1995). In addition, molecular markers will certainly be used to define core collections within genebanks (Gepts 1995). It is perhaps in this molecular area of biotechnology more than any other that critical management decisions must be taken. Applications of molecular biology are certainly in vogue, but that does not mean that many aspects are appropriate yet for all genebanks.

## Managing biotechnology for ex situ conservation

Even though biotechnology has been used effectively for many years to ensure the safe conservation of plant genetic resources in genebanks, several management questions should be addressed before investing heavily in biotechnology:

- Does biotechnology enhance the access to and the management, conservation, and use of genetic resources?
- What alternatives to biotechnology can be used?
- What are the resource implications—human, equipment, or budget—to sustain applications of biotechnology in a genebank?
- What are the trade-offs for not investing in biotechnology?
- Will investment in biotechnology affect resource allocation to other areas of genetic conservation essential for the long-term security of a germplasm collection?

Conservation priorities should shape the strategy for adopting and using biotechnology rather than finding a use for biotechnology under any circumstances. The needs of a genebank with only base collection responsibilities may be different from one that has both active and base collections, distributes germplasm to users, or has a program of germplasm research. Most genetic conservation programs operate with limited financial support. The prioritization of resource allocation across all activities and operations is an essential step to integrate biotechnology successfully into the overall work plan of the genebank. Quite often, different biotechnology tools will be adopted because they are in vogue rather than contributing specifically to the more efficient conservation or exploitation of germplasm. For example, there is the commonly held perception that molecular biology, and particularly molecular markers, will automatically facilitate the development of a core collection or that such markers will help identify traits and their exploitation. Refining these techniques for successful and routine use with diverse germplasm takes time and considerable investment.

## Managing biotechnology and rice genetic resources at IRRI

Located in Los Baños, the Philippines, the International Rice Research Institute (IRRI) holds in trust the world's largest and most genetically diverse collection of rice genetic resources in its International Rice Genebank, which is managed as part of the institute's Genetic Resources Center (Jackson 1997; Jackson et al. 1997). In 1994 the collection was placed under the auspices of the Food and Agriculture Organization of the United Nations (FAO) in the International Network of Ex Situ Collections. Under the agreement with FAO, a material transfer agreement is used to facilitate access to and use of the conserved germplasm. This prohibits IRRI or any other recipient from seeking intellectual property rights (IPR) on the germplasm directly.

The genebank currently maintains a collection of more than 102,700 samples of Asian rice *Oryza sativa* (95%), West African rice *O. glaberrima* (1.5%), and all 21 wild species (3.5%). Since 1991 the infrastructure and operations of the genebank have been upgraded, for example by adding a seed drying room. All these changes were aimed at meeting international genebank standards (FAO/IPGRI 1994), while at the same time increasing the quality of conserved germplasm (defined in terms of seed viability and potential storage longevity). **Our first priority was to ensure the long-term conservation of this strategically important germplasm collection.** This has been achieved by exploiting the seed production environment in Los Baños to achieve maximum seed longevity in storage for all the diverse rice accessions (Ellis et al. 1993; Kameswara Rao and Jackson 1996a, 1996b, 1996c, 1997).

The decision to use biotechnology in various forms for managing and studying the rice germplasm collection was not taken lightly. Our assessment was guided by the need to ensure the safety of the germplasm *per se* and to employ new molecular technologies that would facilitate better understanding of the underlying genetic structure of the collection. IRRI has made considerable investment in biotechnology to support its rice improvement activities, especially through transgenesis and marker-assisted selection. In assessing the molecular marker systems available, we had to determine what level of investment would be appropriate for genetic resource purposes in terms of the overall recurrent costs and infrastructure development in the Genetic Resources Center, as well as safety considerations. Additionally, we felt that rather than using state-of-the-art technology that our partners in the national agricultural research systems (NARS) would not be able to adopt, we should use biotechnology approaches that they might feasibly develop in the foreseeable future.

We were fortunate to establish collaboration with the School of Biological Sciences at the University of Birmingham in the UK in 1993. With funding from the Department for International Development (DfID—formerly the Overseas Development Administration), a research project was initiated to study the diversity of rice germplasm using molecular markers. One of our staff was trained at Birmingham, which put us in a better position to decide what was needed in terms of molecular studies.

As relative costs of molecular techniques fell and their value for the study of germplasm collections was proven, it became clear that we should take the opportunity of adding these to the suite of characterization and evaluation approaches already being used in the genebank. Otherwise, we felt the genebank would be locked in "traditional" approaches and would not take advantage of new technologies in which others had already made the necessary research development investments. We could not hire new staff for this endeavor, but we redeployed existing staff, who were given additional, appropriate training. Biotechnology for genetic resources is supported from the annual budget of the Genetic Resources Center, provided by IRRI from its core budget. We are seeking additional donor support to expand our molecular studies.

By considering these factors and options for the use of biotechnology, we decided that the principal genebank applications of biotechnology would be in vitro culture of seedlings and the study of genetic diversity using a range of biochemical and molecular markers.

## *In vitro culture*

In vitro culture is used to ensure the survival of seed lots with low viability. Seeds may have low viability when they are sent to our genebank for long-term conservation. Sometimes, only a few seeds are sent for conservation purposes. In a 1992 monitoring survey of the International Rice Genebank collection, the viability of about 300 samples (mainly japonica rices) fell into this category. Since such accessions must be multiplied to provide the 500 g necessary for conservation in the active collection and the 120 g for the base collection, it is unwise to plant these seeds directly in the field.

In vitro culture involves germination of hulled seeds (i.e., with the lemma and palea removed) on nutrient agar containing Murashige and Skoog medium (Murashige and Skoog 1962), and a period of growth in culture solution in a phytotron (Yoshida et al. 1976) until vigorous plants are obtained. These plants can then be transplanted to a screenhouse and given more care than is possible in field plots.

### Isozyme electrophoresis

A classification of *O. sativa* varieties into six groups based on the allelic variation at 21 polymorphic loci coding for 14 isozymes (Glaszmann 1987) is an important tool for rice germplasm management, although varieties can be separated quickly using only five loci for two isozyme systems (phosphoglucose isomerase and aminopeptidase). Groups I and VI correspond to the indica and japonica rices, respectively. Furthermore, the javanica rices are included in group VI with the japonica rices. Consequently, they have been renamed "tropical japonicas" and, based on this classification, were selected as germplasm for the development of the so-called "new plant type" (Khush 1995). The remaining groups II to V represent indica varieties found only in the Indian subcontinent, especially in the foothills of the Himalayas, like the floating *rayada* varieties of Bangladesh (group IV) and the *basmati* rices of northern India, Pakistan, and Nepal, prized for their aromatic flavor (group V). Since the crossability barriers between indica and japonica rices affect their utilization in rice breeding, correct identification of these varieties is extremely important. The development of isozyme classification provides an unequivocal biological framework for the use and analysis of diversity patterns of germplasm based on other molecular markers.

### DNA markers

DNA markers such as RFLP, AFLP, RAPD, and SSR are routinely used for the management and evaluation of crop germplasm collections (Westman and Kresovich 1997) for three principal purposes. First, molecular markers may be used to answer so-called forensic questions such as whether two samples are genetically the same. Second, there are questions of location and diagnostics, where the objective is to determine the presence or location of a particular allele or nucleotide sequence, be that in all species and accessions in the genebank, or a population, particularly those related to desirable traits. Such questions are important for monitoring the genetic health and changes of a genebank sample over time and as a consequence of various regeneration procedures. Finally, Westman and Kresovich (1997) highlight the questions of relatedness, of genetic diversity *per se*, and how diversity is distributed in individuals, populations, and species. Such information is useful, perhaps necessary, for adequately targeting areas for germplasm collecting, or designing in situ programs, complementary to ex situ conservation in genebanks. Until relatively recently, such techniques were beyond the means of most genetic conservation programs and may remain for some time to come beyond the immediate resource allocation of many.

In the International Rice Genebank we began using molecular markers after several different approaches had been validated in the joint project with the University of Birmingham (Ford-Lloyd et al. 1997). This collaboration permitted the genebank to take advantage of expertise elsewhere to evaluate various protocols for rice germplasm while facilities for molecular biology were developed and personnel trained.

Initially, our emphasis was on RAPD to understand the diversity in rice landraces (Virk et al. 1995a) and the identification of duplicate accessions (Virk et al. 1995b), but more recently, we started using AFLP. We also established the association between RAPD markers and quantitative variation and were able to predict the performance of rice accessions in the field in Los Baños based on RAPD markers (Virk et al. 1996). We have also used RAPD for taxonomic studies of wild rices, particularly the South American species *O. glumaepatula* (Martin et al. 1997). The results correspond well with taxonomic studies of this species based on morphology (Juliano et al. 1998). AFLP analyses of rice germplasm seem more robust (Zhu et al. 1998) than those based on RAPD, but both correlate well with the isozyme groups referred to earlier. The advantage of both RAPD and AFLP markers is their broad distribution across the rice genome, based on data from a wide diversity of rice varieties. We have been able to use the isozyme classification to validate those based on RAPD and AFLP markers. This gives us confidence that it is not necessary to use only mapped markers, whose distribution and "information" content may be a reflection only of the genetic distance of the original parents of a mapping population (Virk et al. 1999). Choice of markers is important because their position in the genome does affect the analysis of diversity patterns (Parsons et al. 1997).

## Conclusions and management implications

The first priority of a genebank is to ensure the long-term conservation of germplasm with which it has been entrusted. Investments in biotechnology must be made at a level that is consistent with the overall budget and mandate of the genebank and that can be sustained. The development of a core collection using molecular markers is often cited as one activity that many genebanks should initiate. **It is essential that the basic elements of a strong conservation program are in place before taking decisions on developing capability within a genebank to use biotechnology.** Otherwise, the added benefits that biotechnology can bring may not be realized, or germplasm may not be readily available if it has not received proper care in the genebank. We believe that this cautious approach is appropriate for many genebanks where resources are limited.

In the future, molecular analysis of germplasm collections will permit more efficient utilization of wild species in rice breeding (Tanksley and McCouch 1997), and the synteny between cereal genomes (Devos and Gale 1997) presents opportunities to exploit molecular data from one species to search for traits in another. But these are not approaches for a single genebank alone. They will require strong collaboration between different genebanks and molecular biologists worldwide. The investment and resource implications are too great for any one institute alone. Nevertheless, it is necessary to grasp such opportunities in ways that are innovative and that do not compromise the principal purpose of germplasm conservation.

## References

Barlow, B. and G.T. Tzotsos. 1995. Biotechnology. *In* Global Biodiversity Assessment, edited by V. H. Heywood and K. Gardner. Cambridge: Cambridge University Press.

Callow, J.A., B.V. Ford-Lloyd and H. J. Newbury. 1997. Overview. *In* Biotechnology and Plant Genetic Resources: Conservation and Use, edited by J. A. Callow, B. V. Ford-Lloyd and H. J. Newbury. Wallingford: CAB International.

Devos, K.M. and M. D. Gale. 1997. Comparative genetics in the grasses. *Plant Molecular Biology* 35: 3-15.

Ellis, R.H., T.D. Hong and M.T. Jackson. 1993. Seed production environment, time of harvest, and the potential longevity of seeds of three cultivars of rice (*Oryza sativa* L.). *Annals of Botany* 72: 583-590.

Engelmann, F. 1997. *In vitro* conservation methods. *In* Biotechnology and Plant Genetic Resources: Conservation and Use, edited by J.A. Callow, B.V. Ford-Lloyd and H.J. Newbury. Wallingford: CAB International.

FAO. 1996. Global Plan of Action for the Conservation and Sustainable Utilization of Plant Genetic Resources for Food and Agriculture. Rome: Food and Agriculture Organization of the United Nations.

FAO/IPGRI. 1994. Genebank standards. Rome: Food and Agriculture Organization of the United Nations and International Plant Genetic Resources Institute.

Ford-Lloyd, B. and M. Jackson. 1986. Plant Genetic Resources: An Introduction to their Conservation and Use. London: Edward Arnold.

Ford-Lloyd, B.V., M.T. Jackson and H.J. Newbury. 1997. Molecular markers and the management of genetic resources in seed genebanks: a case study of rice. *In* Biotechnology and Plant Genetic Resources: Conservation and Use, edited by J.A. Callow, B.V. Ford-Lloyd and H.J. Newbury. Wallingford: CAB International.

Gepts, P. 1995. Genetic markers and core collections. *In* Core Collections of Plant Genetic Resources, edited by T. Hodgkin, A.H.D. Brown, Th.J.L. van Hintum and E.A.V. Morales. Chichester: John Wiley and Sons.

Glaszmann, J.C. 1987. Isozymes and classification of Asian rice varieties. *Theoretical and Applied Genetics* 74: 21-30.

Jackson, M.T. 1997. Conservation of rice genetic resources—the role of the International Rice Genebank at IRRI. *Plant Molecular Biology* 35: 61-67.

Jackson, M.T., G.C. Loresto, S.Appa Rao, M. Jones, E. Guimaraes and N.Q. Ng. 1997. Rice. *In* Biodiversity in Trust: Conservation and Use of Plant Genetic Resources in CGIAR Centres, edited by D. Fuccillo, L. Sears and P. Stapleton. Cambridge: Cambridge University Press.

Juliano, A.B., M.E.B. Naredo and M.T. Jackson. 1998. Taxonomic status of *Oryza glumaepatula* Steud. I. Comparative morphological studies of New World diploids and Asian AA genome species. *Genetic Resources and Crop Evolution* 45: 197-203.

Kameswara Rao, N. and M.T. Jackson. 1996a. Effect of sowing date and harvest time on longevity of rice seeds. *Seed Science Research* 7: 13-20.

Kameswara Rao, N. and M.T. Jackson. 1996b. Seed longevity of rice cultivars and strategies for their conservation in genebanks. *Annals of Botany* 77: 251-260.

Kameswara Rao, N. and M.T. Jackson. 1996c. Seed production environment and storage longevity of japonica rices (*Oryza sativa* L.). *Seed Science Research* 6: 17-21.

Kameswara Rao, N. and M.T. Jackson. 1997. Variation in seed longevity of rice cultivars belonging to different isozyme groups. *Genetic Resources and Crop Evolution* 44: 159-164.

Khush, G.S. 1995. Breaking the yield frontier of rice. *GeoJournal* 35: 329-332.

Martin, C., A. Juliano, H.J. Newbury, B.R. Lu, M.T. Jackson and B.V. Ford-Lloyd. 1997. The use of RAPD markers to facilitate the identification of *Oryza* species within a germplasm collection. *Genetic Resources and Crop Evolution* 44: 175-183.

Murashige, T. and F. Skoog. 1962. A revised medium for rapid growth and bioassays with tobacco tissue cultures. *Physiologia Plantarum* 15: 473-497.

Parsons, B.J., H.J. Newbury, M.T. Jackson and B.V. Ford-Lloyd. 1997. Contrasting genetic diversity relationships are revealed in rice (*Oryza sativa* L.) using different marker types. *Molecular Breeding* 3: 115-125.

Tanksley. S.D. and S.R. McCouch. 1997. Seed banks and molecular maps: unlocking genetic potential from the wild. *Science* 277: 1063-1066.

Virk, P.S., B.V. Ford-Lloyd, M.T. Jackson and H.J. Newbury. 1995a. Use of RAPD for the study of diversity within plant germplasm collections. *Heredity* 74: 170-179.

Virk, PS., H.J. Newbury, M.T. Jackson and B.V. Ford-Lloyd. 1995b. The identification of duplicate accessions within a rice germplasm collection using RAPD analysis. *Theoretical and Applied Genetics* 90: 1049-1055.

Virk, P.S., B.V. Ford-Lloyd, M.T. Jackson, H. Pooni, T.P. Clemeno and H.J. Newbury. 1996. Predicting quantitative traits in rice using molecular markers and diverse germplasm. *Heredity* 76: 296-304.

Virk, P.S., H.J. Newbury, M.T. Jackson and B.V. Ford-Lloyd. 1999. Are mapped markers more useful for assessing genetic diversity? *Theoretical and Applied Genetics* (in press).

Westman, A.L. and S. Kresovich. 1997. Use of molecular marker techniques for description of plant genetic variation. *In* Biotechnology and Plant Genetic Resources: Conservation and Use, edited by J. A. Callow, B. V. Ford-Lloyd and H. J. Newbury. Wallingford: CAB International.

Yoshida, S., D.A. Forno, J.H. Cock and K.A. Gomez. 1976. Routine procedures for growing rice plants in culture solution. *In* Laboratory Manual for Physiological Studies of Rice. Manila: International Rice Research Institute.

Zhu, J., M.D. Gale, S. Quarrie, M.T. Jackson and G.J. Bryan. 1998. AFLP markers for the study of rice biodiversity. *Theoretical and Applied Genetics* 96: 602-611.

# 10 International Collaboration in Agricultural Biotechnology

*John Komen*

## Abstract

*A wide range of international collaborative opportunities is available for agricultural research organizations that plan or implement research programs in agricultural biotechnology. International collaboration helps research managers monitor new developments and access technologies for agricultural research, facilitate local capacity building, and obtain advice and expertise on research management issues. Starting in the mid 1980s, a number of international initiatives in agricultural biotechnology have been established which provide an important source of new technologies, opportunities for local capacity building, and management assistance. This chapter summarizes various international initiatives in agricultural biotechnology. It analyses the research and training opportunities and management services those initiatives provide to developing-country partner institutions. Finally, it discusses selected management considerations for national organizations participating in international collaborative research.*

## Introduction

In the past 15 years many collaborative opportunities have been created for developing-country agricultural research organizations that are planning or implementing research programs and building national capacity in agricultural biotechnology[1] (see annex 1 for an overview of such international initiatives). In Mexico, for example, the Center for Research and Advanced Studies (Spanish acronym CINVESTAV) conducted large-scale field trials of transgenic potatoes with resistance to potato viruses X and Y in 1994, under a collaborative agreement with Monsanto and the International Service for the Acquisition of Agri-biotech Applications (ISAAA). In Egypt, the Agricultural Genetic Engineering Research Institute (AGERI) conducted field tests of transgenic potatoes with tuber moth resistance in 1996, and in 1997 it did trials for virus-resistant tomatoes and cucurbits. These field trials were done as part of a collaborative program between AGERI and the Agricultural Biotechnology for Sustainable Productivity (ABSP) project.

With the increased incorporation of biotechnology in agricultural research, international collaboration has become more important, as most individual countries are

---

[1] For the purpose of this chapter, international initiatives in agricultural biotechnology are defined as those organizations or programs that conduct, support, or coordinate collaborative biotechnology research that addresses constraints in developing-country agriculture.

© CAB *International*. 1999. *Managing Agricultural Biotechnology—Addressing Research Program Needs and Policy Implications* (ed. J.I. Cohen)

unable to develop all necessary capacity by themselves. At the same time, international collaboration is becoming increasingly complex as the sources for new agricultural technology have become more diverse. The private sector now plays a prominent role in agricultural research and is a driving force in biotechnology R&D. Consequently, technology transfer in agricultural research is changing from a predominantly free exchange of technologies and materials to a process increasingly arranged through formal agreements. Issues such as the management of intellectual property rights (IPR) and biosafety (which are discussed in particular in sections IV and V of this book) have to be addressed and may make collaborative programs more intricate. The existence of adequate local research capacity is increasingly important in a research field that is becoming more and more knowledge-intensive. International collaboration helps developing-country agricultural research institutes to

- monitor new developments and access new technologies for agricultural research
- facilitate local capacity building (through training, information supply, infrastructure development)
- obtain advice and expertise on research management issues.

## New initiatives for agricultural biotechnology

Most international initiatives in agricultural biotechnology began in the second half of the 1980s, when multilateral and bilateral donor organizations and national and international agricultural research institutes began to meet to discuss the potential benefits and challenges of agricultural biotechnology for developing countries (see, for example, IRRI 1985; Cohen 1989; Sasson and Costarini 1989; NRC 1990; Persley 1990). They recognized that applying biotechnology to tropical agriculture and transferring these new techniques to developing countries posed some specific challenges, such as safety reviews of rDNA products and the need to involve the private sector, which had become more involved in agricultural research in general and agricultural biotechnology in particular. A common message from those meetings was that special initiatives and new ways of doing business were required to realize the potential benefits for developing-country agriculture and to prevent the technology gap between industrialized and developing countries from widening. The international donor community responded in various ways:

- It supported the establishment of (1) a specialized international institute for biotechnology, the International Centre for Genetic Engineering and Biotechnology, with locations in New Delhi (India) and Triest (Italy), and (2) specialized national facilities for biotechnology in several developing countries.
- It created special biotechnology initiatives and networks aimed at transferring advanced technology to developing countries.
- It helped develop biotechnology capacity in established international agricultural research programs, including the CGIAR centers.
- It strengthened the international component and outreach activities of advanced national research institutes in industrialized countries.
- It supported special advisory programs on the policy and management aspects of agricultural biotechnology.

This has resulted in a wide range of new programs and activities that are summarized in the annex of this chapter. Brenner and Komen (1994) and Komen (1999) give a detailed discussion of international initiatives in agricultural biotechnology and the donor organizations supporting them.

ISNAR's Biotechnology Service (IBS) set up a database on international agricultural biotechnology, which currently contains data on 43 international programs supporting agricultural biotechnology in developing countries. The information was initially collected through a survey conducted in June–December 1993. The database has been analyzed, updated, and expanded since then. The survey collected information on the following:

- overall goals and priorities
- agricultural research focus in terms of crops, livestock species, livestock diseases, and other
- regional focus and collaborating institutes from developed and developing countries
- type, number, and location of training opportunities provided
- activities and methods developed in program planning, policy, and management
- information products and services
- support for infrastructure development in developing-country programs
- research and development projects, their progress, and technology transfer routes
- funding sources and expenditures

The data collected forms the basis of this chapter.

## International biotechnology initiatives: Opportunities available

The majority of international initiatives discussed in this chapter focus on tropical food crop improvement. The IBS database includes 19 international programs on crop biotechnology, four livestock programs, and two research programs with a mixed focus on crop-related and livestock research (see annex 1). Assessing the different categories of expenditure within the research programs, the IBS survey showed a strong emphasis on R&D, followed at a distance by human resource development (table 10.1). This section of the chapter discusses opportunities available through collaborative research, human resource development, and research-management advice.

**Table 10.1.** Research Programs: Categories of Expenditures (1997)

| Program Element | Share (%) |
| --- | --- |
| Research and development | 50 |
| Human resource development | 18 |
| National program participation | 10 |
| Infrastructure support | 8 |
| Information products and services | 6 |
| Policy and program management | 5 |
| Other | 3 |
| Total | 100 |

*Source:* IBS *BioServe* Database.

## Collaborative research & development

Current plant biotechnology research primarily focuses on developing varieties with enhanced pest resistance and on pest control (diseases, viruses, insects) (see table 10.2); some 60% of the 126 projects recorded in plant biotechnology deal with pest resistance and control. Cereal grains, including rice, maize, and sorghum, are the main crop category in terms of research activities, followed by root crops (potato, cassava, yam, sweet potato) and tropical perennials. By far the most important crop is rice, which accounts for 17% of all crop research projects recorded. The projects in crop biotechnology tend to be at the advanced end of the research spectrum. Around 30% of them involve crop transformation, 29% involve the use of molecular markers, while cell biology (micropropagation and regeneration) is applied in 31% of the total.

Table 10.2. Crop Research Objectives (in Number of Projects)

| Crops | Disease resistance | Insect resistance | Virus resistance | Quality traits | Micro-propag. | All |
|---|---|---|---|---|---|---|
| Cereals | 9 | 13 | 8 | 12 | | 42 |
| rice | 5 | 4 | 6 | 6 | | 21 |
| maize | 1 | 6 | 2 | 3 | | 12 |
| sorghum | 1 | 3 | | 2 | | 6 |
| other | 2 | | | 1 | | 3 |
| Root crops | 4 | 5 | 7 | 2 | 1 | 19 |
| potato | 1 | 3 | 2 | | | 6 |
| cassava | 1 | | 3 | 2 | | 6 |
| yam | 2 | | 1 | | 1 | 4 |
| sweet potato | | 2 | 1 | | | 3 |
| Legumes | 4 | 6 | 4 | 6 | | 20 |
| bean | 1 | 2 | 1 | 2 | | 6 |
| groundnut | 1 | 1 | 3 | 1 | | 6 |
| chickpea | 1 | 1 | | 2 | | 4 |
| other | 1 | 2 | | 1 | | 4 |
| Horticulture | 2 | | 3 | | 1 | 6 |
| Perennial | 2 | 2 | 2 | 2 | 15 | 22 |
| banana/plantain | 2 | | 1 | | 5 | 8 |
| industrial crops | | | | 1 | 4 | 5 |
| coffee | | 1 | | | 4 | 5 |
| sugarcane | | | 1 | 1 | 1 | 3 |
| cocoa | | | | | 1 | 1 |
| Forestry species | | | | 2 | 5 | 7 |
| Miscellaneous | 3 | 3 | | 2 | 2 | 10 |
| All | 24 | 28 | 24 | 26 | 24 | 126 |

Source: IBS Bioserve Database.
Figures are based on information gathered from the 22 international research programs that include activities in crop research. For this table, we used those research activities with a specific applied objective, excluding research activities aimed to generate general technology.

Compared with crop research, livestock research plays a modest role. Projects in this area concentrate on developing new vaccines and diagnostics for tropical livestock diseases, such as trypanosomiasis, tick-borne diseases (e.g., theileriosis and cowdriosis), rinderpest, and foot-and-mouth disease (see table 10.3). The major share of the livestock effort relates to cattle.

### Human resource development

In parallel with their research projects, all crop and animal research initiatives have developed a strong component for human resource development. These training opportunities are primarily found at the CGIAR centers and in advanced research institutions and universities in Europe and the USA. Some laboratories of international private companies have training positions. Table 10.4 gives an overview of these positions available at the time of writing in the 25 research programs. Training activities are concentrated at the postdoctoral and doctoral levels. There are few opportunities for training at the Master's level or for technicians. As indicated in table 10.2, other aspects of capacity building, such as infrastructure development and information services, receive much less attention in international collaboration.

**Table 10.3.** Livestock Programs: Research Focus

| Program | Research focus (examples) |
|---|---|
| CIRAD – Animal Production and Veterinary Medicine Division | • Vaccine for *peste des petits ruminants* (PPR)<br>• Vaccine for *contagious caprine pleuropneumonia* (CCPP)<br>• Vaccine for *contagious bovine pleuropnomia* (CBPP)<br>• Molecular probes and PCR to identify mycoplasmosis<br>• Monoclonal antibodies to identify mycoplasmosis |
| ICIPE – Biotechnology Research Unit | • Development of vaccines for tick-borne diseases |
| International Laboratory of Molecular Biology for Tropical Disease Agents | • Recombinant rinderpest vaccine<br>• Diagnostic kit for rinderpest<br>• Recombinant foot-and-mouth disease (FMD) vaccine<br>• Diagnostic kit for foot-and-mouth disease |
| ILRI – Biosciences Research Programme | • Recombinant vaccine for East Coast Fever<br>• Diagnostic tests for tick-borne diseases<br>• Genetic maps of protozoa<br>• Vaccine development for trypanosomiasis<br>• Molecular analysis of genetic basis of trypanotolerance<br>• Quantification of drug levels using HPLC and ELISA technology<br>• Parasite characterization<br>• Diagnostic tests for trypanosomiasis<br>• Bovine genome mapping |
| Indo-Swiss Collaboration in Biotechnology | • Diagnostic tests for *contagious caprine pleuropneumonia* (CCPP) and *contagious bovine pleuropneumonia* (CBPP)<br>• Foot-and-mouth disease (FMD) diagnostics (development of diagnostic kits for FMD virus and antibody detection) |
| International Program on Vectors and Vector-borne Diseases | • Recombinant vaccine and improved diagnostics for heartwater<br>• Development of attractant decoys for control of ticks |

Source: IBS *BioServe* Databas.e
This list is illustrative, not exhaustive.

**Table 10.4.** Human Resource Development Opportunities Provided by International Biotechnology Programs

| HRD type | Number of opportunities (all programs) |
| --- | --- |
| Postdoctoral | 204 |
| Doctoral | 210 |
| Master's | 82 |
| Technician | 101 |
| Management | 3 |
| Internship | 113 |
| Other (mostly short-term courses) | 146 |

*Source:* IBS *BioServe* Database.
Figures are based on information obtained from 25 international research programs with a research focus on crops and/or livestock.

### Advice on research management

In addition to research and technical training, most international research programs provide advice and training on policy and management aspects of agricultural biotechnology. Biosafety and IPR are priority topics. Advisory services are mostly provided in the form of workshops to raise awareness on priority topics. Some international initiatives concentrate entirely on policy and management issues and play a useful complementary role to the research-oriented programs. For example, the Biotechnology Advisory Center, based at the Stockholm Environment Institute, provides on-request advice on biosafety aspects.

IBS is an advisory service to developing countries on matters of biotechnology research program management and policy, including biosafety and IPR. IBS initiated a series of regional policy seminars that reviewed the issues involved in planning and policy formulation for agricultural biotechnology. Using the findings of the policy seminars it now provides specific in-country advisory services. In addition, IBS has initiated a management training course with support from the Government of Japan, aimed at agricultural research managers who are responsible for implementing biotechnology programs.

## Management considerations: Maximizing the benefits of collaboration

In order to maximize the benefits derived from international collaboration, national organizations participating in international collaborative research should consider a number of management issues, which are discussed in this section: (1) ensuring that research priorities of the international partner or donor agency match national priorities for agricultural research; (2) assessing whether the available local research infrastructure is adequate to become a partner in advanced collaborative research and training; (3) ensuring that research activities lead to actual products, which may involve biosafety review, private-sector partnerships, and dealing with intellectual property.

## Matching research priorities

Through the many activities generated through international collaborative research, resources are spread over many products and different types of research and services. The range of objectives, activities, and regional focus generally reflects the interests of the organizations providing financial support. Few international programs set their priorities in consultation with organizations in the target countries. IBS data shows that funds available for national-program participation are limited.

As international collaborative research presents exciting research and training opportunities for national-program scientists, there's a risk that it takes precedence over other research priorities. Research managers should carefully determine whether the opportunities presented by international collaboration correspond to the national or institutional priorities for agricultural research in general and agricultural biotechnology in particular.

## Assessing institutional capacity

Most international initiatives discussed in this chapter conduct research at the advanced end of the "biotechnology spectrum:" crop transformation, molecular markers, and rDNA vaccines. Similarly, training opportunities are found at the doctoral and postdoctoral levels. This implies that countries with well-established scientific and technological capabilities are the best qualified partners in international programs, which is reflected in their geographic coverage. Many developing countries are taking part in international biotechnology initiatives, and efforts are more or less evenly spread among the different geographic regions. Within each region, however, international collaborative activities are concentrated in a small number of countries: Kenya, Egypt, Zimbabwe, and Côte d'Ivoire in Africa; Indonesia, Thailand, and India in Asia; and Costa Rica, Mexico, and Brazil in Latin America.

But even in these countries, the physical research infrastructure can be a limiting factor in applying the results of international collaboration. Infrastructure development in partner countries is not a priority of most international initiatives. On average, less than 10% of funding is directed towards, for example, the construction of facilities or the acquisition of equipment. However, adequate research infrastructure will help ensure that scientists taking part in international training activities can apply their newly acquired skills when they return.

## Ensuring product development

The ultimate success of international collaboration is reflected in tangible products from farmers' fields. The IBS survey showed that although the majority of the collaborative research activities are in the experimental, laboratory phase, some of them have already resulted in products that are ready for wider diffusion, such as:

- **Disease-free planting material.** Various tissue-culture techniques are applied for the micropropagation of disease-free planting material. They involve mainly export crops such as coffee, cocoa, banana, oilpalm, and sugarcane.
- **Biocontrol agents.** Pilot production of biopesticides is undertaken in India as part of the ISCB program. Products based on *Bacillus thuringiensis* and *B. sphaericus* have been provisionally registered. In livestock research, a pheromone-based attractant

decoy for bont tick vector control was developed at the University of Florida under the International Program on Vectors and Vector-borne Diseases.
- **Transgenic plant varieties**. Virus-resistant potatoes were planted in Mexico as part of the CINVESTAV-Monsanto collaboration. Field trials of virus-free potatoes and transgenic tomatoes and cucurbits are being conducted in Egypt under the ABSP program.
- **New diagnostics and vaccines for livestock diseases.** This appears to be the most significant area for product development to date, with diagnostic tests and rDNA vaccines for rinderpest, cowdriosis (heartwater), theileriosis (East Coast Fever), and foot-and-mouth disease.

Some of these products will require biosafety review, such as transgenic plant varieties and recombinant livestock vaccines. Donor agencies supporting international initiatives in biotechnology often require that the home countries of collaborating institutes have a regulatory structure in place before the transfer of transgenic material can take place. The presence or absence of a national and institutional biosafety review system has therefore become an important issue in international collaboration and technology transfer.

The private sector has been involved in developing products on a modest scale, although some programs actively involve the international private sector for transferring proprietary technology at favorable terms (e.g., ABSP and ISAAA). In addition to such technology "donations," the local private sector may become involved in developing and diffusing products that result from collaborative research. This is a more difficult issue, as markets for those products are usually not very attractive to the private industry. For research managers this means that mechanisms and partners for product development and diffusion should be considered in the early stages of international collaborative projects.

Finally, when proprietary technologies or commercial product development is involved, research managers must address IPR-related issues. Private companies routinely seek patent protection for their inventions, and industrialized-country universities and advanced research institutes increasingly do so as well. Some international initiatives have resulted in proprietary technology. Each collaborating institute in international biotechnology initiatives will need to specify the rights and obligations related to intellectual property.

### *Maximizing the benefits: Lessons from Thailand*

Thailand's experiences in international collaboration were well presented by Grudloyma (1997). Recognizing the need for international research cooperation at an early stage, the Government of Thailand has promoted international collaboration for many years. It assigned its National Center for Genetic Engineering and Biotechnology (BIOTEC) to manage these international collaborative biotechnology programs. The following are examples of successful collaborations in recent years:

- **Asian Rice Biotechnology Network (ARBN).** The ARBN initiative, hosted by the International Rice Research Institute promotes infrastructure and human resource development for rice biotechnology at selected institutes in Asia through joint research activities and training. Thailand is represented in the network through Kasetsart University's Plant Genetic Engineering Unit, one of the specialized laboratories supported by BIOTEC.

- **International Rice Biotechnology Program.** This international Rockefeller Foundation supported program supports research on molecular markers for tagging submergence tolerance genes in rice and research on salt tolerance in rice. In addition, the Rockefeller Foundation supports BIOTEC and Kasetsart University for establishing the "Introduction Station for Genetically Engineered Plants," which will accommodate the testing of transgenic material in Thailand, following international and national biosafety guidelines.
- **Farmer-centered Agricultural Resource program (FARM).** BIOTEC is a member of the FARM advisory board, which is supported by the United Nations Development Programme (UNDP) and implemented by the Food and Agriculture Organization of the United Nations (FAO), with specific responsibility for implementing the subprogram on Biotechnology and Biodiversity. The goal of FARM is achieving sustainable use and management of natural resources in agriculture and household food security through innovative approaches.

The above examples illustrate the various ways in which Thailand benefits from international collaboration, ranging from funding for infrastructure development, to research partnerships, to networking. The Thailand study also revealed the following issues:

1. Areas for collaboration inappropriately defined: Few international programs determine their priorities in consultation with national organizations. This may result in collaborative activities that are undertaken in isolation from local needs. In one case, BIOTEC had to renegotiate a proposal for international collaboration that did not correspond to national priorities.
2. Lack of infrastructure and resources: International collaborative programs generate new tools, but due to a lack of local transfer agents and proper local infrastructure their application is limited. BIOTEC attempts to assess such constraints in advance.
3. IPR: Current international developments regarding IPR and routine use of patents in public and private institutions appear to cause distrust among researchers and negatively affect international collaboration and access to new technologies. BIOTEC addresses this issue by working with international agencies that can negotiate a license for using specific genes.

## Conclusion and recommendations

There are many opportunities for international collaboration in agricultural biotechnology. Collaboration can contribute to developing-country objectives for agricultural and social development. Data presented in this chapter show that research is undertaken on important commodities for tropical agriculture and emphasizes the development of pest-resistant plants and animal health products. The hands-on experience gained by national research managers through their involvement in collaborative research, training programs, and international technology transfer projects are a great benefit as such. Through our research[2] we have identified a number of challenges that research managers are facing with

---

[2] Cf. the data collected by IBS, the findings from Thailand, and the two case studies described in chapters 11 and 12 of this volume.

international collaboration. In order to address these issues effectively, we recommend the following:

- Involve national scientists and decision makers closely early on in the planning stage of international collaborative projects. Different countries have different systems for agricultural research and technology development. For international collaboration to be effectively integrated into the system, detailed knowledge is required of the national priorities for agricultural research, the functioning of the research system, and of scientific capacity.
- Make support for policy and management aspects of biotechnology, such as biosafety and intellectual property, an integral and more prominent component of international initiatives. In most developing countries, the infrastructure and human resources needed to evaluate and eventually conduct field trials of genetically modified organisms are limited. This may be an obstacle to transferring new products and technologies. Similarly, in cases where collaboration from the international or local private sector is sought for product development and diffusion, the arrangements regarding intellectual property should be clearly defined.
- Increase support for developing local research capacity to ensure the impact (i.e., tangible products) resulting from international collaboration. On average, the support for local infrastructure development is low. To pave the way for broader participation, it will be necessary to provide financial support (from national and international sources) and technical assistance for building strong local partners.

## References

Altman, D.W. and K.N. Watanabe (eds). 1995. Plant Biotechnology Transfer to Developing Countries. Austin: R.G. Landes Company.
Brenner, C. and J. Komen. 1994. International Initiatives in Biotechnology for Developing Country Agriculture: Promises and Problems. Technical Paper No.100. Paris: OECD Development Centre.
Cohen, J.I. (ed). 1989. Strengthening Collaboration in Biotechnology: International Agricultural Research and the Private Sector. Washington, DC: Agency for International Development.
Cohen, J.I., and J. Komen. 1994. International Agricultural Biotechnology Programmes: Providing Opportunities for National Participation. *AgBiotech News and Information* 6(11): 257N-267N.
Grudloyma, U. 1997. Thailand and its Experiences with International Biotechnology Collaboration. Paper prepared for the IBS course "Managing Biotechnology in a Time of Transition."
IBS. BioServe Database on International Agricultural Biotechnology.
IRRI. 1985. Biotechnology in International Agricultural Research: Proceedings of the Inter-Center Seminar on International Agricultural Research Centers (IARCs) and Biotechnology. Manila, Philippines, 23-27 April 1984. Manila: International Rice Research Institute.
Komen, J. 1999. International Initiatives in Agri-food Biotechnology. *In* Biotechnology Worldwide, edited by G.T. Tzotzos. Wallingford: CAB International. *Forthcoming.*
National Research Council. Board on Science and Technology for International Development. 1990. Plant Biotechnology Research for Developing Countries. Washington, DC: National Academy Press.
Persley, G.J. (ed). 1990. Agricultural Biotechnology: Opportunities for International Development. Wallingford: CAB International.
Sasson, A. and V. Costarini (eds). 1990. Plant Biotechnologies for Developing Countries. Proceedings of an International Symposium, Luxembourg, 26-30 June 1989. Ede, Netherlands: Technical Centre for Agricultural and Rural Cooperation.

# Annex 10.1. Summary of International Agricultural Biotechnology Initiatives

| Program name and host institution | Priorities | Agricultural focus (crop/livestock) | Region/country focus |
|---|---|---|---|
| **CROP BIOTECHNOLOGY PROGRAMS** | | | |
| Agricultural Biotechnology for Support Project, ABSP (Michigan State University, USA) | • genetic engineering of crops for pest/disease resistance<br>• development of micropropagation systems<br>• integration of biotechnology within a general agriculture and business framework | • maize<br>• potato<br>• coffee<br>• sweet potato<br>• horticultural crops | Indonesia<br>Egypt<br>Kenya |
| Bean/Cowpea Collaborative Research Support Program, B/C CRSP (Michigan State University, USA) | • control of pests and diseases<br>• increase crop yields<br>• increase nutritional quality | • bean<br>• cowpea | international |
| Center for the Application of Molecular Biology to International Agriculture, CAMBIA (Australia) | • novel biotechnologies and methods for agricultural innovation<br>• genetic markers and diagnostics<br>• apomixis | • rice<br>• cassava<br>• bean<br>• agroforestry | international |
| CATIE – Biotechnology Research Unit (Centro Agronomico Tropical de Investigacion y Ensenanza, Costa Rica) | • enhance regional program capabilities<br>• genetic improvement of tropical crops | • banana/plantains<br>• coffee<br>• cocoa<br>• agroforestry | Latin America and the Caribbean |
| CIAT – Biotechnology Research Unit (International Center for Tropical Agriculture, Colombia) | • increasing the efficiency of CIAT strategic research<br>• institutional development in biotechnology | • cassava<br>• common bean<br>• rice<br>• tropical forages | international |
| CIMMYT – Applied Biotechnology Center (International Center for Maize and Wheat Improvement, Mexico) | • enhanced resistance to pests and diseases<br>• enhanced stress tolerance | • maize<br>• wheat | international |

| Program name and host institution | Priorities | Agricultural focus (crop/livestock) | Region/country focus |
|---|---|---|---|
| CIP – Applied Biotechnology Program (International Potato Center, Peru) | • reduce dependence on costly toxic chemical pesticides for potato and sweetpotato production<br>• host-plant resistance in potato | • potato | international |
| CIRAD – Plant Breeding Division (Centre de coopération international en recherche agronomique pour le développement, France) | • develop genetically improved crops | • cotton<br>• rice<br>• sorghum<br>• tropical perennials<br>• tropical fruits<br>• forestry | international |
| DFID Plant Sciences Research Programme (University of Wales, UK) | • genetically improved crops | • cereals<br>• roots and tubers<br>• legumes<br>• oilseeds<br>• fruit and vegetables<br>• fibres | international |
| Feathery Mottle Virus-Resistant Sweet Potato for African Farmers (Agency for International Development, USA) | • human resource development<br>• production of virus-resistant, African varieties of sweet potato<br>• enhance capacity in biosafety regulation of transgenic crop plants<br>• export of transgenic sweet potato to Africa for field testing | • sweet potato | Kenya |
| ICGEB – Plant Biology Programme (International Center for Genetic Engineering and Biotechnology, Italy/India) | • capacity building<br>• genetically improved rice | • rice | international (ICGEB member countries) |

| Program name and host institution | Priorities | Agricultural focus (crop/livestock) | Region/country focus |
|---|---|---|---|
| ICRISAT – Molecular and Cellular Biology Program (International Crops Research Institute for the Semi-Arid Tropics, India) | • support and complement conventional crop improvement programs at ICRISAT | • sorghum<br>• pearl millet<br>• groundnut<br>• chickpea<br>• pigeonpea | international |
| IITA – Biotechnology Research Unit (International Institute for Tropical Agriculture, Nigeria) | • tackle recalcitrant problems in crop improvement<br>• enhance national research capabilities | • cowpea<br>• yam<br>• cassava<br>• banana/plantain | Africa |
| International Laboratory for Tropical Agricultural Biotechnology, ILTAB (Scripps Research Institute, USA) | • genetically engineered food crops with virus resistance | • rice<br>• cassava<br>• tomato<br>• sugarcane | international |
| International Program on Rice Biotechnology (Rockefeller Foundation, USA) | • rice genetic improvement<br>• capacity building | • rice | international |
| International Service for the Acquisition of Agri-biotech Applications, ISAAA (Cornell University, USA) | • transfer and delivery of appropriate biotechnology applications to developing countries and the building of partnerships between institutions in the South and the private sector in the North, and by strengthening South-South collaboration | • vegetables<br>• fruits<br>• field crops (e.g., cotton)<br>• cereals<br>• forestry | international |
| Philippine-German Coconut Tissue Culture Project (Albay Research Center, Philippines) | • micropropagation of coconut | • coconut | Philippines |
| Regional Program of Biotechnology for Latin America and the Caribbean (several UN organizations) | • collaborative research projects<br>• training | • maize<br>• potato<br>• sugarcane | Latin America and the Caribbean |

| Program name and host institution | Priorities | Agricultural focus (crop/livestock) | Region/country focus |
|---|---|---|---|
| Research on the Date Palm and the Arid Land Farming Systems (Estacion Phoenix, Spain) | • in vitro propagation<br>• biological control technology | • date palm | Africa<br>Asia |
| **LIVESTOCK BIOTECHNOLOGY PROGRAMS** | | | |
| CIRAD – Animal Production Division (Centre de coopération international en recherche agronomique pour le développement, France) | • development of heat-stable vaccines through genetic engineering<br>• improved diagnostic tests<br>• determination of genetic resistance to diseases | • cowdriosis<br>• dermatophilosis<br>• rinderpest<br>• peste des petits ruminants<br>• mycoplasmosis<br>• trypanosomiasis | international |
| ILRI – Biosciences Research Programme (International Livestock Research Institute, Ethiopia/Kenya) | • novel, subunit vaccines<br>• improved diagnostics and parasite characterization<br>• development of disease-resistant livestock | • theileriosis<br>• trypanosomiasis<br>• cowdriosis<br>• anaplasmosis<br>• babesiosis | international |
| International Laboratory of Molecular Biology for Tropical Disease Agents, ILMB (University of California, USA) | • live recombinant virus vaccines for animal diseases<br>• technology transfer | • rinderpest<br>• peste des petits ruminants<br>• foot-and-mouth disease | international |
| International Program on Vectors and Vector-borne Diseases (University of Florida, USA) | • development and commercialization of improved vaccines and diagnostic tests | • heartwater | SADC countries<br>Caribbean |

| Program name and host institution | Priorities | Agricultural focus (crop/livestock) | Region/country focus |
|---|---|---|---|
| **CROP / LIVESTOCK PROGRAMS** | | | |
| ICIPE – Biotechnology Research Unit (International Centre of Insect Physiology and Ecology, Kenya) | • biological control of pests (plant protection) and vectors<br>• development of anti-tick vaccines<br>• development of diagnostics tools | • maize<br>• sorghum<br>• cowpea<br>• cattle | Africa |
| Indo-Swiss Collaboration in Biotechnology, ISCB (Federal Institute of Technology, Switzerland) | • research capacity building<br>• human resource development<br>• development, production and commercialization of specific biotechnology products<br>• partnerships between research groups (public and private sector) | • foot-and-mouth disease<br>• contagious caprine pleuropneumonia<br>• plant biopesticides | India |
| **NETWORKS** | | | |
| African Biosciences Network – Sub-Network for Biotechnology, ABN-BIOTECHNET (University of Nigeria, Nigeria) | • genetically improved crops and farm animals<br>• disease control through new vaccines<br>• capacity building | | Africa |
| Asia Network for Small-Scale Bioresources, ANSAB | • plant tissue culture<br>• biopesticides<br>• biofertilizers | • potato<br>• kapok tree<br>• rice<br>• mushroom | Asia |
| Asian Rice Biotechnology Network, ARBN (International Rice Research Institute, The Philippines) | • DNA fingerprinting of pests and pathogens<br>• low-cost marker-aided selection<br>• transgenic rice | • rice | Asia |

# International Collaboration in Agricultural Biotechnology 125

| Program name and host institution | Priorities | Agricultural focus (crop/livestock) | Region/country focus |
|---|---|---|---|
| Bean Advanced Biotechnology Research Network, BARN (International Center for Tropical Agriculture, Colombia) | • constraint identification<br>• technology transfer<br>• information exchange | • beans | international |
| Cassava Biotechnology Network, CBN (International Center for Tropical Agriculture, Colombia) | • stimulate cassava biotechnology research on priority topics<br>• integrate priorities of small-scale farmers, processors, and consumers in cassava biotechnology research planning<br>• information exchange | • cassava | international |
| Technical Cooperation Network on Plant Biotechnology, REDBIO (Food and Agriculture Organization of the United Nations, Regional Office for Latin America and the Caribbean, Chile) | • generation, transfer and application of plant biotechnology<br>• national and regional policies<br>• information exchange | • vegetables<br>• roots and tubers<br>• cereals | Latin America and the Caribbean |
| **DONOR AGENCIES** | | | |
| Australian Centre for International Agricultural Research, ACIAR | • use biotechnology wherever appropriate as a research tool within any of ACIAR's projects | | international |
| DGIS Special Programme Biotechnology and Development Cooperation (Ministry of Foreign Affairs, The Netherlands) | • improve developing-country access to biotechnology, with special emphasis on small-scale producers and women<br>• technical cooperation<br>• international collaboration and coordination | • "orphan" commodities<br>• cassava<br>• maize | Colombia<br>India<br>Kenya<br>Zimbabwe |
| FAO/AGP Programme on Plant Biotechnology (Food and Agriculture Organization of the United Nations, Italy) | • information dissemination and cooperation<br>• advisory services<br>• capacity building<br>• promote research, technology transfer and adoption | • rice<br>• roots and tubers<br>• horticulture<br>• industrial crops | international |

| Program name and host institution | Priorities | Agricultural focus (crop/livestock) | Region/country focus |
|---|---|---|---|
| GTZ Biotechnology in Plant Production (Agency for Technical Collaboration, Germany) | • development of micropropagation systems with diagnostic and pathogen elimination<br>• training and capacity building<br>• integration of biotechnology within BMZ/GTZ supported projects | • potato<br>• cassava<br>• yam<br>• date palm<br>• coconut<br>• banana | Africa<br>Asia |
| Swedish Agency for Research Collaboration with the Developing Countries, SAREC | • plant and forestry genetics<br>• diagnostics and vaccines in veterinary medicine<br>• environment<br>• biosafety<br>• policy research | | Africa<br>Asia |
| United Nations Development Programme | • productive and sustainable agriculture | • food crops<br>• cash crops<br>• livestock | international |
| United Nations Educational, Scientific, and Cultural Organization – Biotechnology Action Council | • human resource development | | international |
| World Bank | • invest in biotechnology as a contribution to economic development in World Bank member countries | | international |
| **POLICY / MANAGEMENT PROGRAMS** | | | |
| Biotechnology Advisory Center, BAC (Stockholm Environment Institute, Sweden) | • provide independent, impartial advice on biosafety development and implementation to developing countries<br>• biosafety capacity building | | international |

# International Collaboration in Agricultural Biotechnology

| Program name and host institution | Priorities | Agricultural focus (crop/livestock) | Region/country focus |
|---|---|---|---|
| Canada-Latin America Initiative on Biotechnology and Sustainable Development, CamBioTec (Center for Technological Innovation, National Autonomous University, Mexico) | • identify opportunities for biotechnology research and applications by tracking technological trends and carrying out priority-setting exercises<br>• strengthen public policies in biotechnology<br>• promote improved management of innovations<br>• foster partnerships between Canadians and Latin Americans | | Latin America |
| ISNAR Biotechnology Service, IBS (International Service for National Agricultural Research, The Netherlands) | • biotechnology research program management and policy formulation<br>• country reviews<br>• identify international program expertise | | international |
| Support to Agricultural Biotechnology Policies (Interamerican Institute for Cooperation in Agriculture, Costa Rica) | • biosafety, IPR<br>• industry development | | Latin America and the Caribbean |

# 11 Public- and Private-Sector Biotechnology Research and the Role of International Collaboration

*Joel I. Cohen*

## Abstract

*International collaboration plays an essential role in building capacity in developing countries for making research, management, and policy decisions for biotechnology. This chapter analyzes the design and implementation of an international initiative in biotechnology undertaken by a bilateral development agency. It describes the project's origins, how priorities were set, innovative mechanisms for supporting commercial research, and the way in which final proposals were solicited and selected. It highlights the associated managerial, oversight, and funding issues faced by the agency during this process with regard to stimulating public-private sector partnerships.*

## Introduction

A major challenge for international development agencies is determining effective means to link the needs of farmers and other beneficiaries with an increasing array of applications in agricultural biotechnology. Assessing new projects raises questions of organizational and departmental priorities as well as agency strategies and funding limitations. More recently development agencies have also reviewed their role in enhancing efficiencies between public and private research organizations. Agencies supporting biotechnology need to consider internal capabilities such as the capacity to provide significant technical, managerial, and regulatory oversight for relevant projects, and the commitment of financial resources once such research projects are approved.

This chapter examines the design and implementation of a collaborative project in biotechnology, undertaken by the United States Agency for International Development (USAID) through its Office of Agriculture. The objective of this paper is to analyze the decision-making process that was used to develop this bilateral initiative in biotechnology. It highlights the following critical steps taken during the process:

- securing USAID's commitment and financing for a priority setting exercise that tapped developing-country and international expertise, and making recommendations from the exercise available for peer review, with final results published for wider dissemination and transparency

© CAB *International*. 1999. *Managing Agricultural Biotechnology—Addressing Research Program Needs and Policy Implications* (ed. J.I. Cohen)

- developing opportunities for funding public and commercial research partnerships with national and private-sector organizations in developing countries
- enhancing USAID's internal capability to manage and provide oversight for a new initiative in plant biotechnology
- integrating the technical dimensions of biotechnology with respective policy considerations regarding biosafety, intellectual property rights (IPR), and internships for training in these areas.

## Project development – lessons learned from project evaluation

### The Tissue Culture for Crops Project

Developing the new biotechnology project began with a review of USAID's first major initiative in plant biotechnology: the Tissue Culture for Crops Project (TCCP). Beginning in 1984 and implemented from Colorado State University, the TCCP sought to (1) develop methodologies for improved stress-tolerant germplasm in several cereal and legume crops, (2) train scientists from developing countries in the project's methodologies, and (3) form an international network of scientists working on plant biotechnology.

The main research methodology applied in the TCCP project involved the in vitro selection of cells tolerant to various stresses (insects, salinity, acid soils, and drought) in rice, wheat, and sorghum. Regenerated lines were tested under both field and greenhouse environments to verify stress tolerance and to determine efficiency of in vitro selection. Teams of scientists from US universities, research centers of the Consultative Group on International Agricultural Research (CGIAR), and developing-country collaborators conducted this research.

A 1998 external review and evaluation of the TCCP project concluded that with the methodology applied and the institutions involved, the ability to successfully derive germplasm tolerant to the selected complex range of abiotic stresses was limited. A few notable exceptions occurred, including the registration and release of sorghum germplasm with improved tolerance to fall armyworm and to acid/aluminum soil conditions. These successes, however, were not on the scale expected, nor did they verify the methodologies being employed. The evaluation further identified management issues as follows:

> "There has been some ambiguity and resultant confusion regarding the purpose of the TCCP, the research hypothesis to be tested, and the methodology to be employed in validating it. This has contributed to some of the management and communication problems that are now being resolved by the concerted efforts of both parties. Less than adequate management performance has been manifested by poor and late reporting to [USAID]; static and untimely work planning; insufficient emphasis on producing outputs; inefficient use of outside advisory expertise; and missed opportunities to involve [USAID] and Colorado State University in joint, substantive decision- making." (USAID 1991).

Other problems with the project's design also became apparent. The project was based on tissue culture methodologies from which agronomic variation was expected. However, the bulk of the financial and human resources in the project focused on laboratory procedures, leaving few resources for field-testing and confirmation of agronomic fitness. Also, the project as conceived relied exclusively on in vitro technologies and did not account for advances in genetic engineering. Finally, difficulties existed in moving from model systems in the lab to germplasm with agronomic importance to developing-country farmers. In summary, the evaluation of the TCCP project highlighted the need for an integrated research initiative, with resources equally distributed between in vitro and conventional agricultural research, emphasizing the need to confirm agronomic performance.

### Considering the context

The evaluation of TCCP was the starting point for developing the new project. However, three trends affecting agricultural research and biotechnology emerged after TCCP had begun. These were incorporated into the context analysis for the new project:

1. **Increased importance of IPR and commercial opportunities in agricultural research.** As project development began, public and private research collaboration in the USA was increasing, responding to national laws regarding availability of patents from federally sponsored research. Universities began filing for patents to effect technology transfer through private-sector product development. Previously, publication of nonpatented information critical for product development and commercialization could result in the loss of (foreign) patent rights and additional revenues. Patents became one tool used to protect university-based research, secure industrial support for university research, and raise awareness of the usefulness of university findings (Nelson 1998).
2. **Need for biosafety review and regulation in developing countries.** The production of transgenic crops called attention to the need for biosafety protocols and regulation in developing countries. As international collaboration increased, enhanced national capabilities in biosafety helped ensure that these countries participated in and benefited from technology exchanges. Special efforts were needed from the international research community to develop this capability and to ensure appropriate regulatory review.
3. **Increased emphasis on sustainable agriculture in USAID projects**[1]. USAID began designing a new project in sustainable agriculture at the same time as it considered biotechnology. The increased emphasis on sustainability, which became more apparent after the 1992 United Nations Conference on the Environment and Development, affected USAID's perceptions for projects using biotechnology. Sustainability discussions raised questions as to whether new technologies contribute to broader views of the agricultural system, such as natural resource management.

## Preparatory steps

After analyzing these trends, USAID sought to benefit from experiences gained from developing projects within and outside the agency. At this point, it had four options:
1. extend the agreement with Colorado State University for another three to five years to address issues coming out of the evaluation, with the project left under prior management

---

[1] Sustainable agriculture, as used here, defines a system that meets rising demands for food at economic, environmental, and other social costs consistent with rising welfare of the people served by the system.

2. undertake a thorough review of the need for alternatives in plant biotechnology, which might lead to advertising for and awarding a new project
3. undertake biotechnology in a more limited manner through ongoing projects, which would mean that an additional project was not justified
4. drop biotechnology in favor of other pressing needs confronting the USAID.

After internal consultation, USAID decided to pursue option 2, raising the first key management challenge. This required securing commitment and financing for a formal priority-setting exercise that would provide for peer review, involvement of international and developing-country expertise regarding plant biotechnology, and the publication of results.

## Setting priorities: consulting with public, commercial, and developing-country representatives

Prior to designing an international biotechnology program, an assessment of the needs, constraints, and priorities and the potential impact of biotechnology on the agricultural system should be conducted. There are many ways of doing this, as shown in chapters 4 and 5 of this book and in other volumes (Cohen and Komen 1994; Toenniessen and Herdt 1989; CTA/FAO 1990). Setting priorities helped ensure that eventual research programs would address relevant productivity constraints in developing countries.

It was suggested that USAID use a more rigorous analysis for setting priorities and determining constraints than it had done in the past. It was also agreed that a panel of experts would undertake an external consultation. Convened by the National Research Council (NRC) and funded by USAID, the panel would analyze innovations in biotechnology expected in the coming three to five years that could benefit developing-country agriculture. The panel included members from national agricultural research organizations in developing countries, private industry, universities, CGIAR centers, the United States Department of Agriculture (USDA), and the Rockefeller Foundation. At its first meeting in September 1989, the panel identified priority areas in biotechnology that are sufficiently advanced to support collaborative initiatives between US and developing-country scientists.

As an initial step in the priority-setting process, panel members received background materials, including summaries of priorities made by some of the panelists, recommendations from two prior meetings on agricultural biotechnology, and a number of related reports. The chairman collected potential research activities on an individual basis, each activity addressing how it would "improve agriculture significantly in developing countries." These were ranked using a modified scoring technique (NRC 1990). The number of initial suggestions was reduced to eight by eliminating duplicates, combining closely related ideas, and disregarding projects that were likely to exceed the three- to five-year target and those that did not fit within USAID's mandate. The final grouping was circulated to panelists unable to attend the meeting and to outside experts who had peer-reviewed the report.

The panel's report (NRC 1990) found a distinct need for a collaborative biotechnology program encouraging developing-country researchers to focus on critical problems of local importance. Institutional priorities that were identified included the following:

- **Biosafety.** USAID should assist developing countries in implementing and monitoring appropriate biosafety regulations.
- **IPR.** USAID should participate in developing policies to promote cooperation in IPR.
- **Human resource development and networking.** USAID should enhance biotechnology capabilities through doctoral and postdoctoral fellowships.

The scientific recommendations for biotechnology were divided into three categories representing the most promising near-term applications of biotechnology to agriculture:

- **Tissue culture, micropropagation, and transformation.** The panel recommended that USAID support the building of developing-country capacity in plant tissue culture technologies, including micropropagation, cell selection, embryo rescue, and haploid techniques to augment conventional plant improvement programs. They also advocated improved ability to use micropropagation for production of virus-free planting material. USAID was also urged to support the development of transformation and regeneration technologies.
- **Plant disease and pest control.** (1) *Bacillus thuringiensis* strain identification to assist developing countries in identifying and cloning *Bacillus thuringiensis* strains effective against insect pests of importance in tropical areas. These bacteria produce a protein crystal that is selectively lethal to foliage-feeding lepidopterous insects but not to others or to animals and humans. (2) Antiviral strategies: support development of antiviral technologies for plant viruses that attack beans, cassava, sweet potatoes, groundnuts, and tropical fruits and vegetables. (3) Pathogen diagnostics and probes: support research to develop DNA probes, as well as antisera and monoclonal antibody probes for plant bacteria, fungi, and viruses that attack crops of importance in the developing world.
- **Genetic mapping of tropical crops.** This category summarized interest in developing specific genetic maps for major crop plants. It included guidance that USAID should assist CGIAR-center and developing-country crop breeders to acquire the capacity to use restriction fragment generated maps in their plant breeding programs.

### Consultations to confirm priorities

USAID favorably reviewed the NRC report and sought broader agreement for the proposed technical objectives and project modalities from national agricultural research organizations in developing countries, in-country USAID offices, and CGIAR centers. Restrictions on available funds eliminated the diagnostic research and genome mapping. The agency agreed on the following:

*Mutual benefit*

The project would seek to derive mutual benefits from research and commercialization programs for both US and developing-country public- and private-sector participants. This would be achieved by encouraging partnerships that allow institutions to disseminate products of research through either the public sector or the commercial sector.

*Proven research, integrated programs, and sustainability*

Project proposals were reviewed with a preference to proven applied research integrated with conventional and sustainable agricultural programs that provide for human resource development. The project would *not* provide funds to create new biotechnology centers or move scientific capability out of ongoing agricultural research programs. The limited funds were not to be used for constructing buildings and laboratories or making renovations that required extensive recurrent funding but that had no proven source of revenue. Support was given to research connecting agronomic research at the cell or molecular level with product development.

Integrating biotechnology with sustainable agriculture was regarded as a priority. Biotechnology research results derived from new cultivars would allow farmers to control weeds, insects, and diseases more effectively and with less dependence on chemicals. Research should strengthen biotechnology capabilities in developing countries through postdoctoral fellowships while supporting appropriate training in biosafety and IPR.

Finally, building an integrated approach maximized efforts to relate the field-testing of genetically modified plants with regulatory needs and conditions. This approach would bring together the technical and policy dimensions of biotechnology, providing a learning opportunity for scientists and policymakers in the participating countries.

*Multiple partnership mechanisms for germplasm development, including the commercial sector*

The project would provide alternative mechanisms for building research partnerships, including public-public, public-private, and private-private (see chapter 22 of this volume). These partnerships provided alternative routes for disseminating improved germplasm, using public-sector institutions such as the national agricultural research system (NARS), university cultivar release programs, international release of germplasm through the CGIAR, and commercial cultivar development. Providing these options increased the means by which products arising from research could reach targeted end users and beneficiaries.

## Project design and management implications

One of the recommendations of the NRC report was that proposed project should address a number of significant technical issues. To achieve effective oversight for such a project required the USAID to increase its internal competency in this area. A science and diplomacy fellow was recruited through the American Association for the Advancement of Science. With this additional expertise, USAID was better able to communicate technical matters to its field missions and stakeholders in the international agricultural research community.

The starting point for conceptualizing the project would be to address perceived technical and institutional inadequacies that affect the application of biotechnology in developing-country agriculture. Less developed countries often lack the technical capability, scientific infrastructure, and financial support to apply biotechnology to constraints affecting agricultural production. Privatization of the technology in developed countries, with applications favoring crops of obvious commercial interest, compounds the problem, limiting the ability to access needed technologies and products.

Increasing technical capability alone does not ensure the sustained participation of developing-country scientists in local and global biotechnology research. Another essential component of the project therefore would be a coordinated and integrated effort to strengthen institutional capacity in the management of science. Additional efforts were required to ensure that the technological capability of the institutions, which may be enhanced through the research and technology transfer component of the project, is fully employed. The project document outlined complementary efforts designed to strengthen capabilities in formulating and implementing regulatory policy, intellectual property protection, and, where possible, downstream product development and commercialization of laboratory research results.

These developments led to another managerial challenge for USAID: it did not have necessary expertise in-house to provide regulatory oversight for issues associated with the development of transgenic crops and the policy and management issues identified for the project. It therefore appointed a Standing Committee on Biotechnology to serve as an "institutional biosafety committee," securing appropriate linkages with official US government regulatory bodies and developing mechanisms to ensure that appropriate regulatory review and approval was undertaken in and by the relevant developing countries (Cohen and Chambers 1990). The committee was a novel development, the importance of which has since been recognized by other development organizations, such as the Department for International Development (DfID) in the UK (DFID 1999).

### Options for research: public and private

The project put equal weight on designing mechanisms that facilitated the participation of the commercial sector and that allowed federal funding of public- and private-sector research addressing productivity constraints in developing countries. USAID had already had a number of discussions, both internally and with the private sector, on possible collaboration with the private sector, but few developing-country-oriented projects actually stimulated private-sector research. It was hoped that providing adequate funding and management mechanisms would stimulate the interest and resources of private-sector research organizations. To encourage the private-sector organizations to participate, they were made equal partners in project implementation from the start and were able to propose activities in agreement with their particular expertise.

Financial commitments from public and private sources would be needed to support this involvement. It was argued that part of the financial burden must be provided from USAID's development budget, and that the commercial sector should be eligible to compete for funding. Getting the private sector to join in research is one means of gaining access to the expertise and resources of commercial agriculture and applying it to the needs of developing countries. Specific activities were then considered that would enhance capabilities in managing biotechnology research and support the transition of basic laboratory results into products. These activities, which were considered innovations at the time, included the following:

1. establish links with the commercial sector for appropriate technologies developed in the project between US companies and universities and developing-country institutions
2. use internships to raise awareness about the protection of IPR and the development of science-based regulatory policy

3. provide consultants for intellectual property agreements, economic impact analyses, and marketing analyses
4. enable developing countries to participate and network in biotechnology trade associations, such as the Association of Biotechnology Companies
5. coordinate with other USAID projects in plant biotechnology and field missions
6. sponsor three industry-based management seminars
7. provide comprehensive environmental and field analysis of proposed larger-scale field tests in both host countries and the USA

The unique design of the project at the time called for a full-time project manager, a position that was eventually advertised as managing director. The position was to be based at the institution that would be awarded the project. In addition to the managing director, other positions were envisioned to address many of the items above, either separately or in a coordinated training function.

Including the above activities in the project required careful budgeting and ensuring that the budget would not be entirely allocated to support ongoing research. This raised some controversy, as some argued that more of the project's budget be attributed purely to research. However, in the final budget the seven points above were supported through separate line items in the budget, with some only becoming available as research neared the product development phase. Thus, it was ensured that researchers could move towards field-testing and product development as rapidly as possible.

## Soliciting and evaluating proposals

The final design of the project was based on the objectives and modalities presented above. Existing international biotechnology programs were analyzed to ensure the current project would be unique and complementary. Once it was established that the proposed initiative would not duplicate other efforts, USAID approved the Agricultural Biotechnology for Sustainable Productivity (ABSP) project. USAID was authorized to provide funding for proposals received from both public and commercial institutions addressing identified constraints on sustainable productivity at the local level in developing countries. Following external peer review of proposals, the project's managing director would coordinate the research.

USAID's call for project proposals had to provide a mechanism supporting alternative routes for technology transfer and delivery of benefits. It was decided to first solicit proposals from public and private nonprofit institutions to address pest- and disease-related productivity constraints and to handle the management and network functions of the project. Following this first round of proposals, commercial institutions were solicited to submit proposals addressing constraints regarding the production of disease-free planting material or the development of germplasm with improved disease or pest resistance, to be distributed through commercial production. After the peer review, USAID would enter into a single cooperative agreement with the institution that submitted the "best" proposal in the first round. USAID would then designate a second award to the commercial sector, to be issued in a subagreement in the following year.

## Monitoring and evaluation functions

The institution selected to implement ABSP would also be tasked with establishing a technical advisory group. This group would advise on technical matters and evaluate progress in meeting the objectives and expected outputs. Specifically, it would assist in the following:

- reviewing annual workplans and quarterly and annual reports, as well as recommending additions to or modification and deletion of project components
- consulting with USAID and the ABSP managing director on the management implications of the technical advisory group's recommendations
- arranging the annual review and in-depth evaluation of each component of the cooperative agreement
- securing appropriate persons and resources, in conjunction with the project's managing director, including a financial audit of project expenditures

Establishing the technical advisory group was of fundamental importance because of the managerial complexities expected for the project, the need to monitor regulatory matters, confidentiality and financing implications arising from working with the commercial sector, and having external evaluators available to provide continuity over time. This group would oversee construction of public and private partnerships and evaluate progress on the technical and policy dimensions regarding the project's research.

## Project budget: Separating public- and commercial-sector research

Once budget categories are established and the budget is approved, USAID's financial procedures prohibit the movement of moneys between items without formally modifying the agreement. To ensure opportunities for public-private-sector collaboration (indicated as the innovative activities of the project), funding was divided between activities before the project began. The second award for research was to address a different productivity constraint, undertaken by a private-sector or commercial organization.

Agreeing to such segmentation in the budget indicated that USAID was able to provide public moneys for collaborative research with the private sector, as well as for subsequent needs for commercialization. It could also contribute to the equitable protection of IPR by providing funds for legal assistance while developing collaborative research agreements. This also meant that total funding available in the cooperative agreement would be divided between the public and private sectors, with innovative funding available depending on research progress and on the needs associated with structuring public-private agreements.

## Award of cooperative agreement

Applications to implement the project were received from public institutions, CGIAR centers, and the private sector. NRC commissioned a number of independent, external consultants to peer-review the proposals and to rate them in terms of technical and managerial merit. The prioritized list of proposals was submitted to USAID, which analyzed the budgets and institutional matching costs. Based on the peer reviews and following agreement from USAID missions and relevant bureaus, the first cooperative agreement for the ABSP project (see figure 11.1) was awarded in September 1991 to

# Public- and Private-Sector Biotechnology Research 137

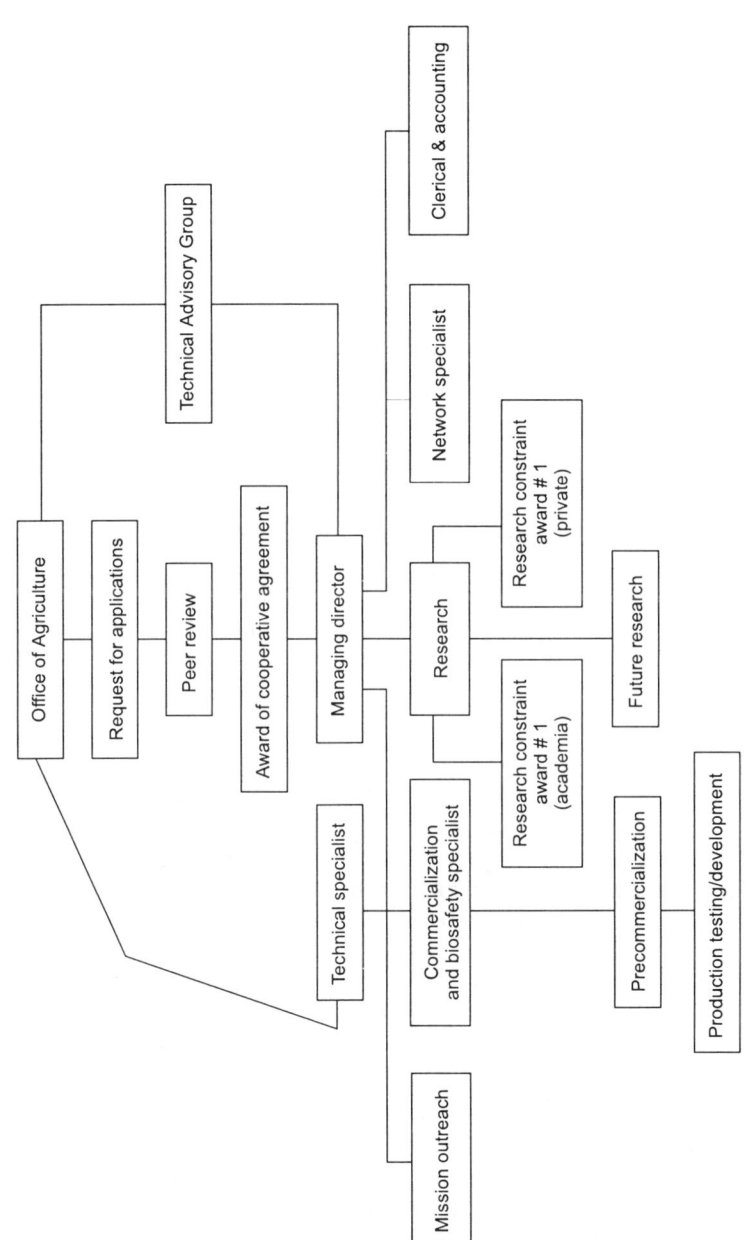

**Figure 11.1.** Design scheme for the ABSP Project

Michigan State University in collaboration with Kenya, Egypt, Indonesia and Costa Rica. The second cooperative agreement, issued at a later date for the commercial sector, went to DNA Plant Technology, Inc in collaboration with Fitotek of Indonesia and Agrobiotechnologia de Costa Rica.

## Conclusions

One of the most important managerial decisions taken was to increase transparency and disseminate USAID's priorities among stakeholders. NRC's review of the panel's report and its publication helped raise awareness and understanding of and reach consensus for the research that the project would undertake. It also laid the foundation for the research to be integrated with relevant policy and managerial capacity building. The time and money invested by USAID for this preliminary work helped build support for the project and its implementation.

The priority-setting process also stimulated discussion on how the project should be awarded. It reinforced the need for consensus and facilitated dissemination of results to developing-country scientists. Its formal publication by the NRC added legitimacy to the project development exercise. It provided confidence in the process to key decision makers, and it provided potential stakeholders from the public and commercial sector with a solid understanding of the proposed objectives.

While developing the project, USAID had to make fundamental choices with regard to its human resources. Managing a long-term biotechnology project implied that it had to enhance its technical expertise in this area. USAID accomplished this by recruiting a senior professional within the Office of Agriculture and using external fellowship programs and interagency exchanges. These individuals helped ensure internal understanding of biotechnology developments within USAID and provide a framework for biosafety and licensing of technology that met with federal guidance.

The project's focus on innovative planning and budget mechanisms helped secure and fund public- and private-sector research partnerships. The mechanisms for including commercial agricultural research institutes have provided many learning experiences, including licensing and IPR agreements (see also chapter 22 of this volume). Public and commercial partnerships allowing for distribution of products to end users were made possible. Both mechanisms are functioning in the current project. Providing such options overcame a major flaw of the previous project: inadequate mechanisms for improved germplasm dissemination. Also, TCCP research was based exclusively at the university and the CGIAR centers, whereas ABSP enriched the supply of available and new technologies by involving the private sector. The project was also designed to ensure that capabilities for biosafety and IPR were developed within the public and private collaborating institutions through a series of internships with regulatory, legal, and university offices.

Applications for project implementation had to be submitted for external peer review. Many private-sector companies, however, are reluctant to share research results with an external review committee. By establishing a long-standing technical advisory committee, thus eliminating the need for one-time or ad hoc reviews by outsiders, USAID helped instill the necessary confidence in companies that research results would be treated confidentially. In this way, external review and guidance could be provided for both public and commercial research activities.

Confidence was also gained with regard to project management, with resources committed to an expanded management and oversight function within USAID, as well as recognition of the need for an overall managing director for the project. Close attention to internal and external management was essential for integrating developments regarding the technical research base of the project with the policy and managerial dimensions posed by such collaborative biotechnology projects. This integration also ensured that related policy matters were not addressed in the abstract, but rather came directly from specific elements of the projects' public and private research partnerships.

## References

Cohen, J.I. and J. Komen. 1994. International agricultural biotechnology programmes: providing opportunities for national participation. AgBiotech News and Information. 6: 257N-267N.

Cohen, J.I. and J.A. Chambers. 1990. Biotechnology and biosafety: perspective of an international donor agency. Pages 378-394. In M.Levin and H.Strauss (eds). Risk Assessment in Genetic Engineering. Environmental Release of Organisms. New York, USA: McGraw-Hill.

CTA/FAO (Technical Centre for Agricultural and Rural Co-operation and Food and Agricultural Organization of the United Nations). 1990. Plant biotechnologies for developing countries. A. Sasson and V. Costarini (eds.). London: Trinity Press.

DFID. 1999. Genetically modified organisms and developing countries. Background Briefing, May 1999. London: Department for International Development.

DGIS (Directorate General for International Cooperation). 1991. Cassava and Biotechnology. J. Komen (ed). Proceedings of a workshop, 21-23 March 1990, Amsterdam, The Netherlands.

Nelson, L. 1998. The rise of intellectual property protection in the American university. *Science Online* 279 (5356): 1460-1461.

NRC (National Research Council). 1990. Plant biotechnology research for developing countries. Washington, DC: National Academy Press.

Toenniessen, G. and R.W. Herdt. 1989. The Rockefeller Foundation's program on rice biotechnology. p. 291-317. In J. Cohen (ed.) Strengthening Collaboration in Biotechnology: International Agricultural Research and the Private Sector. Washington, DC: USAID.

USAID. 1991. Project Paper: Agricultural Biotechnology for Sustainable Productivity (936-4197). April 1991. Washington, DC: USAID.

# 12  Indo-Swiss Collaboration in Biotechnology: Lessons Learned and Future Strategies

*Katharina Jenny and Ernst Schaltegger*

## Abstract

*This chapter discusses the major challenges in managing a bilateral research and capacity-development program, providing a critical analysis of its achievements. In addition, the planning of the next phase of the Indo-Swiss Collaboration in Biotechnology (ISCB) program is outlined, which offers the advantage of integrating findings and experience from the past. The process for planning and project review under the "new" ISCB is discussed, based on the concept of the "integrated value chain." Finally, the main changes in the organizational setup of the program are reviewed. The chapter concludes with recommendations for international collaborative programs, based on prior experiences, a recent external review of the ISCB, and the current planning process for the new ISCB program.*

## Background of the ISCB initiative

### The origins of ISCB (1974–84)

The Indo-Swiss Collaboration in Biotechnology (ISCB) program began with a bilateral agreement between the Government of India (Department of Biotechnology [DBT]) and the Government of Switzerland (Swiss Agency for Development and Cooperation [SDC]) to explore ways of cooperating in various biotechnology-related areas. The origins of the program date back to the early 1970s, when the Institute of Biotechnology of the Swiss Federal Institute of Technology (ETH) and the Biochemical Engineering Research Centre of the Indian Institute of Technology first began to collaborate. In this first stage of close collaboration, major emphasis was put on developing a curriculum, infrastructure, and human resources in the area of biochemical engineering. A comprehensive exchange program helped the participating institutes to achieve the program's goals. The Indian partner institute became a showcase research center and played a key role in developing biotechnology in India. The positive results prompted the Indian and Swiss governments to extend the agreement in 1988.

© CAB *International*. 1999. *Managing Agricultural Biotechnology—Addressing Research Program Needs and Policy Implications* (ed. J.I. Cohen)

## Program expansion (1988–98)

The current phase of the program started in 1988. Based on recommendations by the DBT and the ISCB chair, four Indian research institutions were selected for collaboration. After the project had been reviewed for the first time in 1995, two more partners were added. The main objective of the program was to strengthen the scientific and technological capabilities of R&D institutions in biotechnology in India and to establish a network for product development and technology transfer. The specific sector (agriculture, health, or environment) in which research activities were supported played a minor role in the process of project selection. Projects were selected according to the following criteria:
- scientific quality, significance, and feasibility of the proposal
- provision of joint research activities between Indian and Swiss partner institutes
- potential for future commercialization, chances for technology transfer
- legal and ethical feasibility of proposed activities
- relevance for the development of the institute and the region
- matching guidelines of the SDC and the DBT
- meeting requirements that are not already covered by major biotechnology programs

This explains the broad range of ISCB-supported biotechnology projects, including projects on human health, animal husbandry, and microbial processes and products for agricultural purposes as well as the pharmaceutical industry (see table 12.1 for an overview of the main partner institutes and projects in India).

## Strategy for capacity building

In order to enhance capacity building in R&D in the Indian partner institutes and to help transfer technologies, ISCB supports all the activities required in this process, not just research. Project proposals and annual action plans are assessed in collaboration with the scientific partners and, if required, with external experts. Specialized short- or long-term training events are organized at the Swiss partner institutes to strengthen knowledge and skills in newly emerging technologies and methodologies. Joint research activities between the Indian and Swiss institutes are encouraged whenever possible.

**Table 12.1.** The Main ISCB-Supported Biotechnology Projects

| Institute | Project title |
|---|---|
| Anna University, Center for Biotechnology (Chennai) | Bioprocess development and optimization for the production of biopesticides from *Bacillus thuringiensis* and *B. sphaericus* |
| MS University, Biotechnology Centre (Baroda) | Bioprocess development and genetic manipulation in microbial systems (non-conventional yeasts, new bioactive molecules, bioconversion) |
| MK University, School of Biotechnology (Madurai) | Molecular analysis of *Mycobacterium leprae* and immunology of leprosy |
| Indian Veterinary Research Institute (Bangalore) | Development of molecular and immunological tools for diagnosis of foot-and-mouth disease (FMD) in India |
| BAIF Laboratories (Pune) | Diagnostics for Contagious Bovine Pleuro-Pneumonia (CBPP) |
| National Environmental Engineering Research Institute (Nagpur) | Application of molecular genetics for the management of aromatics in waste |

Relative to the training provided and the research supported, ISCB also supports the purchase of essential laboratory equipment and other research tools. Regular visits of the principal researchers and their Swiss partners contribute to an intensive scientific exchange between India and other countries. The program also encourages Indian scientists to participate at international conferences to discuss their results and make new contacts. Training courses are organized at the Indian partner institutes to transfer the newly acquired know-how. Workshops and seminars with guests invited from public, private, and government institutions enable the researchers to discuss the potential routes for product development and to establish useful contacts with producers.

ISCB not only encourages its partners to establish contacts with private institutions, but it also offers them specific expertise (validation of products such as biopesticides and FMD diagnostic tools). Projects are regularly monitored, which ensures coherence with national technology policies as well as biosafety and risk management aspects. When required, policy questions are discussed at the government level. External consultants help manage project-related issues such as biosafety and risk analysis.

### Organization and monitoring

The ISCB program is managed by the Institute of Biotechnology at ETH, as formally agreed by SDC and ETH. Although the program has steadily grown in terms of financial resources and objectives, the organizational arrangement never changed. At ETH, a small team of one or two scientists and one administrative staff member is full-time responsible for managing the Swiss funds, implementing the objectives, and handling the operational component of the program. An advisory board chaired by a member of the Institute of Biotechnology advises the ISCB management in its activities.

In principle, projects are approved for a period of three years. The Joint Project Committee, with representatives of SDC and DBT and the chair of the ISCB program, meets annually to review progress and approve the annual action plans of each project. ISCB management has developed a semi-annual system to monitor the progress made in the individual projects. ISCB also reports on all training events and visits of scientists from or to India. If necessary, external experts are consulted on specific scientific or management matters. Using the internal monitoring reports, ISCB can decide to amend the action plans. The entire program was subjected to external reviews in 1992 and 1997.

### Funding

Funding comes from DBT, as matching funds for individual projects, and SDC, according to a bilateral agreement. ETH administers and distributes the SDC funds. At present, the Indian contribution to the individual projects is less than 20% of the Swiss contribution. Since the project began, SDC has contributed some 10 million Swiss francs in total, 75% of which has been used directly for projects in India. A large part of these funds (65%) was used for purchasing equipment and chemicals and subscriptions to journals.

### Main accomplishments

The accomplishments of the program (see table 12.2 for a summary) convinced SDC to expand the program, provided that the experiences from the present program and the recommendations of the last external review are used as a basis for the planning process.

**Table 12.2.** Summary of ISCB's Main Accomplishments

| Activity | Specific output |
|---|---|
| Research | • 20 PhD theses in project areas supported by ISCB<br>• 63 scientific papers written, 50 of which appeared in international journals<br>• 12 publications jointly written<br>• two patent applications for processes to produce *Bacillus* based pesticides |
| Human resource development | • 56 Indian scientists trained at Swiss and other European institutes (2–12 months) |
| Infrastructure | • all laboratories at long-term partner institutes upgraded in terms of analytical instruments, bioprocess equipment, electronic communication, and availability of other essential research tools |
| Technology transfer | • 20 workshops and courses held on transferring specific technologies (e.g., ELISA, PCR)<br>• laboratory manuals prepared for wider distribution<br>• eight international seminars held on product development |
| Products | • prototype diagnostic kits for detecting foot-and-mouth disease<br>• effective systems for heterologeous protein expression in nonconventional yeasts and their production<br>• *Bacillus* based pesticides |

## Management challenges in a changing world

Funding agencies are generally reluctant to invest heavily in research that yields results only in the long term. It is therefore most encouraging that SDC has supported this particular project from the outset. However, the project's long-term perspective does not diminish the need to act on and react to emerging developments, not only those regarding the scientific environment but also those in the political and economic context. It is difficult to establish a mode of operation that corresponds to the wishes and needs of the scientific community and that takes into account international cooperation, government policies, and technology development in a highly competitive market.

Although the ISCB initiative has been constantly modified to the changing environment, the general framework of the program has remained the same, which has prevented the program organization and mode of operation from satisfactorily responding to all the challenges. As one SDC officer stated it: "We followed a somewhat 'additive' strategy, starting out with scientific discussions, material, and institutional support. Training has also always played an important role. We later put more emphasis on product orientation and technology development." The implications of this approach—adding new elements to a program in progress to give it a new direction—are discussed below.

### *Program-level challenges: objectives and goals*

At the outset of the program, the socioeconomic status of biotechnology in both Switzerland and India was different from today. Whereas R&D capacity building used to be the major focus of development collaboration, today the development of biotechnology products and processes is regarded as the core of the collaboration. This shift entails a more integrated approach in which partnering within an innovation system and working hand in hand with experts from different sectors is necessary to achieve the defined goals. Such an approach requires collaboration with institutions that not only have a good track record in science but also have good management capacity—an essential factor for a successful partnership.

With the changes in the world economy in the past decade, the concept of international development cooperation has changed. The liberalization of the Indian market exposed the country to global competition, but at the same time it facilitated access to important research tools. In combination with political and economic constraints, SDC's mandate to contribute to poverty alleviation by relevant and need-oriented, high-tech research, poses a permanent challenge. Given the broad range of projects and partners, it is not surprising that difficulties arise in agreeing on a common goal. In addition, not only the projects but also the research objectives and goals are sometimes far from unique. Some projects have an applied nature and a limited time frame and may reach the production stage in a foreseeable time. For other projects the perspective for product development is long term. Here, the building of strong, scientific competitiveness and research capacity has first priority.

### Project level: Partnership issues

The present bilateral agreement does not provide for joint project-implementation contracts between the Indian and Swiss research institutes. Only project activities in India are funded. Real research partnerships have therefore not been formed. Although the situation differs from project to project (depending on the project's history and personal relations), none of the existing projects is formally jointly executed by Indian and Swiss researchers. So far, the Swiss partners have supported project work on a voluntary basis because mutual accountability does not exist within the framework of the ISCB. At present, the ISCB management only makes arrangements for procedural accountability, not scientific accountability.

The building of new partnerships and the networking for the ongoing activities are also hampered by the rigid organizational structures of the Indian project partners. In government institutions, for example, the principal project researchers have very little flexibility in deciding on research partnerships and funding. Scientists have to abide by the rules and regulations of their mother organizations, which in general are very hierarchic systems. Moreover, there are strict rules for international travel, while adequate tools for technology development or ways of transferring results or entering into public-private partnerships are virtually nonexistent.

### Organizational level: Mode of operation and management

Within the current institutional arrangement, the ISCB management at ETH plays a central role: it is the main link between the Indian and the Swiss institutes and coordinates virtually all the activities at all levels of the ISCB program. This includes writing credit proposals; budgeting, transferring, and accounting of the SDC funds; reporting and monitoring of project activities; purchasing of goods for the partner institutes; and networking the interests of Indian and Swiss partners.

No doubt, there were good reasons for such an arrangement in the past and they still present substantial advantages for the Indian partners today. On the other hand, the ISCB management is faced with increasing organizational and administrative bottlenecks. A clear designation of research management tasks within the individual projects could help address these problems.

## Responding to emerging challenges

As ISCB is an ongoing program—there has been no break for a major evaluation or reconsideration since 1988—we could only try to respond to some of the challenges when entering into the third-phase agreement in 1995.

### Program level: Promoting peer review and product development

To improve their ability to respond to the highly competitive environment in biotechnology, both existing and potential partners went through a project review exercise, including scientific peer review. SDC changed its policy in India and limited new collaboration to areas in which SDC was active. For individual project requirements, guidance in biosafety issues (review commission, high-level safety laboratory) was offered, and several workshops were held. Titled "From an Idea to a Product," these workshops addressed the various phases in a product-development process to successfully launch a product on the market—from R&D and technical development to registration and regulation.

However, these activities do not fully address the program goal to support projects to move towards application and product development, which is the explicit aim of the two funding agencies (DBT and SDC) today. There is a need to involve new partners that are either capable of developing the research results and products up to the market level or of giving advice regarding the management of intellectual property rights (IPR), regulatory issues, technology transfer, and biosafety. The ISCB management intended to integrate new partners into projects from the beginning of the project and not to involve them only for specific occasions.

Although the bilateral agreement encourages public-private-sector collaboration, a clear concept for creating these partnerships was never developed, partly due to a lack of rules and regulations at the government level in this regard. ISCB's lack of guidelines on IPR was another problem. To link up with new partners, a small fund was established to support small-scale activities (e.g., specific training abroad or support for short-term research). This resulted in contacts with some new institutes. Incorporating this new initiative into the existing agreement was very difficult, as it was not based on existing mechanisms. Nevertheless, even though this small project fund was implemented slowly, some positive feedback on the initiative has been received.

### Project level: Promoting partnerships

As far as "gaps" in the partnerships between the Indian and Swiss institutes are concerned, the ISCB management tried to fill them as much as possible. However, it is clear that these activities cannot compensate for the lack of established partnerships between individual Indian and Swiss researchers. As there was no joint project planning, it was increasingly difficult to find motivated Swiss partners, because their research interests often do not relate to Indian needs. The problem was not a lack of interest, but the Swiss partners would have liked to see a mutual benefit in participating in joint projects.

Under the agreement the project goal and objectives could not be redefined at this stage, nor could the subjects be changed. Neither did the agreement allow management to attract new Swiss partners by offering to fund joint research activities, which would have helped improve the networking. To better coordinate and monitor the activities, ISCB recruited a scientist with a background complementary to that of the program manager. This strengthened the partnership between ISCB management and the Indian scientists, which helped keep deadlines and focus on implementing planned activities more clearly.

To address the partnership problem, scientific contacts with more than one Swiss partner per project were established, based on Indian needs and requirements. To foster mutual scientific exchanges, technicians, PhD students, postdocs, and faculty members were invited to spend a longer period at the Indian partner institutes. For workshops and training courses, many Swiss scientists visited India, which enabled everybody to forge closer contacts and learn from each other, not only on scientific issues but also with regard to the different cultures. Networking of the activities and access to new Swiss partners was facilitated through the presentation of the program on various occasions.

### Organizational level: Mode of operation and management

The ISCB management team at ETH currently plays a rather unsatisfactory role. It has been forced into the position of troubleshooter, thus being *de facto* responsible for solving all organizational and administrative problems. Instead, all project partners should be made jointly responsible not only for the scientific and technical content of their projects, but also, as much as possible, for the management and administration of the projects. An attempt was made to strengthen the capacity in fund management in each project (including purchases, travel, and workshops) through separate agreements with the partners in each project.

This was a very time-consuming and inefficient exercise, however, for a number of reasons: (1) The Indian government initially did not approve direct transfers of funds to partner institutes, which led to delays in setting up the agreements. (2) The Indian partner institutes have limited access to foreign exchange accounts. They lose funds through currency exchange, and they must use the funds following the rules and regulations of the Indian institution (overhead charges, bookkeeping, purchase system). (3) Although most international companies have agencies in India, they usually do not accept Indian currency (each order still has to be reconfirmed by the ISCB). (4) Budget calls were often late and unrealistic. As a result of these constraints, the Indian partners preferred to have ISCB do their administrative work.

The monitoring system outlined in the program description remained unchanged. Experience shows that outputs can be measured in terms of time and magnitude using the project action plans that are submitted each year. The internal review process, however, could be improved by introducing standard formats and mechanisms.

## Main findings of the external evaluation

The general consensus emerging from a 1997 external program evaluation was that ISCB created considerable goodwill between the participating Indian and European scientists, which demonstrates how international cooperation in biotechnology can benefit both developing and developed countries. However, in order to make the program more effective, self-propelled, and sustainable, the evaluation team made a number of recommendations, the main ones being the following:

- To make collaboration in biotechnology between the two countries more sustainable, the project has to respond to the needs of both countries and must relate to the

changing context in India and Switzerland. In a global competitive environment, a high quality and high standards of organizational style and scientific integrity are required.
- The goal and objectives of the ISCB should be examined and redefined. The overall goal, which can be defined as "strengthening India's capacity for self-development," should be set on a long-term basis.
- Interactions between the research projects and better networking between them should be endeavored.
- Future thrusts should respond, on the one hand, to Indian needs and, on the other, to research interests and capacities in both India and Switzerland and the guiding principles of both governments.
- Genuine research partnerships should be supported, sharing of responsibilities among researchers should be determined clearly, and the framework for joint R&D should allow all partners to equally benefit from the collaboration, both scientifically and financially.
- Improved monitoring mechanisms should be developed for projects and the program, as well as for internal and external monitoring. Consequently, a new organizational structure should be established.
- At the program level, the Joint Program Committee should be strengthened, and the task of the program management should be made more meaningful and more relevant.
- Regarding the major topics that are to be addressed and considered when planning a new program, the review team recommended taking enough time for planning and also inviting the potential key actors to participate in it.

## Management implications: Towards a new ISCB Program

The implications for management are best described by the review comments that contributed to formulating a new perspective for the ISCB. A fundamental guiding principle for future changes in the project is that negotiating and planning should be done jointly with research institutions as well as with potential future users and producers, regulating state agencies (or other official institutions), authorities dealing with safety matters, and organizations or individuals dealing with ethical questions. Moreover, project preparation must focus on the managerial capacities and flexibility of the concerned research institutions.

## The planning process

In August 1997 planning for a renewed program began. In March 1998, a task force comprised of representatives of DBT and SDC, an expert from ISNAR's Biotechnology Service (IBS), as well as scientific resource persons from India and Switzerland was charged with designing the new ISCB. A consultant was hired to moderate the exercise.

To set priorities among research sectors, two surveys were conducted in Switzerland. In a first step, more than 140 biotechnology R&D institutions were approached to make suggestions for a future Indian-Swiss biotechnology program. The responding institutions were invited to provide some background on their capacities and expertise and to indicate whether they were interested in collaborating in the project. The first survey indicated that the Swiss biotechnology institutions ranked health, agriculture, and environment as priority areas. (SDC's guiding principles in India, however, excluded human health as a potential thrust area.) A major focus was put on R&D and technology transfer through joint research and training modules. Using the surveys, the task force identified potential Swiss partners and their specific interest in collaboration.

On a subsequent exploratory mission to India, the task force outlined the elements for a planning platform, including a list of Indian and Swiss resource persons for a two-day planning workshop. Held in Delhi, the workshop helped the participants reach a consensus on the logical framework of the new ISCB and its thematic areas, list key guiding principles, and sketch a preliminary organizational framework for the program. At a first task force meeting, guidelines for identifying potential project partners were established and invitations for project "preproposals" were distributed. Also, SDC and DBT were invited to indicate preferences regarding ISCB's new organizational structure. At a second task force meeting, more than 100 project preproposals were reviewed, and potential project partnerships were shortlisted. The key elements for a call for project proposals were also laid out. Three "information days" were held to present the objective and scope for ISCB and to encourage networking among Swiss and Indian research institutes. On these days, a "first call for project proposals" document was handed out to all invited institutes.

### Defining the premises

At an early stage of the preparation process, a set of premises was defined, the most important of which was that the new ISCB should address fundamental development concerns in India, including poverty alleviation for rural and urban populations and sustainable management of natural resources. This paved the way for the second premise: the adoption of the concept of an "integrated value chain" in biotechnology R&D. The value chain concept is widely used for analyzing trends in the biotechnology industry, but it can also be applied as a planning and management instrument to move research activities to product development and diffusion. The concept is best understood as a chain of events that starts with a problem definition and ends with sustained market penetration of a new product, process, or service (see figure 12.1).

The integrated value chain is particularly important to a system where needs-based and product-oriented R&D is carried out. The reason to apply the chain is to clearly locate a given project along the chain and to identify commitments required in terms of time and resources to achieve a given milestone. It also requires a clear idea upfront of the expected

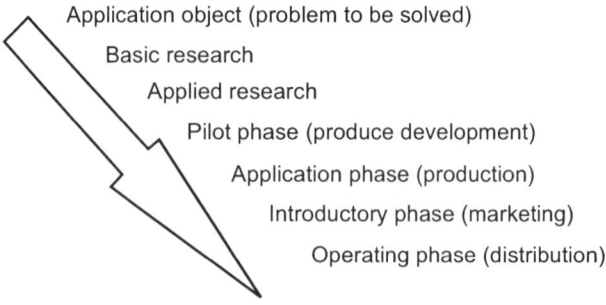

**Figure 12.1.** General sequence of the integrated value chain

final products and processes, the intermediary stages to be implemented, and the project's monitoring and potential for success.

A third important premise during program preparation was that although it ultimately addresses some fundamental development goals, the new ISCB focuses on certain niches in the thematic areas of "agriculture" and "environment" while supporting research and training on "transsectoral" topics. Consequently, transsectoral topics such as technology assessment and transfer, biosafety, ethics of biotechnology, and intellectual property management are to be given due importance in an area that is as innovative as it is controversial.

Last, the planning process also considered the needs and interests of the project partners. The initiative to foster collaborative R&D partnerships, with at least one Indian and one Swiss partner, is a new feature of the ISCB, allowing for funding of Swiss partners involved in the projects.

## Initial selection of projects

In a first selection round, after analyzing the preproposals, a crop-based approach was chosen for a call for project proposals. Selected candidates were invited to submit joint proposals on the production of wheat and pulses in semi-arid and rainfed agricultural systems. At the time of writing, the project proposals were to be peer-reviewed by independent scientific advisors. After a shortlist of proposals is prepared, a second call for transsectoral topics will be announced. The Joint Apex Committee, which is currently being formed, will then select the projects.

All partners in the "old" ISCB program were invited to participate in the project-submission- and -review process. It was clearly stated that their prior involvement does not guarantee future participation. The competitive approach adopted for the new program will show whether the collaboration so far has been successful. Partnerships that had been formed in thrust areas that are no longer included in the new program were terminated by March 1999.

## Program organization and management

The essential functions of the new program are highlighted in table 12.3. In the five-year bilateral agreement between the two governments, which defines the framework for the collaboration, the Indian government designated DBT as the funding agency for the program and Switzerland appointed SDC. A Joint Apex Committee will assume overall responsibility for program implementation. This eight-member committee will consist of representatives of SDC, DBT, universities, nongovernmental organizations, and industry. The committee monitors the program and defines its strategic orientation, approves projects, and (re)designates its members as well as the program management units. Scientific review and advisory functions are carried out by a pool of peer reviewers and advisors.

An important change in the program is the creation of a Program Management Unit based in India, complementing the Program Support Liaison Unit in Switzerland. These units will share management functions, and they are both accountable to the Joint Apex Committee.

Third, the ISCB organization and management will be shaped significantly by the fact that projects are jointly defined and managed by Indian and Swiss institutes or companies.

The project will be the reference point in scientific terms (leadership, coordinated implementation, and accountability). This is important for ownership and accountability at project level and will adequately be reflected in the project implementation contracts, which will also cover IPR, biosafety, and ethical issues, if applicable.

Table 12.3. The Former and New ISCB Programs Compared

| Elements | Former ISCB (1988–1999) (Phases 1–2: April '88–March '95 phase 3: April '95–March '99) | New ISCB (1999–2004) |
| --- | --- | --- |
| Planning process | Phase 1-3: Without prior identification of research priorities and participation of interested groups. | Participatory, with representatives from India and Switzerland from public and private research institutions, government agencies, and consultants. |
| Program goal & objective | *Goal:* ISCB has developed sustainable and effective R&D capacities in India. | *Goal:* ISCB has developed capacity and partnership along the entire value chain with strong economic, social and ecological relevance. |
| | *Objective*: Project partners have achieved a strong scientific track record and have established useful contacts with potential new partners and/or funding institutions. | *Objective:* Collaborative partnerships have produced verifiable progress in the value chain. |
| Thrust areas | Microbial biotechnology; veterinarian diagnostic; human health; waste management. | Main area: Environment, agriculture and transsectoral issues such as need assessment, biosafety, IPR, technology transfer. |
| | Phase 3: New areas for research collaboration restricted to SDC's policy in India. | |
| Project partners | Indian institutes. | Indian institutes, private firms, and their respective counterparts in Switzerland |
| Project review process | Phases 1–2: No project reviews. | Project review process and competitive selection based on quality, partnership, approach, and problem orientation. |
| | Phase 3: Project reviews and selection of final projects by JPC of a limited number of candidates. | |
| Organizational set-up | Phases 1–3: Overall responsibility with the funding agencies represented in the JPC (SDC, DBT). | Overall responsibility remains with the funding agencies. Extended Joint Apex Committee (SDC, DBT, academia, NGOs, industry), and additional Research Advisory Pool. |
| | Phase 3: Advisory group for project implementation and strategic support in Switzerland only. Flexible pool of scientific advisors and consultants. | |
| Management | Central position of the management team at ETH, Switzerland. | A new Program Management Unit in India complementing the liaison in Switzerland. |
| | | Designation of scientific and operational leadership to the projects. |

| Elements | Former ISCB (1988–1999) (Phases 1–2: April '88–March '95 phase 3: April '95–March '99) | New ISCB (1999–2004) |
|---|---|---|
| Funding | Phases 1–3: DBT funds directly to the projects. Phases 1–2: Use of Swiss funds through the management unit for the Indian projects. Phase 3: Indian institutes with "Terms for the use of the Swiss funds." 10 years: around 10 million SFr from SDC; matching funds by DBT with less than 20% for Indian projects. | Direct funding by DBT to the projects in India. Swiss funds for the Indian projects through the PMU and for the Swiss partners through the PSLU. Accounting within the projects. Swiss funds for specific research activities in Switzerland. 5 years: around 10 million SFr from SDC; matching funds by DBT with an estimated 50 % for Indian projects. |
| Agreements | Bilateral program agreement (SDC, DBT) for 3 years. Bilateral agreement between SDC and ETH for program implementation. No project implementation contracts. | Bilateral program agreement (SDC, DBT) for 5 years. Tripartite agreements (PMU, SDC, DBT) and bilateral agreement (PLSU, SDC) for program implementation. Project implementation contracts for 3 years that will also cover project and financial organization, project evaluation and review, biodiversity, IPR, and ethical issues (when applicable). |

# SECTION IV

# Ensuring Environmental Responsibility

Biotechnology can have a positive impact on food security and can contribute to the sustainability of modern agriculture. It may be able to reduce some of the negative impacts of current agricultural practices, and it may help stave off impending agricultural and environmental crises. However, the use of biotechnology to address constraints in agricultural production brings with it questions regarding the potential of genetically modified organisms (GMOs) to cause unacceptable impacts on human health and the environment. Use of biotechnology and its products must therefore be practiced in a safe and sustainable manner that minimizes the possibility of adverse effects.

As explained in chapter 13 ("Biosafety Management: Key to the Environmentally Responsible Use of Biotechnology"), a biosafety system can help ensure environmentally responsible use of biotechnology and its products. Biosafety **guidelines** are the most visible part of such a system. Equally important are the **people** involved, the biosafety **review process,** through which potential risks are assessed and management strategies identified, and **feedback mechanisms** to incorporate new information into the system. These four elements are interdependent. For example, people who conduct biosafety reviews do so within the regulatory framework articulated in the guidelines. The extent to which the guidelines are transparent, science based, and flexible is reflected in the quality of the decisions that biosafety reviewers take. These decisions, in turn, are a product of the competence and confidence of reviewers and applicants. All are strengthened by accumulated experience and new information that is fed back into the system.

In a well designed biosafety system, the dynamic nature of the whole system is also characteristic of each individual element. In formulating guidelines for field-testing, the Philippines took great care to make their procedures adaptable to the changing backdrop of scientific accomplishment and environmental and ecological insight (chapter 14, "Formulating Guidelines for Field-Testing in the Philippines"). That adaptability provides ways to correct unintended and unwanted barriers to the use of biotechnology that may arise in the course of implementing biosafety systems.

Importantly, the Philippines' biosafety guidelines are structured to elicit information needed for sound decision making. They promote a regulatory atmosphere in which people learn from their own experiences and are open to lessons gleaned from other countries. Finally, they use a conservative step-by-step approach to accumulate the information needed. Some of this information can be gathered from experiments conducted under confinement. However, limited field evaluations are essential to provide other, additional information.

Ultimately, the fate of agricultural biotechnology is in the hands of the public. It is the public's acceptance, or refusal, of GMOs and geneticaly modified foods that will determine whether or not the potential benefits of this technology are realized. The ongoing contentious debate in Europe has dramatically increased global awareness of biotechnology and the use of GMOs. Public acceptance, however, has not followed the same course. Farmers and consumers in Europe and a growing number of other countries have expressed serious concerns about the long-term safety of GMOs and genetically modified foods. High profile media stories have reported organized boycotts and demonstrations, filing of lawsuits, vandalized field test sites, and disruptions in international trade. Public demand for labeling, which would allow consumers to choose whether or not they buy genetically modified foods, is growing. Efforts to advance international cooperation in biosafety stalled in late 1998 due to the failure of 138 signatories to the Convention on Biological Diversity to ratify a legally binding Biosafety Protocol. These recent developments signal an urgent need for countries seeking to use GMOs to mount responsible initiatives for communicating with the public.

Public acceptance of biotechnology depends on the extent to which consumers feel they have the information they need to make informed decisions. Building an understanding based on science, rather than speculation, is the fundamental goal of programs that address public health and safety concerns. A variety of approaches, ranging between crisis management, persuasion, facilitation, and education, have been used. Crisis management is a reactive approach often seen when opposition groups provoke the public with half-truths and unscientific, emotionally-charged campaigns. Caught unprepared, biotechnology proponents are likely to deliver an ineffective response. More proactive, but slower facilitation and education methods have longer-lasting impacts on public acceptance. Each approach has its own management implications in terms of time frame, cost, and impact.

Efforts underway in Japan, which are reviewed in chapter 15 ("Addressing Public Acceptance Issues for Biotechnology: Experiences from Japan"), provide a good example of a multifaceted approach to informing and educating the public. Information is disseminated in a variety of formats—classes, symposia, brochures, and Internet resources—using understandable language. Managers therefore need to be proactive in building public acceptance based on sound, well articulated, scientific principles.

Biotechnology applications can be evaluated for potential negative and positive impacts on health and the environment even at the planning and research stage. Chapter 16 ("Balancing Needs for Productivity and Sustainability—Genetic Engineering of Rice at IRRI") illustrates how principles embodied in the biosafety system can be translated into sound technological decisions. DNA constructs to be inserted in crops are scrutinized for biosafety consequences arising from the choice of gene, promoter, selectable marker, transformation method, and other factors. Technical options that provide for sustainability may feature polygenic strategies, use of tissue-specific promoters, or inducible expression systems.

Technical options can also influence appropriate deployment strategies. For many transgenic traits, inappropriate deployment may reduce opportunities to apply biotechnology to agricultural crops. Further, large investments of human, financial, and other resources are squandered if the technology is deployed in a way that brings about its own demise. The utility of engineered pest- or disease-resistant crops is lost when pests and pathogens resistant to the control mechanism emerge. Chapter 17 ("Managing Target Pest Adaptation: The Case of Bt Transgenic Plant Deployment") appraises responsible deployment principles for crops modified to express Bt genes. Applying these principles requires detailed knowledge of the target organism and its capacity to adapt, as well as knowledge of the gene's non-target impacts.

Resistance management programs should be an integral part of the deployment strategy for crops engineered to resist disease or insect attack. Given the lack of motivation for end users to implement resistance management plans, it falls to the GMO producer to find ways to incorporate resistance management into the product, for example by engineering time- and tissue-specific expression of the transgenic control mechanism. Alternatively, producers can look for ways to increase incentives for resistance management planning and adoption and monitor subsequent compliance with the prescribed program.

## Recommendations

Individual chapters in this section address the management challenges associated with the topics introduced briefly above. Based on personal experience and insight, the papers present lessons learned, strategies and possible solutions, as well as examine some of the costs. The main recommendations from these accounts are summarized here:

- Establish an integrated, responsive and flexible biosafety system, ensuring functionality of all parts.
- Train people at all levels so that they are competent in their duties and confident in their abilities to implement biosafety policy.
- Formulate biosafety guidelines that are science-based and dynamic.
- Seek public input in the design and operation of the biosafety system.
- Use proactive, targeted measures to build public acceptance.
- Respond to public concerns with rational and scientifically sound information.
- Incorporate biosafety principles into research planning and technical decision making.
- Protect biotechnology research options and investments by sound and responsible field deployment strategies.
- Incorporate pest resistance management strategies in the deployment of pest- or disease- resistant crops.

# 13 Biosafety Management: Key to the Environmentally Responsible Use of Biotechnology

*Patricia L. Traynor*

## Abstract

*Agricultural crops improved through genetic engineering raise expectations of enhanced agronomic, nutritional, and marketing qualities, as well as numerous other benefits. These same crops also raise concerns about their potential long-term effects on human health and the environment. Such concerns have prompted developed and developing countries alike to implement guidelines for the safe use and handling of genetically modified organisms (GMOs) in the environment. Appropriate oversight of GMO releases in the environment is achieved through the establishment of a biosafety system in which guidelines are the most visible part. Equally important in a biosafety system, however, are three other elements – the people involved, the biosafety review procedure, and mechanisms for feedback. These four elements function together to produce environmentally responsible decisions. Establishing and maintaining a functional, effective biosafety system presents management challenges at every step. It requires education and coordination across government ministries, universities and research institutes, private-sector interests, individual scientists, and the public. Significant investments may be needed in training and human resource development, information and communications systems, facilities, and follow-on activities. In facing this task, it is worth bearing in mind that the elements in a biosafety system, and the challenges they present, are interrelated and interdependent. Therefore efforts to strengthen any one part will ultimately strengthen the entire system and the biosafety decisions coming from it.*

## Introduction

Biosafety is associated with the use of genetically modified organisms (GMOs) and, more generally, with the introduction of nonindigenous species into natural or managed ecosystems. A relatively new concept in agricultural research, it tempers the adoption of a new technology by carefully considering its potential effects on human health and the environment.

Biosafety emerged as a global priority in chapter 16 of Agenda 21, adopted during the United Nations Conference on the Environment and Development, and in Articles 8(g) and 19 of the Convention on Biological Diversity. In a series of regional policy seminars organized by ISNAR's Biotechnology Service (IBS), biosafety was designated a high

© CAB *International*. 1999. *Managing Agricultural Biotechnology—Addressing Research Program Needs and Policy Implications* (ed. J.I. Cohen)

priority for countries in Southeast Asia, Africa, and Latin America, regardless of their level of indigenous technical capacity (Cohen et al. 1998). Building competence in biosafety is thus strategically important to successfully integrating biotechnology into agricultural research.

Biosafety is achieved by assessing and managing environmental risks, evaluating potential ecological consequences, and weighing these against potential benefits. A step-by-step, case-by-case approach fosters a deliberate, informed, and environmentally responsible use of biotechnology. Policymakers begin by naming a lead agency under which biosafety policy is to be implemented. The next step is usually to draft biosafety guidelines that provide a framework for administrative procedures and decision making regarding the appropriate use of GMOs.

Managers charged with implementing biosafety policy can do so more effectively if they think in terms of instituting a biosafety system, rather than simply writing a document. Even though the guidelines are the most visible part of all biosafety systems, three other components are equally important: (1) the people involved, (2) the review process they implement, and (3) mechanisms for feedback. All well-functioning biosafety systems have these four elements in common, though they may take different forms and have different degrees of importance in implementation. Taking this broader view, a system must be put in place in which the four elements function together to produce environmentally responsible decisions. Each element presents its own management challenges:

- The **guidelines** formulated should be transparent, science-based, and flexible.
- Competence and confidence must be built in the **people** who are involved.
- The biosafety **review process** must be appropriate.
- There should be means of ensuring **feedback** so that the system improves through experience.

These elements and the challenges that they present are related and interdependent. People who conduct biosafety reviews do so within the regulatory framework articulated in the written guidelines. The extent to which the guidelines are science based and clear is reflected in the quality of decisions that they take. These decisions, in turn, are a product of the competence and confidence of reviewers and applicants. All are supported by accumulated experience and new information that is fed back into the system. Efforts to strengthen any one part ultimately strengthen the entire system and all the biosafety decisions coming from it. In the same way, neglect of any one part will weaken the whole system, hampering sound decision making.

## The four elements of a biosafety system: Challenges and recommendations

### Guidelines

Biosafety guidelines set forth policies and procedures for ensuring the safe use of biotechnology and its products. Ideally, the document clearly defines the objectives and scope of regulatory authority, specifies the allocation of responsibility within government agencies, lists the duties and membership of national and institutional biosafety review committees, and describes application and review procedures. Guidelines should stipulate that scientific principles are the basis for risk assessment. They should be flexible so that recommendations for risk management are commensurate with the level of identified risk.

The inherent flexibility of biosafety guidelines determines the extent to which the review process can, over time, be increasingly focused on cases that may pose some risk. As experience and familiarity increase, the same flexibility would allow a gradual relaxation of regulatory oversight for those releases that are found to pose little or no risk to human health or the environment. To draft guidelines that meet these standards, managers must consider a number of factors, including the statutory form, scope of oversight, objectives, and the stakeholders.

## *Statutory form*

Biosafety policy can be implemented in the form of new legislation, adaptation or extrapolation of existing legislation, or through nonlegislative means such as a ministerial decree. In deciding which form to use, managers must weigh the merits and drawbacks of each, for example, in complexity, timeliness, flexibility, and cost. While a new law carries the weight and enforcement power of government regulatory authorities, it is difficult to amend, and a lengthy and expensive effort is needed to enact and implement it. A ministerial decree, in contrast, is faster and simpler to issue, and it is more readily amended or replaced. Without regulatory authority to enforce compliance, however, it may lack substance. Adaptation of existing laws, as was done in the USA, avoids the necessity of drafting new laws but can lead to redundancy and delays, for example, where a product such as Bt potatoes cannot be marketed until cleared by three separate regulatory agencies.

A biosafety system needs to be compatible with other regulatory authorities and procedures. For example, importation of seed, commodities, embryos, livestock, or biocontrol agents may be subject to plant protection and quarantine rules or regulations on animal health or food safety. Administrative jurisdiction, application procedures, and record-keeping requirements pertaining to biosafety should be coordinated with existing authorities in order to minimize duplication and conflicts.

Some countries, Australia being a good example, have developed guidelines in a form that serves a practical purpose in helping researchers and reviewers in risk assessment and risk management procedures (GMAC 1998). Applicants requesting approval for a GMO release are prompted to answer comprehensive and detailed questions about the proposed release, thereby providing the information necessary for evaluating potential risk.

## *Scope*

Guidelines should clearly articulate what is subject to biosafety review and what is not. Does the system apply to products of modern molecular techniques such as chromosomal manipulations and embryo rescue? Or does it apply only to research involving recombinant DNA methods? Are all agricultural plants, animals, and microorganisms considered? Are laboratory and greenhouse experiments included, or only releases into the environment? Will the guidelines address vaccines, pharmaceuticals, or industrial feedstocks such as lubricating oils produced in recombinant animals or plants? What about nonagricultural applications such as industrial fermentation technologies? Will there be provisions for evaluating commercial-scale releases? Will the safety of genetically engineered foods be evaluated the same way as nonbiotech foods? Answering these preliminary questions enables managers to determine the size of the task and estimate the resources they need to complete it.

## Objectives

A careful analysis of objectives can be one of the most useful preliminary steps in designing a biosafety system. By definition, the primary objective of biosafety systems is to ensure the environmentally responsible use of biotechnology in order to reap its benefits without causing significant, unintended negative effects on people or the environment.

There may be secondary objectives, however, which are more immediate. Having a policy and procedures in place may be seen as a way for developing countries to secure access to biotechnology products. When a multinational company is seeking to field-test GMOs, the initiative to develop guidelines may stem from a pragmatic decision to quickly "get something on the books" so that an opportunity for technology transfer will not be lost. Instituting a biosafety system may also serve political objectives. Examples are asserting national autonomy, facilitating international trade, and participating in global initiatives such as the Convention on Biological Diversity. The recent failure of the open-ended ad hoc working group on biosafety under the Convention on Biological Diversity to reach agreement on an internationally binding biosafety protocol puts added pressure on individual countries to develop their own guidelines.

## Stakeholders

Biosafety policies and procedures affect a diverse group of stakeholders. Scientists and project leaders engaged in field research are directly subject to biosafety regulations. The private sector also has a stake in how the guidelines address the use of GMOs. Local as well as multinational companies want a regulatory environment that is transparent, reasonable, and consistent. They have a stake in how commercial use of GMOs will be regulated in terms of application and approval procedures, environmental safeguards, and consumer issues such as labeling, in addition to matters of trade and importation.

For biotechnology regulators, biosafety is typically an added responsibility that must be integrated into their other duties. National biosafety committee members and technical reviewers, in particular, are invested in the biosafety system. Their decisions may be subject to local, national, and even international scrutiny. Government agencies concerned with biotechnology applications, typically including ministries of environment, health, and science and technology, and regional or local authorities have vested interests in how the system is established and run.

The public's concerns about biotechnology's environmental effects and food safety are based in part on past experience with scientific advances whose immediate benefits were trumpeted before longer-term negative effects were realized. This and concerns raised by environmental groups and consumers' organizations need to be addressed. One way to secure public input is to have a respected member of the community (e.g., a local official, medical specialist, teacher, or extensionist) serve on the local institutional biosafety committee. A system that actively seeks public input and openly addresses the risk-benefit issues is, in the long run, one of the most effective ways to build public acceptance of biotechnology and its products.

## People

The people working in a biosafety system are the public- and private-sector scientists seeking to test GMOs and members of the review committees who decide on proposed releases. The applicant and members of the biosafety review committee seem to have somewhat different roles, but they are engaged in a cooperative effort with a shared goal:

- The research scientist prepares a proposal to test transgenic plants in the field.
- Members of the institutional biosafety committee review the researcher's application.
- Members of the national biosafety committee are responsible for granting approval.

All those involved need to be familiar with the environmental issues associated with biotechnology products and have a working knowledge of the biosafety review process. They need to recognize what constitutes a potential risk and what risk management strategies may be applicable. Because advances come quickly and biosafety concepts are in constant evolution, all these people need ready access to the latest scientific, regulatory, and biosafety information.

## The review process

A biosafety review systematically evaluates a GMO, the site where it will be released, and the conditions under which the release will be conducted (see, for example, UNEP 1996; USDA 1994; and OECD 1992). The process is governed by the idea that, just as with any technology or human endeavor, there is no such thing as zero risk. Once the potential risk is identified, appropriate management procedures are specified to reduce it to an acceptable level. The emphasis of most biosafety reviews is on potential risk. However, applicants and biosafety review committee members also need to consider the potential benefits of allowing use of the GMO. Here they must recognize that *not* using a particular GMO may imply risks as well.

The review process begins with a proposal to introduce a GMO into the environment. (In some countries, greenhouse experiments with GMOs are also subject to biosafety committee review and approval.) The applicant has primary responsibility for assessing possible impacts or environmental consequences of a proposed release by comparing the GMO with the nonmodified or parental organism. This evaluation takes into account

- the nature of the organism that receives the new trait
- the donor organism from which the trait was derived
- the vector or mechanism used to transfer the trait
- the nature of the introduced trait, including potential toxicity of a gene product
- characteristics of the site or environment into which the GMO will be introduced
- elements of the release plan that provide containment or control

If a risk of sufficient probability and having consequences of sufficient magnitude be identified, the applicant determines suitable risk management measures that reduce risk to an acceptable level. This evaluation, conducted by the person most familiar with the GMO and documented in the release proposal, constitutes the first and most important level of biosafety assessment.

An *institutional* biosafety committee conducts the second level of review. Generally, it evaluates the proposal for scientific content and verifies that the proposed release conforms to institutional and national guidelines. Institutional biosafety committee members collectively should have sufficient expertise in the relevant scientific disciplines to conduct an independent risk assessment and make appropriate risk management recommendations, if needed. Final review and approval by a *national* biosafety committee may encompass nonscientific issues, such as accordance with quarantine regulations, international trade and treaty considerations, and socioeconomic impact.

Not surprisingly, there are ongoing discussions over what constitutes an environmental risk and what is the limit for acceptable risk. Currently, biosafety reviews generally focus on a limited number of environmental issues associated with the release of transgenic crops (see, for example, Snow and Palma 1997; Kendall et al. 1997). Two of these that concern the possibility that crops or their relatives may invade new territory, displace existing plant communities, or reduce species biodiversity. They may have added importance in regions that are centers of origin or diversity for the crop.

Weediness—the potential for a crop to become established and to persist and spread into new habitats as a result of newly introduced genes—is an issue when there is scientific evidence that acquisition of the new genes is sufficient to convert a domesticated species into a successful weed. Geneflow—in which new genes are spread by normal outcrossing to wild or weedy relatives of the engineered crop—becomes an issue if the new trait(s) confers a fitness advantage and becomes stably introgressed into the recipient genome. Toxicity is an issue associated with human health concerns over allergenicity and the safety of biotechnology foods and potential negative effects on nontarget organisms, especially beneficial species. Pest and pathogen effects include a range of possible consequences such as the generation of novel viruses by molecular exchange within a transgenic plant, or emergence of target pest populations resistant to an engineered control mechanism.

For many biotechnology products, none of these concerns is relevant. For example, based on long experience with conventionally bred "paste" tomatoes grown for processing, there is no evidence that tomatoes that are genetically engineered to have a higher solids content are more likely to become weeds than regular tomatoes. Also, there is no ecological evidence that increased solids confer a fitness advantage, and there is no reason to assume that pests or pathogens will be altered by their normal interactions with these GMO tomatoes. Issues of toxicity or allergenicity are not raised by increased amounts of normal fruit constituents. If, however, a new marker gene or other gene product is involved, toxicity can be readily tested prior to release. It is incumbent upon applicants and reviewers to recognize those cases where a particular combination of crop, gene, and environment may present legitimate questions that need to be addressed before the GMO is grown in certain areas or on a large scale.

Many "decision support tools" have been developed to guide reviewers through the biosafety review process (see, for example, Tiedje et al. 1989; Persley et al. 1993; Rissler and Mellon 1994). These charts, tables, decision trees, flow diagrams, etc. are not intended to supply yes-or-no answers, rather they can be used as roadmaps or checklists to ensure that all issues have been considered. An ongoing ISNAR research activity is also geared to help in this regard (see box 13.1).

## *Feedback mechanisms*

New information and accumulated experience are essential inputs that allow biosafety systems to evolve. Feedback mechanisms are the built-in supply lines through which guidelines, people, and the review process keep up with the rapid pace of scientific advance and an accelerating rate of small- and large-scale releases of GMOs worldwide. Access to

> **Box 13.1.** Biosafety Research at ISNAR
>
> ISNAR's Biotechnology Service (IBS) is expanding its biosafety activities to support the appropriate application of biotechnology methods to enhance agricultural productivity and the environmentally responsible use of its products.
>
> A two-year collaborative research project has been set up to (1) assess the impact of a genetically engineered crop being commercially released in two partner developing countries, and (2) review biosafety policies and procedures associated with the introduction, in order to assess the efficacy of their biosafety systems. This work will lead to a set of recommendations to address identified gaps in the technical, human, and information resources needed to strengthen biosafety capacity.
>
> Recommendations may address how, for example, to clarify application and review procedures; to facilitate consensus on risk assessment/risk management issues; to share biosafety data and pool information among agencies, institutions, and regional neighbors; and to identify opportunities for cooperative use of human and financial resources. The same strategy may be applied in subsequent years to other crops, other countries, and other regions, as IBS client needs dictate.

technical information and data gathered from prior releases is essential to support and strengthen subsequent biosafety decisions. At the same time, the accumulated experience of biosafety reviewers constitutes another source of feedback that can be used to improve oversight procedures. In a well-designed system, scientific and procedural feedback operate simultaneously to improve the quality of biosafety decisions.

*Scientific feedback*

Scientific feedback comes from external as well as internal sources. In evaluating a proposed release, biosafety review committee members may benefit from the experience of other countries by considering the acceptability of data from similar releases conducted elsewhere. Databases of field test information and biosafety reviews from the USA, Europe, and Australia can be useful to developing nations, so long as potentially significant differences in the environment, affected ecosystems, and agronomic practices are recognized.

Where there are specific gaps in knowledge concerning potential risks, issue-oriented meetings can be convened to clarify what is known and what management practices are appropriate. For example, there is some uncertainty about the potential to generate new plant viruses in transgenic plants expressing viral genes that confer virus resistance. Biotechnology regulators at the Animal and Plant Health Inspection Service of the United Stated Department of Agriculture (USDA) invited more than 20 prominent virologists to discuss the issue and propose experiments that would provide data to support more accurate risk assessment (AIBS 1995).

Review committees anticipating a need for data may specify reporting and record-keeping requirements as a condition of approval. Where appropriate, monitoring procedures specified as a condition of approval can be tailored to generate useful information. This type of feedback allows institutional and national biosafety review committees to gather information and identify emerging issues of local or regional importance. The relevance and usefulness of such feedback, however, will depend on the quality of data submitted. If, for example, monitoring to determine survival of a GMO or to detect transgene flow via pollen movement is required, it should be conducted in order to provide scientifically valid data. Subjective observations or anecdotal evidence are of limited value, but appropriate monitoring and reporting data can support subsequent decisions on proposed releases.

*Procedural feedback*

A periodic re-evaluation of guidelines and implementation procedures gives applicants, reviewers, regulators, and the public an opportunity to assess how well the system is working. Are applications reviewed on a timely basis? Is the burden of paperwork excessive or redundant? Can the review process be streamlined without compromising environmental responsibility? Are there new factors to consider, such as an administrative reorganization or a change in the budget? Is the public satisfied that their health and safety concerns are being properly addressed?

Implementing biosafety policy does not end once guidelines are written, people are trained, and reviews are conducted. It is a dynamic process that evolves through mechanisms for incorporating new information. As reviewers' experience and familiarity with introductions of GMOs increase, flexible regulations allow a gradual relaxation of oversight for those releases that pose little or no risk to human health or the environment. Limited biosafety review resources can then be focused on cases that do present some element of risk.

## Human resource development and other costs

Establishing a biosafety system imposes substantial human resource development (HRD) costs on the designated organizations or institutions. The major HRD cost is for various kinds of biosafety training, as discussed below. Loss of productivity can impose a substantial cost when researchers and managers are diverted from their regular jobs to draft guidelines and conduct biosafety reviews. Site visits, facility inspections, and monitoring activities usually fall to regulators or institutional biosafety committee members who still have all their regular duties. As the number of proposals increases, additional personnel may be needed to serve as liaison to potential applicants, to handle applications and decision documents, to process reports, and to perform other information management tasks. Last, and perhaps most important to ensure long-term access to the benefits of biotechnology products, designated people will be needed to engage and inform the public.

### Training

Biosafety training builds the competence and confidence of scientists, biosafety reviewers, and regulators alike. It enhances their awareness of environmental issues and potential consequences, and it provides a systematic approach to the evaluation of proposed GMO releases. A number of international agencies and programs support various forms of introductory and technical biosafety training at the institutional, national or regional level. Among these are the United Nations Industrial Development Organisation (UNIDO), the United States Agency for International Development (USAID), the Rockefeller Foundation, and the United Nations Environment Programme (UNEP). Many government agencies are also active in biosafety education and training.

Short and long training programs should be accessed on a regular basis. Biosafety, like biotechnology, is dynamic; new products, new technologies, and new concerns will continue to appear, presenting a moving target for regulators as well as research managers. There are one- or two-day programs that introduce current and future agricultural biotechnology applications, their potential benefits and associated risks, and how the biosafety system operates to guard against harm. These are useful to raise awareness and broaden the perspectives of senior people who influence policy decisions, manage research, or communicate with the public. Longer, more intensive technical training gives members of institutional and national biosafety committees, regulatory officials, and ad hoc reviewers a thorough grounding in how to apply science to the risk assessment/risk management process. It affords practical experience in reviewing field-test proposals in accordance with national guidelines.

In selecting candidates for technical biosafety training, managers must make the difficult choice between exposing many people to at least one biosafety course versus providing recurrent training to fewer people. There is value in providing initial training for as many people as resources will allow, as it tends to create the "critical mass" needed to keep biosafety under discussion once the workshop is over. Further, it is only realistic to acknowledge that conflicts of interest can arise when a shortage of technically trained people results in the same few individuals being involved with both the research and the review.

On the other hand, becoming competent and having a sense of confidence in making biosafety decisions comes with practice. Providing more training to fewer people means that when there are field-test applications to evaluate, those who have had opportunities to review a variety of proposals under the guidance of experienced instructors will bring more skill and confidence to the review committee.

In selecting candidates to serve on biosafety review committees, managers need to identify people with suitable expertise who can bring good analytical skills to the job. The ability to discriminate, to discern what is relevant and what is tangential; the capacity to recognize what is scientifically credible and what is merely anecdotal; the competence to judge the applicability and acceptability of data; all of these skills are crucial to making sound biosafety decisions.

## *Other costs*

It takes time and money to establish and maintain a biosafety system. A partial list of commonly encountered costs includes the following:

- **Meetings.** Biosafety review committees and teams that draft guidelines are usually composed of people from different institutions, often in different cities.
- **Consultations with international experts.** Outside expertise used in the early stages allows managers to capitalize on what has been learned elsewhere.
- **Access to information.** People in the biosafety system and outside stakeholders need to be informed of national and international biosafety information.
- **Monitoring.** Costs depend on who follows up, what they look for, and what they report.
- **Record keeping/information management.** Paperwork needs to be centrally organized and accessible. Documents and procedures need to be distributed to interested parties.

- **Administrative overhead.** Telephone, fax, photocopying, e-mail, and Internet access charges are recurring expenses.

Anticipating these costs, and knowing that there will likely be others, gives managers a more realistic basis for establishing a biosafety implementation plan.

## Managing biosafety: Measures of success

An effective biosafety system fosters the judicious use of biotechnology to improve agricultural productivity and food quality and secure economic benefits, while protecting human health and the environment. Such a system is one in which the four elements are balanced and effective:

- The **guidelines** clearly define the structure of the biosafety system, the roles and responsibilities of those involved, and how the review process is to operate.
- The **people** are knowledgeable and well-trained, confident in their ability to make decisions, and supported by their institutions.
- The **review process** is based on up-to-date scientific information; focuses on specific combinations of crop, gene, and environment; promotes appropriate risk management practices; and balances risks against benefits;
- **Feedback** mechanisms are used to incorporate new information and revise the system as needed.

Criteria such as these are open to interpretation and not always measurable. More objective measures of success can provide a better assessment of how well the biosafety system is working. Objective criteria to use could include the following:

- The system is operational: applications are being submitted, reviews are being conducted, and decisions are being made.
- Decisions have substance and weight; there is no need to seek repeated consensus on the same issues.
- Proponents, reviewers, and regulators have participated in appropriate training courses; training is an ongoing effort.
- Compliance is the norm; enforcement measures are seldom if ever needed.
- Stakeholders remain informed and involved by attending regular update and advisory meetings.
- Stakeholders collectively view the system as adequate; a consistent message is communicated to the public.

Biotechnology is one of the tools that will help solve our shared and increasingly critical problem of ensuring an adequate and sustainable food supply for all people. It opens up new possibilities for improving crops and livestock for higher yield and quality, enhanced nutritional content, improved storage and processing characteristics, and many other desirable traits. It also provides a means for developing alternative uses for agricultural crops, and even remedying environmental problems through bioremediation. Agricultural research managers charged with implementing biosafety policy play a major role in ensuring the environmentally responsible use of this powerful technology.

# References

AIBS. 1995. Transgenic Virus-Resistant Plants and New Plant Viruses. Proceedings of an American Institute of Biological Sciences workshop sponsored by the USDA Animal and Plant Health Inspection Service and the Biotechnology Industry Organization. Washington, D.C.: American Institute of Biological Sciences.

Cohen, J.I., C. Falconi, and J. Komen. 1998. Strategic Decisions for Agricultural Biotechnology: Synthesis of Four Policy Seminars. ISNAR Briefing Paper 38. The Hague: International Service for National Agricultural Research.

GMAC (Genetic Manipulation Advisory Committee). 1998. Guidelines for the Deliberate Release of Genetically Manipulated Organisms. Canberra: Ministry for Industry, Technology, and Commerce, Commonwealth of Australia.

Kendall, H.W., R. Beachy, T. Eisner, F. Gould, R. Herdt, P.H. Raven, J.S. Schell, and M.S. Swaminathan. 1997. Bioengineering of Crops. Report of the World Bank Panel on Transgenic Crops. Washington, D.C.: The World Bank.

OECD. 1992. Safety Considerations for Biotechnology. Paris: Organisation for Economic Co-operation and Development.

Persley, G.J., L.V. Giddings, and C. Juma. 1993. Biosafety: The Safe Application of Biotechnology in Agriculture and the Environment. ISNAR Research Report 5. The Hague: International Service for National Agricultural Research.

Rissler, J. and M. Mellon. 1994. Perils Amidst the Promise. Cambridge, MA: Union of Concerned Scientists.

Snow, A.A. and P.M. Palma. 1997. Commercialization of Transgenic Plants: Potential Ecological Risks. *BioScience* 47:86-97.

Tiedje, J.M., R.K. Colwell, Y.L. Grossman, R.E. Hadson, R.E. Linski, R.N. Mack and P.J. Regal. 1989. The Planned Introduction of Genetically Engineered Organisms: Ecological Considerations and Recommendations. *Ecology* 70:298-315.

UNEP. 1996. International Technical Guidelines for Safety in Biotechnology. Nairobi: United Nations Environment Programme.

USDA. 1994. Guidelines for Research Involving Planned Introduction into the Environment of Genetically Modified Organisms. Supplement to Minutes, Agricultural Biotechnology Research Advisory Committee. Washington, D.C.: Office of Agricultural Biotechnology, U.S. Department of Agriculture.

# 14 Formulating Guidelines for Field-Testing in the Philippines

*Emerenciana B. Duran*

## Abstract

*The Philippines was one of the first countries in Asia to develop a biosafety system. This chapter provides an overview of the management challenges and lessons learned in formulating guidelines for field-testing and for implementating biosafety guidelines in general in the Philippines. The most important message is that biosafety guidelines should be based on sound science and management, implementable within the conditions prevailing in the country, and structured in a way that enables rational risk analysis. Biosafety guidelines provide the framework for ensuring the safe use of and benefits from the products of modern biotechnology.*

## Introduction

In developed countries, modern agricultural biotechnology involving genetic engineering has advanced rapidly; many products are being field-tested and some are being grown commercially. But also in developing countries, which rely heavily on agriculture to feed to their rapidly growing population and to fuel their economy, genetic engineering is being regarded as a powerful tool to increase the potential and capabilities of breeding practices for crop improvement.

But genetic engineering, like any other technology, also has risks. In the mid-1980s, concern among Filipino scientists about risks associated with genetic engineering prompted the creation of an Ad Hoc Committee on Biosafety. Composed of representatives of the University of the Philippines in Los Baños, the International Rice Research Institute, and the Department of Agriculture, the committee published a report in 1987 with several recommendations, the most important of which were to create a national biosafety committee and to formulate national policies and guidelines on biosafety. The committee based its proposed biosafety guidelines on those of Australia, USA, and Japan.

© CAB *International*. 1999. *Managing Agricultural Biotechnology—Addressing Research Program Needs and Policy Implications* (ed. J.I. Cohen)

## Biosafety system in the Philippines

### National Committee on Biosafety of the Philippines

In October 1990, then Philippine President Corazon C. Aquino issued Executive Order No. 430 to create the National Committee on Biosafety of the Philippines (NCBP). While a presidential executive order does not give the authority to make policy, an executive order is easier to amend than a legislative act. This was considered an advantage as the Philippines had little or no experience with biosafety regulations at the time.

The NCBP is composed of 10 members: the undersecretary for research and development of the Department of Science and Technology (chair), one biological scientist, one environmental scientist, one physical scientist, one social scientist, two members of the public, and one representative from each regulatory agency: the Department of Agriculture, the Department of Environment and Natural Resources, and the Department of Health. The NCBP is aided in its functions by these regulatory agencies, the institutional biosafety committees (IBCs) and the ad hoc technical subcommittees (see figure 14.1).

The NCBP provides scientific expertise and regulatory experience and ensures community representation within it. The four scientist members must have a minimum of seven years of academic and post academic training (degree and/or nondegree). The term of office of the chair coincides with his/her appointment as undersecretary. The term of office of the scientist members and the community representatives is three years, renewable for another term or more under exceptional circumstances. Representatives of the regulatory agencies are designated by the heads of their respective agencies.

The mandate and functions of the NCBP are the following:

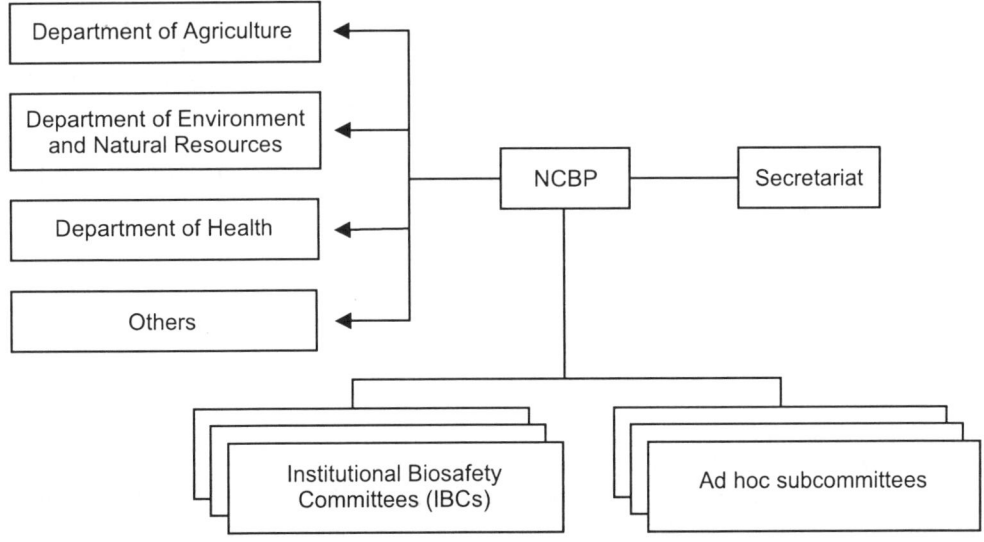

**Figure 14.1.** How the NCBP coordinates with existing regulatory agencies

- identify and evaluate potential hazards involved in initiating genetic manipulation experiments or introducing new species and genetically modified organisms (GMOs), and recommend measures to minimize risks
- formulate and review national policies and biosafety guidelines
- develop working arrangements with the government quarantine services and institutions for the evaluation, monitoring, and review of projects vis-à-vis adherence to national policies and guidelines on biosafety
- hold public deliberations on proposed national policies, guidelines, and other biosafety issues
- provide assistance in formulating and amending relevant laws, rules, and regulations

### Institutional biosafety committees

An IBC is composed of at least five members. Three may be staff of the institution and must be able to assess the safety of research in general and of the specific area of specialization of their organization in particular. In multidisciplinary institutions, such as academic institutions, IBC members must represent the different areas of research specialization in the organization. At least two members must not be affiliated with the organization but represent the interest of the surrounding community with respect to health and protection of the environment. IBCs may have consultants on-call to assist in assessing and managing risks. At the time of writing, 54 IBCs had been established in the country.

### Regulatory agencies cooperating closely with the NCBP

The NCBP also relies heavily on existing regulatory agencies to enforce the guidelines and to avoid duplication of functions. In addition, these agencies have policy-setting authority, an established network, and the necessary personnel to monitor releases as covered by the guidelines. These agencies are the following:

- **Department of Agriculture.** The Bureau of Plant Industry and the Bureau of Animal Industry within the Department of Agriculture play a crucial role in enforcing the guidelines through their quarantine services. Through these agencies, the movement of GMOs and pathogenic exotic organisms is regulated and monitored. The NCBP also interfaces with the Fertilizer and Pesticide Authority in registering biological control agents.
- **Department of Health.** A representative of the Bureau of Food and Drugs of the Department of Health is a member of the NCBP. The Bureau is responsible for registering processed food products, pharmaceuticals, and animal feed. The Bureau will play a crucial role in ensuring biosafety in the very near future, as imported genetically modified food, products, drugs, and animal feed enter the domestic market. Similarly, GMOs produced in the Philippines may also reach the commercial stage at some point in the future. At present, BFAD has no specific guidelines or policies regarding GMOs. It refers matters with biosafety concerns to the NCBP.
- **Department of Environment and Natural Resources.** The Department of Environment and Natural Resources is the focal point for the Convention on Biological Diversity. One of the risks associated with GMOs is their potential adverse impact on genetic resources and biodiversity. Hence, the Department of Environment and Natural Resources is also represented in the NCBP. The NCBP will also refer to

the Department any proposed work with GMOs that requires an environmental impact assessment and monitoring.

## The Philippine Biosafety Guidelines

Published in 1991, the first Philippine Biosafety Guidelines evolved from the 1987 report of the Ad Hoc Committee on Biosafety. The original guidelines covered GMOs as well as exotic organisms, although the latter are not genetically modified. They include procedures for evaluating proposals with biosafety concerns, procedures for the introduction, movement, and field release of regulated materials, and descriptions of physicochemical and biological containment.

Feedback from users and research proponents indicated that the guidelines were difficult to understand, the numbering system for physical and biological containment was not uniform and caused confusion, and data requirements were the same irrespective of scale and type of organism (microorganism, plant, or animal). Some proponents criticized the NCBP for being too strict, while some nongovernmental organizations (NGOs) contended that the NCBP was lax to allow the importation of Bt rice, even for research purposes under contained conditions. But the NGOs shared the general perception that the guidelines were among the strictest in the world. Under the 1991 guidelines, the NCBP assessed nearly 80 proposals, all of which were for contained work.

In the course of the first half of the 1990s, feedback from proponents about shortcomings in the guidelines prompted the NCBP to charge a technical working group, consisting of the four scientist members of the NCBP, to review the existing guidelines and draft new ones.

## Management challenges in formulating and implementing biosafety

### National level

Due to NCBP's limited budget, members of the NCBP typically hold full-time jobs in other organizations and serve on the NCBP in addition to their regular duties. The committee was confronted with two major constraints in executing the functions mandated by the 1990 executive order:

1. *How can a committee of only 10 members with a limited budget perform all of its mandated functions?* The NCBP addressed this constraint by requesting all institutions and organizations that would conduct work covered by the guidelines to establish their own institutional biosafety committees. Functioning as the direct link between the NCBP and the institution or organization, these committees help to closely monitor any activities falling under the guidelines. The NCBP also formulated a policy that the primary responsibility for enforcing the guidelines rests with the institution's committee.
2. *How can a committee with no power to impose stiff penalties give substance to its policies and guidelines?* The NCBP addressed this by interfacing closely with existing regulatory agencies that are empowered by law to enforce the implementation of pertinent provisions in the guidelines. To ensure that there is close coordination between the NCBP and these agencies, at least one designated

representative of each of the agencies participates as a full member of the NCBP. In case of violations, the NCBP can thus impose sanctions and penalties through the relevant regulatory agencies in accordance with existing rules and regulations.

## *Existing regulatory agencies*

Knowing the strong and weak points of the Philippine administrative system, the technical working group drafted practical and implementable guidelines. It also carefully examined the feasibility of any changes that needed to be made to improve enforcement of the guidelines. Interactions among regulatory agencies and the NCBP during the first six years were assessed. The working group concluded that the following was needed:

- **Capacity building for biosafety regulators.** The NCPB's regulators are well-versed in plant pathology, animal diseases, etc., but many are unfamiliar with the fundamentals of molecular biology and related fields. Although the institutional biosafety committees and the NCBP are expected to examine the interplay of GMOs with other organisms in the open environment as part of risk assessment, regulators in the field who may be involved in monitoring (such as quarantine officials), should understand the nature and management of potential hazards associated with GMOs.
- **Review of gaps in the regulatory infrastructure.** Existing biosafety-related rules and regulations had to be reviewed so that gaps could be identified and remedied. The laws that created relevant regulatory agencies date back to the 1960s. The rules and regulations promulgated by these agencies through administrative orders also needed updating as no provisions are made for GMOs.
- **Expansion of NCBP membership.** The committee should be expanded to 15 members to include one representative at the level of each bureau, not just department. They would be involved in monitoring research that presents biosafety concerns. However, this change involved amending the executive order, which required another presidential decree. A quicker alternative would be for heads of relevant regulatory agencies to designate one staff member to coordinate with NCBP on biosafety-related matters. At present, the committee's designated department representatives are expected to coordinate with concerned officials in other agencies within their own department. Given the large size of the bureaucracy, the working group believed that for field-testing, a more direct link between the agency concerned and NCBP will ensure effective monitoring and control.

## *Institutional level*

The requirement that at least two members of an institutional committee must not be affiliated with the institution has been a problem for several institutions, particularly commercial companies that are concerned about confidential business information. On the other hand, NGOs and other citizen groups doubt that institutions are able to police themselves, and they are concerned about conflicts of interest among institutional committee members. The NCBP carefully scrutinizes the biodata of institutional committee members, particularly the community representatives. Community members on institutional committees preferably include parish priests, educators or school administrators, and elected government officials serving the local community.

## Decision to revise field-test guidelines

Some NGOs contended that field-testing should be prohibited altogether. The NCBP, however, argued that the loss to public welfare of not realizing the benefits from modern biotechnology would be tremendous. During intense discussions, many of the risks and benefits of genetic engineering were analyzed and debated. The risks of geneflow and weediness potential were discussed, as well as the poor record of the technology to deal with acquired pest immunity and the perceived accelerated evolution of "super pests" due to the selection pressure exerted by crops that have enhanced pest-resistance. The benefits to be derived from agricultural biotechnology included food security, at least in part, for a population of approximately 70 million Filipinos, by using genetic engineering to reduce losses due to pests and to improve product quality.

As transgenic crops developed abroad are bound to enter the domestic market eventually, the NCBP recognized that having some regulatory framework in place would provide greater safety than not having any guidelines at all. After a series of discussions, the committee arrived at a consensus to allow field-testing subject to compliance with the new guidelines. Until the new guidelines for field-testing were finalized, NCBP decisions on proposals for field-testing were to be deferred. The NCBP decided at the outset to make the following changes:

- divide the guidelines into a series of monographs based on the scale of work: small scale, large scale (including greenhouse trials), field-testing, and possibly commercial or general use. Breaking up the guidelines has several advantages: (1) proponents do not have to go through the entire guidelines document but can refer to the monograph applicable to them based on the scale of work, and (2) modifications can be easily incorporated into one monograph without necessarily affecting other monographs in the series, thus providing for greater flexibility.
- draw up completely new guidelines for field-testing, taking into consideration the experience in other countries where field tests have been conducted

## Limiting factors for the technical working group

The technical working group that was assigned by the committee to draft revised guidelines was confronted with several problems:

### Time and logistics

All members of the working group had regular jobs in addition to their many NCBP duties, which included participating in evaluating the proposals and revising the guidelines. Also, the limited NCBP funds did not allow members to meet for longer periods of time, which would have helped the continuity of the discussions. The fact that the members of the working group were personally committed beyond the call of duty and against all odds has been instrumental in achieving the accomplishments.

## Expertise

The NCBP members had experience in biosafety regulations but not in all of the disciplines needed to draw up the guidelines for field-testing. The working group therefore consulted known experts in various areas, who were fortunately more than willing to help.

## Basis

The working group had to decide whether to pattern the revised guidelines on those existing in other countries. If so, it had to decide between (1) taking the guidelines of only one country that conducted field tests in order to ensure consistency and (2) using the guidelines of several countries whose conditions were similar to those of the Philippines. After reviewing the existing guidelines of other countries, the working group selected the guidelines published in 1996 by the Genetic Manipulation Advisory Committee of Australia. The main reason for choosing these guidelines was that they provide a good basis for rational risk analysis while their data requirements are reasonable.

After a slow start a rough working document for the entire field-test guidelines was prepared. This first draft was based on discussions held, drafts of various aspects of the guidelines that had been submitted earlier, inputs from proponents, NCBP's experience in monitoring projects during the previous six years, the existing regulatory infrastructure and system in the Philippines, and other countries' guidelines.

## Multisectoral consultation on the guidelines

Biosafety involves several stakeholders. The NCBP recognized the need to consult these various stakeholders before adopting the working group's draft guidelines on field-testing of GMOs and potentially harmful new species. The NCBP distributed the document widely and solicited comments, in particular from NGOs, professional societies, academic and public research organizations, and regulatory agencies. Some NGOs suggested consulting the public through a forum in which ideas could be exchanged. With this suggestion, the NCBP organized a public consultation to allow the different stakeholders to participate. The NCBP extended the deadline for submission of comments or suggestions on the guidelines, and in response to a request of NGOs that additional consultations be held, the National Academy of Science and Technology held three public consultations throughout the country.

During these public consultations, the following important issues were raised:

- **Public notification of planned releases.** The guidelines provide for notices of field trials to be posted in conspicuous places and published in newspapers of general circulation, allowing one month for public comment. The NGOs demanded that a public hearing be conducted for every field trial application. The NCBP surveyed the system of public consultation in neighboring countries' guidelines and observed that the system prescribed in the Philippines was comparable to, if not more stringent than, the requirements in other countries.
- **Selection of community representatives on institutional biosafety committees and the NCBP.** The NCBP publishes an invitation for nominees for membership of the committee and invites NGOs to nominate community representatives. Public interest is taken into consideration in biosafety reviews, and the multilevel review

process ensures that sufficient views are taken into account; it is conducted at institutions, by an ad hoc Scientific and Technical Review Panel (composed of experts in the relevant field of specialization), and by the NCBP.
- **Moratorium on field-testing of GMOs.** The NCBP could not find a compelling reason to impose a longer moratorium on field-testing GMOs. There is a wealth of experience not only in field trials but also in the commercial growing of GMOs in more advanced countries. Also, there is a lack of clear, substantiated, scientific evidence of adverse effects.
- **Monitoring capability of regulatory agencies.** The capability of regulatory agencies to monitor field triels has been recognized as a weakness in the regulatory infrastructure. The NCBP has already taken steps to strengthen this capability. Monitoring is also the responsibility of institutional biosafety committees as well as the NCBP.
- **Liability and compensation for any damage to health or the environment arising from a field trial.** The guidelines stipulate the conditions under which NCBP approval can be withdrawn. Under these circumstances, sanctions can be imposed through the regulatory agencies. The guidelines also provide that expenses for clean-up in the event of any accident or unforeseen adverse effects of field trials are borne by the proponent.

## Conclusion and recommendations

After a series of revisions, the revised guidelines for field testing were adopted by the NCBP. The new guidelines became effective in September 1998. Ensuring environmental responsibility through a responsive biosafety system, including reviewing and formulating guidelines, involved several management challenges. These included optimizing limited resources, selecting the right people, remedying flaws in the existing regulatory infrastructure, improving the system through feedback, and harnessing communication and negotiation skills in dealing with various stakeholders.

Based on the Philippine experience in formulating and implementing biosafety guidelines, the following major recommendations can be made:

- **Design dynamic guidelines, not static ones.** Guidelines should be reviewed periodically and updated as guided by experience and newly available scientific information. In the absence of such experience or information, a conservative step-by-step review approach should to be considered. Both overprotection and slackness can be costly.
- **Base guidelines on sound science and implement them with competent, knowledgeable people.** Biosafety reviewers should have, or have access to, appropriate scientific expertise and should conduct their reviews on the basis of scientific principles and knowledge.
- **Recognize and respond to the needs and concerns of all stakeholders.** At the same time, do not impose unnecessary burdens that will delay or prevent the benefits of modern biotechnology from reaching the general public.
- **Learn and benefit from the experience in countries that are advanced in this field.** Build on what is known and adapt it to local conditions, rather than waste limited resources on re-inventing the wheel.

# 15 Addressing Public Acceptance Issues for Biotechnology: Experiences from Japan

*Yutaka Tabei*

## Abstract

*The Government of Japan has declared 22 genetically modified crops safe as food and feed. However, some Japanese consumer and environmental organizations have doubts about the safety of these crops and have protested their import and consumption as human food and livestock feed. Consumer anxiety about these issues is due partly to a lack of reliable information. The government, which regulates genetically modified organisms, plays a role in providing the public with accurate information about genetically engineered crops and their risk assessment. This paper describes Japan's experience with public acceptance activities carried out by the Ministry of Agriculture, Forestry, and Fisheries.*

## Introduction

Developing and using biotechnology for food and agriculture is an important goal of the Government of Japan. Genetically modified foods and their by-products have been available to consumers in Japan for more than five years. As of June 1999, the Food Sanitation Committee of the Ministry of Health and Welfare had declared 22 genetically modified crops safe as food (table 15.1).

As has been the case in many European countries (Gaskell et al. 1999; Cantley et al. 1999), consumers in Japan can be uneasy about genetically modified organisms (GMOs), most often because they are unfamiliar with modern biotechnology, especially transformation technology. Some consumers and environmental organizations in Japan have doubts about the safety of genetically modified food and have protested against the import and consumption of genetically modified crops for human food and livestock feed. Moreover, some antibiotechnology groups have caused public concern with statements that genetically modified foods have allergic and toxic effects.

In the past 10 years, biotechnology companies have actively targeted shareholders, consumers, governmental institutions, and interest groups to convince them that their products are safe for human consumption (Ranchhod and Gurau 1999). Government regulatory agencies, which are responsible for ensuring the safety of biotechnology products, also have a role in providing meaningful, objective information that addresses the public's questions and concerns. In this way, governmental agencies serve as an additional source of information beyond that provided by industry and advocacy groups.

© CAB *International*. 1999. *Managing Agricultural Biotechnology—Addressing Research Program Needs and Policy Implications* (ed. J.I. Cohen)

Table 15.1. Current Status of Commercialization of Transgenic Crop Plants in Japan (as of December 1998)

| Organism | Phenotype | Introduced gene | Company | Purpose |
|---|---|---|---|---|
| 1. Herbicide-Tolerant Soybean (40-3-2) | Glyphosate tolerance | EPSPS | Monsanto | Food, feed |
| 2. Herbicide-Tolerant Canola (HCN92) | Gluphosinate tolerance | Phosphinothricin acetyltransferase NPT II | AgrEvo | Food, feed |
| 3. Herbicide-Tolerant Canola (PGS1) | Gluphosinate tolerance male sterile, fertile restration | bar NPT II, barnase, barstar | Plant Genetic Systems (PGS) | Food, feed |
| 4. Herbicide-Tolerant Canola (GT73) | Glyphosate tolerance | EPSPS, Glyphosate Oxidoredactase | Monsanto | Food, feed |
| 5. Lepidopteran-Resistant Corn (Event 176) | Lepidopteran resistance | Bt (CryIA (b)), PAT | Ciba-Geigy | Food, feed |
| 6. Lepidopteran-Resistant Corn (Bt11) | Lepidopteran resistance | Bt (CryIA (b)), PAT | Northrup King | Food, feed |
| 7. Coleopteran-Resistant Potato (New Leaf Potato) | Lepidopteran resistance | Bt (CryIIIA), NPT II | Monsanto | Food |
| 8. Lepidopteran-Resistant Corn (Yield Guard Corn: MON810) | Lepidopteran resistance | BT (Cry IA (b)) | Monsanto | Food, feed |
| 9. Coleopteran-Resistant Potato (New Leaf Potato) | Lepidopteran resistance | Bt (CryIIIA), NPT II | Monsanto | Food |
| 10. Lepidopteran-Resistant Cotton (Ingard Cotton) | Lepidopteran resistance | Bt (CryIA (c)) | Monsanto | Food, feed |
| 11. Herbicide-Tolerant Corn (T14, T25) | Gluphosinate tolerance | Phosphinothricin acetyltransferase (pat) | AgrEvo | Food, feed |

*(continued on next page)*

| Organism | Phenotype | Introduced gene | Company | Purpose |
|---|---|---|---|---|
| 12. Herbicide-Tolerant Hybrid Canola (PHY14, PHY35) | Gluphosinate tolerance male sterile, fertile restration | bar, NPT II, barnase, barstar | PGS | Food, feed |
| 13. Herbicide-Tolerant Canola (PGS2) | Gluphosinate tolerance male sterile, fertile restration | bar, NPT II, barnase, barstar | PGS | Food, feed |
| 14. Herbicide-Tolerant Hybrid Canola (PHY36) | Gluphosinate tolerance male sterile, fertile restration | bar NPT II, barnase, barstar | PGS | Food, feed |
| 15. Herbicide-Tolerant Canola (T45) | Gluphosinate tolerance male sterile, fertile restration | pat, barnase, barstar | AgrEvo | Food, feed |
| 16. Herbicide-Tolerant Cotton (Roundup Ready Cotton) | Glyphosate tolerance | EPSPS, NPTII | Monsanto | Food |
| 17. Herbicide-Tolerant Cotton (BXN Cotton) | Bromoxynil tolerance | BXN, NPTII | Calgene | Food |
| 18. Herbicide-Tolerant Canola (MS8RF3) | Gluphosinate tolerance male sterile, fertile restration | pat, NPTII, barnase, barstar | PGS | Food, feed |
| 19. Herbicide-Tolerant Canola (MS8RF3) | Gluphosinate tolerance | pat, NPTII | AgrEvo | Food, feed |
| 20. Ripening-Delayed Tomato | Ripening Delayed | ACC, NPTII | Calgene | Food |
| 21. Herbicide-Tolerant Canola (MS8) | Gluphosinate tolerance | pat, barnase | PGS | Food |
| 22. Herbicide-Tolerant Canola (RF3) | Ripening Delayed | pat, barstar | PGS | Food, feed |

Consumer concerns are due partly to insufficient and incomplete information about genetically modified food and biosafety. Providing the public with additional accurate information about biotechnology crops is essential for building public acceptance. Consumers must be well informed about how new crop varieties are developed and how the food and/or environmental safety of GMOs is ensured. The introduction and explanation of risk-assessment procedures in Japan is the most important part of an information program. In addition, it is essential that the public understands the benefits of biotechnology, why it is being used now, and how it may be used in the future.

## Japan's risk-assessment guidelines

In Japan, risk assessment of GMOs is done using the guidelines issued by the Science and Technology Agency, the Ministry of Agriculture, Forestry, and Fisheries (MAFF), and the Ministry of Health and Welfare. There is no legislation for a safety assessment scheme, but the Japanese government requires that all producers and trading companies handling transgenic crops evaluate the safety of their products. Risk assessment is based on two important principles: "substantial equivalence" and "familiarity," as defined by the Organisation for Economic Co-operation and Development (OECD 1994).[1]

### *Safety assessment with respect to environmental effects*

Environmental risk is evaluated in four stages of research: closed greenhouses, semiclosed greenhouses, isolated fields, and open fields. In the experimental phase, the Science and Technology Agency conducts risk assessments for research in closed and semiclosed greenhouses. MAFF does risk assessments for isolated and open-field experiments. If the environmental safety of a particular GMO is confirmed, then the transgenic crop is suitable for cultivation under the same conditions as the unmodified crop.

One example of such an assessment in Japan is the recent cultivation of transgenic melon. Procedures and tests were devised for evaluation in a closed greenhouse and in an isolated field. Risk assessment was determined by comparing characteristics for six traits between transformed and nontransformed plants. The six traits used for this study include the morphology and maturation period required for the melons, their reproductive traits (pollen dispersal, fertility, and longevity), potential harmful impact on other plants, effects of overwintering in relation to weediness, influence of melon cultivation on soil microflora, and potential for residual *Agrobacterium* vector to accumulate on transgenic melons.

The transgenic melon received approval for cultivation in ordinary fields, with subsequent field growth showing no difference in the six traits between transgenic and ordinary varieties (Tabei 1999). As will be stressed later, testing of transgenic organisms for safety-related parameters stimulates public confidence in review procedures and provides a cornerstone for building public acceptance of biotechnology.

---

[1] Familiarity is defined as having enough information to be able to judge the safety or risks of a GMO. It can be used to indicate ways of handling risks. Familiarity is not synonymous with safety. A relatively low degree of familiarity may be compensated for by appropriate management practices. Familiarity can be increased as a result of a trial or experiment. This increased familiarity can then serve as a basis for future risk assessment. Substantial equivalence embodies the idea that existing organisms used as food, or as a source of food, can be used as the basis for comparison when assessing the safety of human consumption of a food or food component that has been modified or is new.

## Safety assessment of food and feed

The Ministry of Health and Welfare is responsible for assessing the safety of genetically modified food, based on guidelines formulated by the Committee of Food Sanitation. It does so exclusively on the basis of data provided by the applicant, which is compiled in the country of origin. The committee can request additional or complementary data if the first set of data submitted does not satisfactorily show the food is safe.

The Livestock Industry Bureau of MAFF established the Agricultural Material Council for assessing the safety of genetically modified crops as feed. The council assesses feed safety according to the "Guidelines for Safety Assessment of Feed Produced by Recombinant DNA Techniques." Most of the tests are similar to those used for determining food safety.

As of July 1999, the minister of MAFF approved 75 applications (covering 123 transgenic lines) for field trials to be conducted at isolated fields and 45 applications (covering 70 lines) for nonregulated cultivation. Twenty-two transgenic crops have already been confirmed as safe for human food; of these, 19 are also confirmed safe for use as feed (see table 15.1).

# Responding to the challenge: Getting the message to the public

MAFF's public acceptance project, begun in 1995, was established to provide accurate information about biotechnology and genetically modified crops to the public. Initial funding was US $200,000. Responsibility for planning the project was given to a section chief in the Innovative Technology division of MAFF and a division director in the Society for Techno-Innovation of Agriculture, Forestry, and Fisheries (STAFF), an extra-departmental body of MAFF. STAFF has two major missions: to enhance research and development of biotechnology and to promote a foundation for the use of biotechnology by supporting, for example, projects to build public acceptance.

Due mainly to stagnation in the economy, the Japanese government, like many other governments, is in a process of saving on personnel costs. This has made it difficult to staff the units that provide information and training in building public acceptance. External organizations like STAFF are therefore essential actors in the government's public acceptance project.

## Seminars, symposia, and educational activities

In collaboration with STAFF, various seminars were organized to establish direct contact with consumers and to provide them with accurate information about genetically modified crops. MAFF allocated $900,000 for a series of one-week seminars and one of two-day seminars, and $950,000 for a series of public biofora.

*One-week seminars*

Intensive one-week seminars were conducted for leaders from various professions, high-school teachers, dieticians, local government employees, representatives of the private sector, and farmers. The purpose of the seminars was to train people to serve as mediators between the government and consumers. Mediators should be able to confidently explain the characteristics of genetic engineering and the merits of the technology.

The seminars were held in Tuskuba twice a year in the summers of 1996 and 1997. For each seminar, the organizers selected a group of participants with a regional and professional balance. The 46 participants attending the 1997 seminars studied current developments in agricultural biotechnology and learned about risk assessment procedures regarding the environment and food safety. Participants could also do some elementary lab work in molecular biology. As most participants had no prior experience in working with DNA, they initially felt that it may be mysterious or dangerous, but they were surprised and pleased to see they could isolate total DNA from rice tissue. Working in the laboratory helped increase their understanding of DNA and related research methodologies.

Using the participants' feedback on the seminar, MAFF and STAFF improved the content and organization of subsequent seminars. To encourage and support the participants of the one-week seminars to be good mediators, MAFF and STAFF regularly send them new information on government policy and product development.

*Two-day seminars*

The two-day seminars, which basically serve the same purpose as the one-week seminars, are aimed especially at high-school teachers and regional leaders who cannot attend or are not selected for the one-week seminars. Participants in the two-day seminars, which are also held twice per year, attend various lectures regarding the Japanese risk assessment system for genetically modified crops, and they conduct simple experiments in molecular biotechnology. Based on participants' feedback, MAFF and STAFF are planning to organize a symposium program targeted specifically at high-school students.

*"Bioforum" sessions*

Organized once a year, biofora consist of a series of lectures and a display of products made from GMOs. The purpose is to give a large number of ordinary consumers a chance to see products of biotechnology and listen to presentations by scientists and government employees. Biofora are announced through STAFF and the MAFF Regional Agricultural Administration Office. Biofora have been held in Tokyo and cities such as Nagoya, Kyoto, and Fukuoka; they will continue to be held in other major cities.

*Information brochures*

Due to budgetary and organizational constraints, only a few seminars can be held per year, and only a limited number of participants can attend them. Therefore, to educate a broader audience about biotechnology, MAFF has prepared two information brochures. Published in August 1995, "Biotechnology in Everyday Life" shows applications of biotechnology such as embryo culture, cell fusion, and genetic transformation of crops, as part of a wide spectrum of techniques for genetic variation. "Quick Guide to Recombinant Crops," published in September 1996, explains the basic mechanisms and benefits of genetic transformation and answers some frequently asked questions and concerns of the public. The brochures are frequently used at seminars and symposia to introduce biotechnology and raise consumers' awareness of genetic engineering. Consumers and environmental groups have shown a great deal of interest in the brochures; as of May 1998, 40,000 copies of "Biotechnology in Everyday Life" and 45,000 copies of "Quick Guide to Recombinant Crops" had been distributed free of charge.

*Information on the Internet*

A growing number of public organizations, private companies, and consumers have access to the Internet. To provide information quickly to this fast-growing audience, MAFF has developed a site on the World Wide Web in both Japanese and English.[2] The Innovative Technology division also operates a web site[3] that provides information about the development of GMOs and the use of the biosafety guidelines. The division adds relevant new information regularly and updates the site about once a month. The Ministry of Health and Welfare provides food-safety-related information on its web site.[4]

*Books and reports*

MAFF invested $25,000 in 1997 to build a small, special library with materials related to ways of building public acceptance. The library project is administered by STAFF, which prepared summaries in Japanese of 21 key books and reports published overseas. STAFF also translated reports, such as *A New Green Revolution* of the Food and Agriculture Organization of the United Nations (FAO), *Codex Committee to Rethink Food Labeling*, and papers from the journal *Nature* that summarize scientific and regulatory trends. All of these summaries and reports have been collected in one volume, which has been distributed to all 21 national MAFF institutes and to STAFF members.

*Talking to consumers*

As genetically modified crops are becoming increasingly available in Japan (particularly canola, corn, and soybean), primary buyers and food-processing companies are becoming more concerned about reactions from consumers and customers. Many companies as well as consumer groups have therefore begun to organize public meetings, panel discussions, and symposia on genetically modified crops. Members of the Innovative Technology Division of MAFF participate in these activities to provide appropriate information. In 1997, there were about 90 of these participatory events. MAFF's lectures usually include the following elements:

1. **An introduction to genetic engineering technology in plant breeding.** Most consumers are unfamiliar with the concept of plant breeding and plant improvement methods. Any explanation of genetic engineering therefore begins by explaining conventional plant breeding and the use of hybridization to combine preferred traits from different genetic sources. The basics of molecular mechanisms through which traits are combined in progeny by meiosis are explained, as are naturally occurring gene transfer mechanisms (such as *Agrobacterium*). It is essential that consumers understand that transgenic technology was developed from research about natural biological processes.

   Various forms of biotechnology, such as cell and tissue culture and cell fusion, are discussed, and it is shown why and how they were developed to circumvent the

---

[2] http://www.maff.go.jp/eindex.html

[3] http://ss.s.affrc.go.jp/docs/sentan

[4] http://www.mhw.go.jp/english/index.html

current limitations in genetic resources available to breeders. In this context, all forms of biotechnology, including genetic engineering technology, are shown to be new tools for plant breeding.
2. **An explanation of risk assessment procedures for environmental and food safety.** Before explaining risk assessment procedures, the history of risk assessment must be explained. Once consumers understand that risk assessment has always been an essential part of biotechnology in every country that is active in genetic engineering, then consumers' confidence in the efficacy of the present risk assessment system is increased significantly. In describing risk assessment procedures, two points are emphasized: (1) any potential risk in a new technology should be assessed by the developer, and (2) risk assessment of GMOs should be done in a step-by-step manner for each level of research and release.
3. **An introduction to commercialized transgenic crops.** Japan's commercial transgenic crops include herbicide-resistant soybean and canola, insect-resistant corn and potato, and blue carnations. Lectures on the merits of transgenic crops use visual materials to convey effectively the rationale for introducing herbicide and insect resistance into crops. These lectures explain why specific transgenic crops have been developed and how they contribute to Japan's welfare. Summaries of global data on commercial releases help consumers realize the extent of worldwide interest in using transgenic crops.

*Discussions with the audience*

After the lectures, the presenters take time for discussions with the audience. While the lectures will have helped many in the audience gain a better understanding of the merits and necessity of transgenic crops, others will still have questions and misgivings about the safety of the GMOs. Typical questions include: How can the safety of foods from these crops be ensured? Will herbicide-resistant crops become weeds or will crossing between weeds and herbicide-resistant crops produce new herbicide-resistant weeds? Are the antibiotic resistance genes in transgenic crops safe for human health? Responding to consumers' concerns in this personal way, as well as through the other mechanisms described in this paper, shows that MAFF tailors its information dissemination to the needs of consumers.

*Use of questionnaires*

Surveys previously conducted by consumers' and public interest organizations interviewed their own members or other small groups of people about biotechnology. Results from these limited efforts, however, cannot be considered to represent opinions of the general public. Moreover, the groups sampled were typically biased against the use of GMOs, yet the results were sometimes cited as showing the general public's opposition to the commercialization of GMOs.

The MAFF team responsible for conducting risk assessments of GMOs needed a more accurate picture of consumer opinion about biotechnology, and a better understanding of what the public was concerned about and what they wanted to know. Consequently, in 1997 MAFF's Innovative Technology division developed a questionnaire that first explained present conditions for development and commercialization of GMOs, and then asked 20 related questions. STAFF sent the questionnaire to 3,000 randomly selected people. A total of 657, mainly housewives, completed and returned the questionnaire.

The results showed that 60% of respondents thought that GMOs would improve life, while 7% did not find them useful. This finding indicates that most Japanese consumers do not necessarily object to commercialization of GMOs and that they may have some understanding of the necessity and benefits of GMOs. Over 90% of the respondents wanted to have additional detailed information about food safety. About 70% were interested in the potential influence of GMOs on the environment. One out of three respondents wanted information summarizing the benefits and quality of GMOs. When asked whom they trusted most to provide information, about 70% of the respondents said they trusted researchers. Consumer and environmental organizations were trusted by about 65%, and administrative organizations by about 60% of respondents. The questionnaire also showed, however, that Japanese consumers do not have the same degree of confidence in manufacturers of GMOs and private companies using GMOs as food material.

The survey revealed issues that Japanese consumers are most concerned about. The results are consistent with other findings regarding information for building public acceptance. Ranchhod and Gurau (1999) emphasize that public acceptance depends mainly on providing the public with accurate information about particular products in a timely manner. They also emphasize that the public needs to understand the "added benefits" that come from biotechnology-derived products, compared with those from conventional research. Finally, responsible parties should address perceived risks posed by a product at the earliest stage possible.

Results of the survey suggest that the direction and method of MAFF's public-acceptance-building activities may be having a positive effect. The collected information will be used to help plan new seminars, symposia, and lectures that will place greater emphasis on explaining Japan's procedures for ensuring food and environmental safety.

## Conclusion and closing remarks

This paper summarizes approaches taken by the Government of Japan to communicate with the public on issues regarding public acceptance and biotechnology. Efforts have included translation and production of user-friendly material in the national language to ensure wide readership and use. As in other countries, communication with the public will continue to be a challenge. In Canada, for example, it was found that consumer confidence regarding acceptance of transgenic products does not rest solely on understanding the details of genetic engineering. It is also based on the public's need to trust that genetically modified foods are as safe and nutritious as non-genetically modified, traditional foods (Badani 1998).

In the past, the Japanese public felt uncertain about whether foods made from GMOs were safe and whether the cultivation of genetically modified crops would impact on the environment. Consumers are demanding more and better information about the safety and merits of GMOs. However, some consumers remain concerned about the safety of genetically engineered food, and they claim the right to have a choice as to whether or not to buy genetically modified foods. It is also important to note that the public can distinguish between the various applications of genetic modification. In Europe, for example, the public is less concerned about plant biotechnology then about biotechnology applications to animals. They accept and welcome medical applications much more than applications to food, and they recognize the ambivalence between their differing attitudes regarding food and medicine (Millar 1999).

As the number of field trials and commercial products grows, the Japanese government must continue to use scientific research to confirm the safety of newly developed genetically modified crops. Communicating this information to the public, combined with the use of educational activities, is one of the keys to raising public awareness and building acceptance.

In addition, meetings held in various areas of the country have targeted relevant stakeholders such as leaders in various professions, teachers, local district officials, private-sector representatives, and farmers. Seminars have been used to train members of the public to serve as mediators between the government and consumers. These activities to build public acceptance in Japan have ensured an important place for the government in debates and discussions regarding biotechnology.

They are a source of information, in addition, to industry and special interest groups, and they provide mechanisms for ensuring public participation in and ownership of the decisions and responsibilities that come with Japan's participation in global and national developments in agricultural biotechnology.

## References

Badani, B. 1998. Public acceptance of genetically modified products: Challenge and response. Paper prepared for the ISNAR course "Managing Biotechnology in a Time of Transition". Haikou, China, November 2-13, 1998.

Cantley, M., T.Hoban and A. Sasson. 1999. Regulations and consumer attitudes toward biotechnology. *Nature Biotechnology* Vol. 17, Supplement BV37-40.

Gaskell, G., M.W. Bauer, J.Durant, and N.C. Allum. 1999. Worlds apart? The reception of genetically modified foods in Europe and the U.S. *Science* 285: 384-387.

Millar, H. 1999. Consumer expectations and openness. *Nature Biotechnology* Vol. 17, Supplement BV39-40.

OECD. 1994. Safety Considerations for Biotechnology: Scale-up of Crop Plants. (92 95 01 1) ISBN 93-93-081. Paris.

Ranchhod, A. and C. Gurau. 1999. Looking good: Public relations strategies for biotechnology. *Nature Biotechnology*, Europroduct Focus Summer 1999: 5-7.

Tabei, Y. 1999. Environmental risk assessment of transgenic melon in Japan. *Plant Biotechnology* 16(1): 65-68.

# 16 Balancing Needs for Productivity and Sustainability: Genetic Engineering of Rice at IRRI

*John Bennett*

## Abstract

*To achieve the productivity gains required to feed the growing population of rice consumers, IRRI scientists use genetic engineering to address a range of problems. Prominent among these problems is the need to enhance resistance to biotic and abotic stresses. Research focuses on finding solutions that are effective, durable, environmentally friendly, and acceptable to consumers. Achieving this balance stimulates research on (1) the isolation of genes and promoters to ensure effectiveness of traits, safety as human food, and friendliness to nontarget organisms, (2) cultivar-independent transformation methods to maintain biodiversity in transgenic rices, (3) environmentally friendly selectable marker genes, (4) deployment strategies favoring durability of transgenic traits, and (5) biological confinement to prevent the spread of certain transgenes to related plants. Several areas of uncertainty may delay the implementation of this valuable technology. This paper explores some of these issues with regard to managing biotechnology for rice improvement at IRRI.*

## Introduction

Over the last 30 years, the introduction of high-yielding semi-dwarf cultivars and allied technologies doubled rice production and halved the world price of rice (Khush 1995a). To feed the growing population of rice consumers over the next 30 years, similar gains in production will be required, but they will have to be achieved with less land, water, and labor and with more pollution than in the past. The Philippines-based International Rice Research Institute (IRRI) is dedicated to attaining this goal through scientific research ranging across many disciplines from biology to economics.

## Rice improvement and biotechnology

Improved seed is one of the most effective ways of delivering the benefits of biological research to farmers and consumers. IRRI's rice improvement programs are developing new cultivars for irrigated, rainfed, and flood-prone ecosystems. The programs aim to achieve significant gains in yield ceiling, yield stability, nutrient use efficiency, human nutrition, and grain quality (Khush 1995b). These ambitious programs are underpinned by the development of new breeding tools, especially those of biotechnology. The biotechnology tools of greatest relevance are anther culture, wide hybridization, DNA marker technology, DNA fingerprinting, and genetic engineering. With the exception of genetic engineering, each of these tools can be used to increase the efficiency of conventional breeding programs.

The choice between conventional breeding and genetic engineering as the preferred route to trait improvement depends on several factors. One factor is the availability of rice cultivars that already possess the desired trait or its component parts. If suitable donor cultivars are available, conventional sexual hybridization may be used to transfer the trait into the desired genetic background. If no suitable donors are available, as in the case of sheath blight resistance, the justification for investing in genetic engineering is enhanced. A second factor is the depth of molecular knowledge about the trait. If knowledge is scant and appropriate genes are not available, the case for genetic engineering is weakened. Resources may then be devoted to either conventional breeding or the basic science required to understand the trait. A third factor is the grain quality or environmental adaptation of the recipient cultivar. The difficulty of recovering genetically complex traits after sexual hybridization provides an argument for direct addition of DNA by genetic engineering. For this reason, stemborer resistance was transferred to a high-quality aromatic rice by genetic engineering (Ghareyazie et al. 1997).

## IRRI's program in genetic engineering

Genetic engineering by protoplast transformation began at IRRI in 1991. It was followed by microprojectile bombardment in 1993 and *Agrobacterium*-mediated transformation in 1996. The first agronomically important trait enhanced at IRRI by transformation was resistance to striped and yellow stem borers. Subsequently, resistance to bacterial blight and sheath blight and tolerance of submergence were enhanced (see table 16.1). The transgenic plants generated in these four projects have been evaluated in IRRI's transgenic greenhouse for at least two generations and in one case for eight generations. They are homozygous for the genes of interest and carry both the antibiotic resistance genes *bla* and *hpt*. They now are ready for field-testing to determine (1) whether the transformation process has affected any aspect of plant growth and reproduction, including yield potential, (2) whether trait enhancement is seen under field conditions, (3) whether there is any adverse impact on nontarget organisms, and (4) whether expression of the trait is stable through multiple generations.

Table 16.1. Traits Introduced into Rice by Transformation at IRRI

| Trait | Gene | Promoter |
| --- | --- | --- |
| Bacterial blight resistance | Rice *Xa21* | Rice *Xa21* |
| Sheath blight resistance | Rice chitinase | CaMV 35S |
| Stemborer resistance | *Bacillus thuringiensis* synthetic truncated *cry*IAb | Maize $C_4$ phospho*enol*-pyruvate carboxylase |
| Submergence tolerance | Rice pyruvate decarboxylase | CaMV 35S |

Since 1995, IRRI and its national partners have followed a strategy to encourage field-testing of rice in Asia. The strategy is based on the assumption that public acceptance of field-testing is more likely when the following conditions are met:

- the trait is important, novel, and without perceived risk
- the trait has been introduced into a locally popular cultivar
- the trait is achieved by transformation of rice with a rice gene under the control of a rice promoter
- scientists in national agricultural research systems (NARS) have played a key role in the genetic engineering program

IRRI's leading candidate gene is the *Xa21* gene of *O. longistaminata*, a wild species from Africa. This gene is highly effective in providing resistance against a range of pathovars of the causal agent of bacterial blight, *Xanthomonas oryzae*. The *Xa21* gene is being introduced into popular Asian cultivars with the active participation of national scientists (Tu et al. 1998). The transgenic products of this research should receive rapid approval for field-testing, because they retain in full the high-quality characters of the parental lines.

Deployment of other rice genes, such as chitinase for sheath blight resistance and pyruvate decarboxylase or alcohol dehydrogenase for submergence tolerance, will follow soon. At a later date, when regulatory experience has been gained by pursuing the above applications, more complex cases such as Bt rice can be considered.

## Management challenges

IRRI's genetic engineering program presented several management challenges relevant to balancing productivity and sustainability. The challenges arose in making technical decisions in the following areas: (1) plasmid constructs to be used in transformation, (2) cultivar to be engineered, (3) DNA delivery method, (4) selectable markers, and (5) physical and biological confinement.

### 1. Plasmid constructs for transformation

The balance between productivity and sustainability in the genetic engineering of rice often comes down to choices made at the time of plasmid construction. For a food crop, public perception is an especially important issue and so IRRI scientists are concerned that a gene should be both effective in enhancing a key trait (high benefit) and acceptable as a component of the diet (low risk). In general, plant genes are used wherever possible, which may require special auxiliary programs in gene isolation and evaluation. The protein should be nontoxic and nonallergenic at concentrations manyfold above those encountered in the diet. If possible, the parts of the plant consumed as human food (e.g., the grain of rice) should be made free of the transgenic protein by appropriate choice of gene promoters. It may not always be possible to find promoters that express to high levels in the correct tissues but are inactive in the rice grain, and if we were targeting post-harvest insects, we would want expression in the grain. In such cases, the whole burden of ensuring safety would fall on the choice of gene.

In addition to concern about human consumers, care should also be taken to evaluate transgenic plants in relation to nontarget organisms. IRRI has a program to study the effect of insecticidal transgenic proteins, such as crystal proteins from *Bacillus thuringiensis*, on friendly insects. Another environmental question, increasing the weediness of the crop's wild relatives, is considered later in this chapter.

A fundamental concern is whether a short-term productivity gain will put at risk the long-term effectiveness of a transgenic trait. IRRI is undertaking collaborative studies with Ohio State University, USA, to determine which crystal protein genes from *B. thuringiensis* can be combined to give the most durable resistance. The effectiveness of promoters such the maize $C_4$ PEP carboxylase promoter throughout the entire growing season (in photosynthetically active tissue and senescent tissue) is being examined (Ghareyazie et al. 1997).

## 2. Choice of cultivar

Critics of genetic engineering claim that this technology will erode the biodiversity in farmers' fields because of the limited range of transformed cultivars likely to be available. To solve this problem, IRRI emphasizes the development of cultivar-independent methods of transformation. Protoplast transformation, microprojectile bombardment, and *Agrobacterium*-mediated transformation are all used at IRRI. Together, these methods provide a high probability that any cultivar may be transformed, including traditional cultivars of high quality.

## 3. DNA delivery method

Of available methods for DNA delivery, microprojectile bombardment appears at this moment to provide the best route to cultivar-independent transformation protocols. Regeneration from protoplasts is inefficient for most cultivars, and *Agrobacterium*-mediated transformation is more efficient for japonica lines than for indica lines. On the other hand, *Agro*-transformation has two major advantages. First, unlike protoplast and biolistic transformation, *Agrobacterium* does not transfer its ampicillin resistance selectable marker gene (*bla*) to plant cells. Second, compared with other methods, the pattern of gene integration is generally simpler for *Agrobacterium*-mediated transformation, and so the chances of gene silencing are correspondingly diminished (Kampatla et al. 1997). However, microprojectile bombardment could be made more attractive if (1) the *bla* gene is excised from plasmids by use of a suitable restriction endonuclease, and (2) the 50–70% of transformants containing more than three copies of the transgenes are discarded.

## 4. Selectable markers

Certain selectable marker genes, such as some that confer resistance to an antibiotic or herbicide, work effectively in rice. Given the controversy surrounding the presence of such genes in genetically engineered crops and foods, however, it may be best to find ways to address the issue before the final transformed cultivar is released. Several solutions, in the form of methods for removing or inactivating marker genes by recombination, segregation, or intron insertion, are available. At the same time, new selection systems are being developed that do not present the concerns posed by the use of antibiotic or herbicide resistance genes (table 16.2).

Removal of selectable markers by recombination can be accomplished by use of one of several molecular systems. In both the *cre*-lox system of phage P1 (Bryant et al. 1992) or

**Table 16.2.** Alternatives to the Currently Used Antibiotic Resistance Marker Gene (*hpt*) in Rice Transformation.

| Gene | Selection agent | Advantages | Disadvantages |
| --- | --- | --- | --- |
| *hpt* removed by recombinase | Hygromycin B | Good selection | Requires crossing with recombinase$^+$ line |
| *hpt* in super-binary vector | Hygromycin B | Good selection | Must use large *Agrobacterium* plasmid |
| *hpt* in separate plasmid | Hygromycin B | Good selection | Must use two *Agrobacterium* plasmids |
| *hpt* interrupted by plant intron | Hygromycin B | Excellent selection | Antibiotic resistance gene still involved |
| Herbicide resistance | Basta (*bar*) glyphosate (*epsps*) | Excellent selection | Spread to weedy or wild relatives |
| Mannose-6-P isomerase | Mannose | No environmental problems | Selection efficiency unclear |

the *FLP*-FRT system of the yeast 2µ circle (Lyznik, Rao, and Hodges 1996), the selectable marker gene is flanked in the transformation plasmid by direct repeats of target sequence (the lox or FRT recombination site). In the presence of the corresponding recombinase enzyme (encoded by the *cre* or *FLP* gene), the repeat sequences are joined by excising the intervening sequence containing the selectable marker gene. These systems function with reasonable efficiency in plants but their use must be planned in advance; they cannot be used to remove a gene unless the flanking recombination sites are already in place. A rather different form of the recombination approach uses a maize transposable element system that results in the loss of both the element and the marker gene that had been inserted into it.

When the marker gene and the gene of interest integrate into separate sites in the recipient genome, as can happen when they are introduced on separate plasmids or on an *Agrobacterium* superbinary plasmid containing two distinct T-DNA regions, normal segregation can result in the loss of one. Second-generation transformants can be screened by PCR to identify progeny containing the gene of interest but lacking the selectable marker. However, recombination between separate plasmids prior to integration can eliminate the utility of this option.

A simpler approach is to use a plant intron to break up the coding sequence of an antibiotic resistance gene into two exons. Exons are readily spliced together at the RNA level in plants but would act as two separate and nonfunctional coding regions if transferred to bacteria. This approach gives highly effective selection when the intron that is introduced contains enhancers of gene expression. The one negative aspect to the intron approach is that the antibiotic resistance gene is still present. The public might easily be persuaded by critics to regard the insertion of an intron as an insufficient response to their concerns.

Researchers at IRRI are also investigating the use of the mannose-6-phosphate isomerase gene as a selectable marker (Joersbo et al. 1998). This enzyme catalyzes the reversible interconversion of two forms of six-carbon sugar. Calli of some plant species cannot exploit one of the forms as a carbon source because of inadequate levels of the enzyme. Expression of a foreign mannose-6-phosphate isomerase gene in such cells converts the one form into the other, useable form and thus stimulates growth. The effectiveness of this approach in rice is unknown and may depend on the role of the enzyme in other biosynthetic pathways.

## 5. Physical and biological confinement

The first edition of the Philippine Biosafety Guidelines, published in 1991, prohibited field-testing transgenic plants that could cross with native or agricultural plants grown in the country, including rice. The second edition, published in 1998, allows field-testing but under closely regulated conditions. In moving from extreme physical confinement to eventual field release, IRRI has made a series of management decisions related to the balance between productivity and sustainability. The following first addresses when and why containment is needed and then discusses containment options for transgenic rice.

### When and why containment is needed

The region from central India to southern China, including northern parts of Southeast Asia, is recognized as the center of origin of cultivated Asian rice (*Oryza sativa*) (Vaughn 1994). There are several special concerns for growing transgenic rice in Asia that are not present for transgenic crops grown away from centers of origin, e.g., corn, soybean or potato grown in the USA or Europe (De Kathen 1996; Rissler and Mellon 1996):

- contamination of wild germplasm in native habitats
- contamination of in situ germplasm conservatories
- the loss of biodiversity when plots for deployment of transgenic rice are cleared of wild relatives
- increased weediness of wild relatives of rice after introgression of certain transgenes that might give the recipient extra vigor

However, if suitable precautions are being taken against the cross-pollination of wild germplasm and in situ conservatories by nontransgenic rice, then these same precautions will be effective against transgenes. Suitable precautions taken to collect and conserve wild relatives from areas where nontransgenic rice is cultivated will be applicable for areas where transgenic rice is cultivated. IRRI will draw on its experience in hybrid rice production and germplasm conservation to provide further guidance in this regard (Virmani 1994; Vaughan 1994).

Many rice cultivars are naturally stress-tolerant, and farmers in rainfed areas grow them as part of a risk-avoidance strategy, even though they are often low-yielding. These cultivars are not shunned as a danger to wild relatives or as possible progenitors of superweeds. By contrast, the introduction of stress tolerance by a transgenic mechanism is often met by calls for confinement until studies can be conducted to assess the probability of these scenarios (Kareiva 1993; King 1996; Rissler and Mellon 1996). Conventional and transgenic plants are treated differently in this context. The main concern is that transgenic stress-tolerance mechanisms are usually discussed in terms of single dominant genes. It is thought that single genes are more likely to spread to weedy relatives than are natural stress-tolerance traits conferred by several unlinked genes.

There are at least two responses to this argument. First, transgenes do not "escape" to weedy relatives. They and about 30,000 other genes in the cultivated rice plant combine with 30,000 genes from the weedy relative to form a hybrid. Although weediness is notoriously difficult to define and measure, the hybrid is likely to be intermediate in weediness between the two parents. It is doubtful that one stress-tolerant transgene will increase weediness significantly, when the many allelic differences between the cultivated rice and the weedy relative will tend to reduce weediness in the hybrid compared with the weedy relative.

Second, claims of significant improvements in stress tolerance through insertion of a single transgene may be valid under laboratory conditions, where stresses are applied singly and for a brief period, but they may not hold up under field conditions. Transgenic stress-tolerant traits are likely to be built up from several genes. An adequate response to the above criticism is to ensure that these genes are unlinked. The molecular breeder therefore sacrifices ease of trait transfer in breeding programs for the sake of sustainability in agriculture. The sacrifice will in fact be quite minor when all the genes contributing to a trait can be separately evaluated for contribution to phenotype and will be easily tracked by existing DNA-based technology that is cheap and robust.

*Containment options for transgenic rice*

Construction of a containment greenhouse at IRRI was preceded by lengthy discussion. The earliest decision was to construct a transgenic greenhouse to the strictest standards (CL4), even at the risk of giving the impression that the plants in the greenhouse were dangerous. As a guest institute in a host country that was regional leader in biosafety, IRRI wanted to establish clearly from the very beginning its commitment to biosafety. Furthermore, in a country that experiences about 30 typhoons a year, only an extremely well-constructed greenhouse could assure adequate confinement. The construction, maintenance and running costs of the 400 $m^2$ facility have been high, but operations have been largely trouble-free, thanks to excellent maintenance and adherence to standard operating procedures by IRRI staff. Since its opening in October 1994, the greenhouse has allowed the growth of 10 generations of transgenic rice plants.

The IRRI transgenic greenhouse should not necessarily be duplicated elsewhere in Asia. Few national institutes in Asia can afford to construct a confinement facility of such size, and few would have the need. If countries devise realistic guidelines that allow environmentally friendly transgenic plants to be grown in ordinary greenhouses, screen houses, and fenced fields during the evaluation phase, strict confinement can be limited to the few controversial plants that need special evaluation. Countries should spend their resources on training regulators to make decisions that are responsible, realistic and timely, rather than on facilities that are overly complex, unnecessarily large and expensive to operate.

In one recent attempt to create a system of biological confinement, a herbicide resistance gene was inserted into the plastid genome using microprojectile bombardment (Daniell et al. 1998). As the plastid genome is usually inherited maternally in most crops, including rice, pollen released by plastid transformants would not contain the herbicide resistance gene. Thus the transgene would not be transmitted to other plants through pollen. However, this argument fails to take into account the fact that pollen flows from cultivated rice to wild rice and vice versa (Virmani 1994). If a plastid transformant is the female parent in an outcrossing event, the resulting hybrid would still contain the transgene.

A preferable approach would be to prevent outcrossing in either direction by some form of mutation that causes failure of the palea and lemma to open at anthesis; such forms or mutations are known in several cereals. It would be necessary, however, to ensure that grain size and quality are not adversely affected by this change.

## The special case of Bt rice

The crystalline δ-endotoxins of the soil bacterium *Bacillus thuringiensis* (Bt) are highly toxic to specific classes of insects. Bt sprays, containing bacterial proteins or spores, have been used with some success for over 30 years to control many coleopteran, dipteran, and lepidopteran insects on contact. These sprays are less effective, however, against insect pests that are not exposed for most of their life cycle, such as rice stemborers hidden in the stem. An alternative strategy is to produce the toxin inside plants at the covert places where the insects feed. Hence the popularity of the concept of transgenic plants containing genes specifying Bt toxins. The low mammalian toxicity of Bt toxins and the low risk to friendly insects are added attractions of this approach.

In recent years, insects with resistance to one or more of the δ-endotoxins have emerged, raising questions about the durability of the transgenic approach. There is also concern that some friendly insects might be at risk. Deployment strategies should be as theoretically and practically sound as we can make them, within our current knowledge of the genetics and behavior of target and nontarget insects. We are likely to learn a great deal by careful evaluation of Bt rice under field conditions. In particular, we shall be able to determine whether transgenic rice plants highly resistant to stem borers in the greenhouse are equally effective against the diverse populations of stem borers found across Asia and under the varied climatic and nutritional conditions found in farmers' fields. Sustainability of the Bt rice approach may depend on deploying multiple insect-control mechanisms (Bt genes and non-Bt genes) under the control of different promoters to ensure adequate levels of gene expression in appropriate parts of the plant (table 16.3). An emerging trend in genetic engineering is away from a single-gene approach and towards a multigenic approach that more closely resembles the genetic complexity of durable natural defense mechanisms.

## Concluding remarks and future challenges

Genetic engineering of crops offers one means of achieving safe and sustainable solutions to many long-standing problems of food production in Asia, and rice farmers and consumers will be among the major beneficiaries of this new technology. There are several areas of uncertainty that may delay the implementation of this valuable technology. These

**Table 16.3.** Strategies for Sustainable Deployment of Bt-Rice.

| Strategy | Rationale |
| --- | --- |
| Use tissue-specific promoters to control Bt gene expression | Minimize access of non-target insects to Bt protein  Exclude Bt protein from grain |
| Use several different Bt genes targeted at different receptors | Multiple targets should create more durable defense |
| Combine Bt genes and other insecticidal genes | Multiple mechanisms should create more durable defense |
| High-dose plants | Kill insects heterozygous for gene conferring resistance to Bt |
| Refugia | Maintain Bt sensitivity in insect population |
| Needs-only basis for deployment | Limit exposure of insect population to Bt genes |

include: (1) regulatory environment, (2) need for field-testing, (3) freedom to operate, and (4) transgenic rice as a food.

1. Until the national biosafety guidelines are formulated and implemented, it will be difficult for scientists to be sure that they are moving on the right track with respect to meeting regulatory requirements. This situation is particularly difficult for IRRI, which traditionally shares its breeding lines with many countries and proposes to do the same for transgenic rice. IRRI is therefore being proactive in ensuring that its genetic engineering program will meet the highest standards of public acceptance at the regulatory level.
2. Many questions about the productivity and sustainability of transgenic rice cannot be settled from an armchair or even from experiments conducted in transgenic greenhouses. Transgenic rice must be evaluated in the field. The yield potential of transformants must be checked, and the efficacy of transgenes must be evaluated under different environmental conditions and against different populations of pests, diseases, and friendly insects. It may be necessary to use several genes or promoters to obtain the required performance in terms of the enhancement and durability of the trait.
3. Intellectual property protection on transformation protocols, genes, cultivars, and allied materials has created an environment where even large companies with substantial legal staffs find it difficult to guarantee their freedom to operate. Public-sector institutes such as IRRI and the NARS, which have few legal resources and little protected intellectual property to use in cross-licensing agreements, are in an even more difficult position. Considerable resources may have to be spent (1) licensing the intellectual property required to maintain the balance between productivity and sustainability or (2) inventing alternative approaches that will be available to rice breeders in the public sector.
4. As rice is a major component of Asian culture, the introduction of transgenic rice into the Asian market place should be handled sensitively. Farmers will not grow transgenic rice if consumers are not willing to buy it. The general public must have confidence in the regulatory process and its implementation and in the value of transgenic rice to farmers. IRRI is developing a program to explain genetic engineering to farmers, extension agents, nongovernmental organizations, and educators. It is important that by the time transgenic rice is ready for general release, both farmers and consumers should regard these new seeds as safe and valuable additions to our tools for sustainable food production.

As a disinterested evaluator of the productivity and sustainability of new technologies, IRRI welcomes responsible, well-documented comments on the application of genetic engineering to rice improvement. The institute encourages open debate that creates accurate public perceptions on key issues, defines new research agendas, and builds useful new partnerships.

## References

Bryant, J., S. Leather, E.C. Dale, and D.W. Ow. 1992. Removal of selectable marker genes from transgenic plants: Needless sophistication or social necessity. *Trends in Biotechnology* 10:274-275.

Chandra, D. 1995. Farmers' strategies for minimizing risk in rainfed lowland rice farming systems in eastern India. *In* Proceedings of the International Rice Research Conference, 13-17 February 1995. Los Baños: International Rice Research Institute.

Christou, P., P. Vain, A. Kohli, M. Leech, J. Oard, and S. Linscombe. 1996. Introduction of multiple genes into elite rice varieties-evaluation of transgene stability, gene expression, and field performance of herbicide-resistance transgenic plants. *In* Rice Genetics III. Proceedings of the Third International Rice Genetics Symposium. Los Baños: International Rice Research Institute.

Daniell, H., R. Datta, S. Varma, S. Gray, and S.B. Lee. 1998. Containment of herbicide resistance through genetic engineering of the chloroplast genome. *Nature Biotechnology* 16:345-348.

De Kathen, A. 1996. The impact of transgenic crop releases on biodiversity in developing countries. *Biotechnology and Development Monitor* No. 28, pp.10-14.

Ghareyazie, B., F. Alinia, C.A. Menguito, L.G. Rubia, J.M. de Palma, E.A. Liwanag, M.B. Cohen, G.S. Khush, and J. Bennett. 1997. Enhanced resistance to two stem borers in an aromatic rice containing a synthetic *cryIA(b)* gene. *Molecular Breeding* 3:401-414.

Joersbo, M., I. Donaldson, J. Kreiberg, S. Guldager Petersen, J. Brunstedt, and F.T. Okkels. 1998. Analysis of mannose selection used for transformation of sugar beet. *Molecular Breeding* 4:111-117.

Kampatla, S.P., M.B. Chandrasehan, L.M. Lyer, G. Li, and T.C Hall. 1997. Genome intruder scanning and modulation systems and transgenic silencing. *Trends in Plant Science* 3:97-104.

Kareiva, P. 1993. Transgenic plants on trial. *Nature* 363:580-581.

Khush, G.S. 1995a. Modern varieties – their real contribution to food supply and equity. *GeoJournal* 35(3):275-284.

Khush, G.S. 1995b. Breaking the yield frontier of rice. *GeoJournal* 35(3):329-332.

King, J. 1996. Could transgenic supercrops one day breed superweeds? *Science* 274:180-181.

Lyznik, L.A., K.V. Rao, and T.K. Hodges. 1996. FLP-mediated recombination of FRT sites in the maize genome. *Nucleic Acids Research* 24:3784-3789.

Rissler, J., M. Mellon. 1996. The Ecological Risks of Engineered Crops. The MIT Press, Cambridge, pp. 168

Tu, J., T. Mew, Q. Zhang, N. Oliva, G.S. Khush, and S.K. Datta. 1998. Transgenic rice variety IR72 with Xa21 gene resistant to bacterial blight. *Theoretical and Applied Genetics* (in press).

Vaughan, D.A. 1994. The wild relatives of rice: A genetic resources handbook. International Rice Resarch Institute, Philippines, Los Baños, pp. 136.

Virmani, S.S. 1994. Heterosis and hybrid rice breeding. Berlin: Springer-Verlag.

# 17 Managing Target Pest Adaptation: The Case of Bt Transgenic Plant Deployment

*Mark E. Whalon and Deborah L. Norris*

## Abstract

*Pest resistance is the evolutionary process whereby pests and pathogens adapt to the strategies, tactics, and tools used to suppress target pest populations. "Resistance management" is the science of ameliorating or preventing resistance from developing. Since genetically engineered plants will deliver highly effective means for reducing pest populations, most pest management experts agree that pests will develop resistance to transgenic plants unless appropriate resistance management strategies are undertaken. This chapter seeks to relate various approaches that research managers and directors of research institutes can use to insure the durability of pest resistance of transgenic plants. In the end, implementing either polygenic traits or the high-dose, large-area susceptible pest refugia is the most likely strategy to yield long-term use-life of transgenic plants once they are released into the field. No transgenic plants should be commercially released into the field unless the project is implemented in the context of integrated pest management.*

## Introduction

Insect pests, weeds, and diseases are a major cause of economic loss to farmers throughout the world. Conventional host-plant resistance cultivars have been pursued since the dawn of civilization. These elite selections have contributed immensely to society's welfare. Yet, there are many examples of insects, mites, nematodes, and plant pathogens that have adapted to these conventionally derived cultivars. Theoretically the best pest-resisting cultivars are those that do not cause selection of the target pest(s). In practice, this is an elusive goal and has yet to be achieved.

Widespread availability of synthetic insecticides, herbicides, and fungicides together with elite cultivars have allowed most countries to manage plant pests. However, reliance on chemical controls has led to new problems, including: (1) pest resistance, (2) health impacts on farmers and farm workers, (3) pesticide residues in food, (4) destruction of beneficial organisms such as pollinators, (5) pollution of surface and groundwater, and (6) reduction of biodiversity, and impact on nontarget species, including some mammals, birds, and fish (Agne et al. 1995). In fact, one of the strongest justifications for the development and deployment of transgenic plants is the amelioration of pesticide effects on humans, the environment, and nontarget organisms.

© CAB *International*. 1999. *Managing Agricultural Biotechnology—Addressing Research Program Needs and Policy Implications* (ed. J.I. Cohen)

"Resistance" is the biological process through which a pest population is selected and, after adaptation, may survive a pesticide, elite cultivar, or transgenic plant. Pests are much more difficult and costly to control once they evolve resistance, and in some instances, they may no longer be managed after resistance is acquired. Since resistance has a genetic basis and the environmental and ecological factors that affect its development are known, it can be managed. "Resistance management" is the process of ameliorating resistance development in insect, disease, and weed pest populations.

One of the most significant developments in agricultural biotechnology has been the engineering of various biological pesticide genes into crop plants. Over 100 transgenic plant species have been experimentally or commercially released worldwide, and this number is expected to grow dramatically within the next few years. There is growing concern, however, that resistance may develop to these transgenic plants once they have been released.

Biotechnology is a relatively new field, and much about the interaction of living genetically modified organisms and their effects on agricultural ecosystems is unknown. Therefore, many developed and developing countries have adopted a conservative stepwise process of introducing transgenic plants into cropping systems. It is interesting that several countries have started from radically different philosophical biosecurity positions on transgenic plants, yet after several years, their biosecurity plans have converged into strikingly similar documents. The biosafety guidelines in these countries often state clearly that various safety measures are to be applied carefully as transgenic plants are deployed into agriculture. Plants are usually to be secured in specially equipped greenhouses, confined field plots, and, finally, in small commercial-scale fields before full-scale release. This process should be followed by larger-scale releases together with ecosystem-level monitoring to determine if any biosafety, biodiversity, or resistance problems are developing. In theory these guidelines are conservative and follow the conservative logic of a planned release process.

However, the likelihood is very low that any developing country will actually ever have an elaborate "ecosystem-level" monitoring system in place. Such a system is just too expensive. Even developed countries cannot always amass the personnel, capital, and political will to carry this out. Therefore, what are the best two or three strategies that would give maximum assurance of durable transgenic plants in the field that developing countries could adopt without elaborate monitoring and management plans?

In the USA, many corn, cotton, and potato varieties engineered with insect control proteins derived from *Bacillus thuringiensis* (Bt) are now grown commercially. The Environmental Protection Agency has allowed up to 80% of a given production field to be planted in transgenic plants. This decision has been forged in the heat of broad public debate spanning more than half a decade. Recent work now suggests that it may not be necessary to deploy large percentages of a transgenic Bt variety within a given field to achieve pest control (Alstad and Andow 1996, Riggin-Bucci and Gould 1997). These scientists are exploring whether smaller percentages of highly effective transgenic plants can be enough to suppress pest populations below damaging levels over large areas.

This chapter will focus primarily on the issues of resistance management, transgenic plants, and the development of policies that enhance the long-term use-life of transgenic plants. Specifically, the two main objectives of this chapter are to (1) present a rationale for preserving transgenic plants as important pest management tools using Bt technology as an example, and (2) discuss the necessary role of resistance management in any deployment scheme for transgenic or conventionally derived pest resistant cultivars, and to offer several strategies for appropriate release of pest-resistant transgenic plants.

## The resistance problem

The introduction of synthetic chemical control agents in agriculture put an enormous selection pressure on target pests. The first case of herbicide resistance was reported in the 1960s. By 1990, there were an estimated 84 cases of weed resistance to at least one herbicide and reported cases of broad-spectrum herbicide resistance in some weeds of Australian wheat (Green et al. 1990). Since then, resistance has continued to rise in weeds globally. The 1960s also saw the first case of fungicide resistance in pathogenic fungi. Over the next 20 years, over 100 species of fungi had developed resistance to at least one fungicide (Green et al. 1990). Insects are probably the best example of resistance selection by pesticides; in only 40–50 years, over 530 species acquired some form of resistance (Georghiou and Lagunes-Tejeda 1991). Insects have many adaptation strategies; they can adapt their digestive processes to escape toxic secondary chemistry in host plants, develop new feeding behaviors or taste preferences, relocate to areas with new plant species on which to feed, or develop detoxification mechanisms and target-site changes to neutralize toxins.

## Bt insecticides

The primary response from agricultural scientists to pest resistance has been to develop new chemical tools for controlling pests. At the same time, a global rise in environmental and consumer activism has appropriately spurred public- and private-sector researchers to seek less toxic, more environmentally safe alternatives. The discovery that proteins from the soil bacteria Bt could be used as an insecticide was a dramatic step toward an environmentally safe insecticide with no impact on humans and limited impact on pollinators and beneficial organisms.

Recently, Bt toxin genes have been genetically engineered into plants such as potato, corn, cotton, tomato, rice, tobacco, clover, fruit trees, various forest tree species, and numerous others. These plants express the Bt toxin within their own tissue, thus protecting the plant from certain insect pests without resorting to pesticide sprays. In addition, plant tissues traditionally difficult to protect with conventional pesticide delivery systems, like roots, tubers, husks, and bolls., can now be protected.

However, field and laboratory data indicate that insects have the potential for resistance to Bt, especially to those transgenic plants that maintain a constant killing dose throughout the season. Therefore, unlike Bt sprays that attenuate quickly, the selection pressure that transgenic Bt plants elicit on susceptible pest populations may be much higher and for a much longer period of time. The basic axiom in understanding resistance development is "the higher the selection pressure on a pest population, the more rapid the resistance response." Fifteen insect species have already developed resistance to Bt sprays under laboratory selection. Diamondback moth was the first species to develop resistance in the field as a response to continual exposure to conventionally sprayed Bt (Tabashnik et al. 1990).

The technology associated with engineering Bt transgenic plants is advancing rapidly. While transgenic crops are being registered and commercialized in northern hemisphere countries, developing countries are beginning to import or develop, evaluate, and commercialize this technology too. A rapid adoption of Bt transgenic plants into agriculture is thus occurring worldwide. Deployment of these plants will very likely lead to the development of resistance if steps are not taken to prevent it from occurring.

## Why should Bt be preserved?

Transgenic technology may help reduce reliance on chemicals, reduce environmental contamination, and reduce human health impacts by conventional pesticides. For example, it is estimated that the introduction of Bt cotton could decrease overall insecticide use worldwide by at least 10–15% (Roush 1994). In some of the cotton-producing areas of Latin America, where insect resistance to chemical pesticides is severe, Bt cotton could provide a much needed economic boost. In the case of potatoes, transgenics will offer an alternative to conventional insecticides for control of Colorado potato beetle, the most notorious resistant pest in the world. In just one state in the USA, pesticide use could be reduced by 538,000 pounds annually with transgenic potatoes (Grafius 1996). This technology is therefore very appealing to developing countries that lack effective pesticide safety systems, because transgenic plants do not incur the same level of human and environmental risks that modern conventional pesticides do.

It has been estimated that a 10% penetration of the global insecticide market by transgenic Bt plants could lead to an 8% reduction in conventional insecticide use (see table 17.1). In addition, Bt transgenic seed has an estimated economic gain exceeding 35% over conventional nonpest-resistant seed. Liberally estimated, this would save approximately US $668 million globally every year. An 8% reduction translates into approximately 109 million fewer pounds of insecticide (active ingredient) introduced into the environment each year (NRDC 1996 and US EPA)[1], which can be quantified as an estimated environmental gain of US $632 million annually. Furthermore, a 10% reduction could result in an 8% reduction in insecticide-related illnesses, which, in terms of an estimated human health gain, could be translated as a $37 million savings in health care costs annually. While these values are only estimates, they indicate a strong economic, environmental, and human health rationale for managing Bt resistance in transgenic plants in a sustainable way.

## Implementing resistance management

History and research have shown that no single pesticide or pest management tool will escape resistance forever. The question is not whether resistance will develop but when and in what form (Green et al. 1990). However, the longer it takes for resistance to develop, the longer one can rely on a given pest control strategy. Resistance management is a way of sustaining the effectiveness of a pest control tool or tactic. It is an effort to delay or prevent adaptation in pest species by managing the factors that may contribute to resistance development.

---

[1] Unpublished EPA data obtained by the NRDC indicate that worldwide conventional pesticide use reached 4.7 billion pounds (active ingredient) in 1995. Approximately 29% of this amount is attributed to insecticides. An 8% reduction would result in approximately 109 million fewer pounds of insecticide used worldwide.

**Table 17.1.** Estimated Global Economic, Environmental, and Human Health Benefits of Bt Transgenic Plants

|  | % | | | |
| --- | --- | --- | --- | --- |
| Global insecticide market penetration of transgenic plants (in %) | 1 | 10 | 25 | 50 |
| Estimated reduction in conventional insecticide use (actual market penetration of transgenics, in %)[1] | 0.8 | 8 | 20 | 40 |
|  | US $ millions | | | |
| Estimated global savings from reduced use of conventional insecticides.[2] | 67 | 668 | 1,670 | 3,340 |
| Estimated Environmental Gain[3] | 63 | 632 | 1,580 | 3,160 |
| Estimated Human Health Gain[4] | 3.6 | 36.5 | 91.2 | 182.4 |

[1] Reduction in use due to the introduction of transgenic plants is not likely to be 1:1. Since narrow-spectrum transgenic plants will replace many broad-spectrum insecticides, the continued use of some insecticides will be required.

[2] 1995 global expenditures on all pesticides (insecticides, herbicides, fungicides, and others) were approximately US $28.7 billion (NRDC 1996). Global expenditures on insecticides alone are estimated at $8.35 billion (29% of the total pesticide market). Estimated Global Savings = (% reduction in insecticide use) x ($8.35 billion).

[3] Estimated Environmental Gain = (% reduction in insecticide use) x (global insecticide-related environmental costs, or $7.9 billion). Based on estimates by Pimentel et al. 1992 and unpublished US EPA data (NRDC 1996).

[4] Estimated Human Health Gain = (% reduction in insecticide use) x (global insecticide-related human health costs, or $456 million). Based on estimates by Pimentel et al. 1992 and unpublished US EPA data (NRDC 1996).

There are three groups of key players in implementing resistance management strategies: producers of Bt seed products, distributors and advisors, and end users (Kennedy & Whalon 1995). A combination of commitment, participation, and cooperation at all levels is important for successful resistance management. There may be a temptation for seed and biotech companies (producers) to maximize profits to recoup the costs of bringing their product to market. However, companies have much to gain by increasing the use-life of Bt products including transgenic plants.

Distributors and advisors are a critical link between seed producers and growers. Public and private institutions alike can serve as advisors by educating at all levels to help prevent emergence of pest resistance through detection and proactive management. Education and awareness about resistance and resistance management could help convince both biotech companies and farmers to sacrifice short-term profit for long-term markets and benefits.

At the same time that transgenic plants are becoming available, societal values and consumer demands have put a greater focus on the environment and food safety. Political pressure is high in many countries for adoption of agricultural policies that decrease chemical inputs (Zadoks 1992). This puts pressure on the farmer to use more environmentally safe pesticides, such as Bt products including transgenic plants. Since Bt products are relatively nontoxic as compared with chemical insecticides, farmers may not realize the necessity to manage their use. Therefore, there is a need to educate and raise awareness among end users in Bt resistance management strategies. An awareness and appropriate regulatory focus by policymakers is especially important for the development of an appropriate management policy.

## Resistance management strategies

Resistance management programs rely on the following five key strategies (McGaughey and Whalon 1992):

### *Diversification of mortality sources*

The premise here is that insects will not adapt if they are faced with more than one mortality mechanism. Several tactics can be used in both conventional (spray) and transgenic Bt programs. Bt toxins can be rotated or alternated with other chemicals, toxins, or other control strategies. Similarly, two or more toxins could be mixed and introduced at once. With transgenic plants, mixtures could be achieved by mixing seeds of different genetic lines, each engineered to express a different toxin, or one plant variety could be engineered to express multiple toxins (Gould 1988). Many biotech companies have experimentally developed transgenic plants with multiple genes (multigenetic, stacked or pyramid) for insect resistance. Other companies have developed new genetically engineered Bt spray products. It will not be long before companies develop and market multigene insect-resistant transgenic plants. However, the development of cross-resistance to Bt toxins is a possibility that could preclude the long-term success of this approach. Some pests have demonstrated the ability to develop resistance to a wide variety of Bt toxins after initial selection by only one toxin (Gould et al. 1992; McGaughey and Johnson 1994; Tabashnik et al. 1990). The best strategy for maintaining transgenic Bt plant success therefore is not to overuse this technology. Integrated pest management (IPM) programs are the best context into which transgenic Bt plants can be introduced, because they rely on multiple strategies, tactics, and tools to manage pest populations below economically damaging levels. Thus IPM is probably the best safeguard for transgenic Bt resistance management.

### *Maintenance of a susceptible population*

Assuming that there is a fitness cost associated with resistance, a reduction in selection pressure may help the population revert back to a more susceptible state. Fitness costs also help maintain the existing susceptible population, thus preserving the important susceptible genes. Refuges may be an effective way to reduce selection pressure by providing an area for habitation and immigration of susceptible insects. In a transgenic deployment scheme, this can be achieved by providing a refuge of nontransgenic plants in one or more ways:
- a seed mixture of transgenic and nontransgenic plants within a field or fields
- a spatial mixture, or field-to-field mosaic, that results in a patchwork of completely transgenic and completely nontransgenic plots within a field or fields
- a temporal mixture, or season-to-season sequence that alternates between transgenic and nontransgenic plantings

Tissue- or temporal-specific gene expression (i.e., limiting where or when Bt toxins are expressed) in the plant are also possible refuge options.

Preserving and managing genetic resources, i.e., susceptible genes, is the key goal of resistance management. Maintaining a susceptible pest population means that resistance genes, when they occur in the pest, will be "diluted" by mating with susceptible insects. The first step in resistance management, then, is to realize the importance of and insure the survival of susceptible individuals into the next generation. The challenge is to reduce only that portion of the population that will result in economical control, rather than attempting to kill 100% of a pest population (McGaughey and Whalon 1992).

### Gene strategies and tactics

A major advantage in resistance management is killing the heterozygote offspring of susceptible X resistance matings. This dramatically reduces the probability of development of a resistant population. Probably the best current strategy to accomplish this is to express Bt genes at a high enough level in the transgenic plant to kill the heterozygote (table 17.2). Other gene strategies include pyramiding, stacking or polygenic expression of multiple mortality mechanisms in a single plant. This latter point is often raised, but is much more difficult to deliver; most current commercial Bt transgenic plants are single-toxin expressers.

**Table 17.2.** Gene Strategies and Tactics for Deploying Insecticidal Genes in Plants

| Category | Possible tactics | Function |
|---|---|---|
| Gene options | • Single gene<br>• Multiple genes<br>• Chimeric genes (genetically altered from original) | Deployment of one or more control agents (insecticidal proteins, insect growth regulators, proteinase inhibitors, etc.) in a single plant, thereby diversifying mortality sources and potentially decreasing the rate of resistance development. |
| Gene promoter | • Constitutive (continual dose in all plant tissues)<br>• Tissue-specific (in specific plant parts, e.g. fruit, roots, pollen)<br>• Inducible (spray a chemical or damage physically to turn on) | Control of toxin dose by regulating where and/or when toxin is released in the plant. |
| Gene expression | • High dose<br>• Low dose<br>• Mixtures | Control of toxin dose by regulating the actual quantity and/or quality expressed in the plant tissue. |
| Field tactics | • Uniform single gene<br>• Mixture of genes (seed mix)<br>• Gene rotation or sequence (time mix)<br>• Mosaic planting (area or region mixtures)<br>• Refuges (spatial or temporal areas untreated) | Regulation of toxin expression by manipulating planting patterns in the field. |

Based on McGaughey and Whalon 1992.

## Prediction and monitoring of resistance

Waiting until resistance occurs before implementing a resistance management program is ineffective (Hoy 1992). A successful program must include tactics for monitoring, predicting, and evaluating resistance as it progresses. Sampling of insect populations at regular intervals is a good approach for monitoring resistance progress, but it requires either a simple, low-cost diagnostic tool for use in the field (especially in developing countries) or more-advanced facilities where insect specimens can be sent for diagnosis. In either case-resistance monitoring requires a high level of commitment from farmers and extension personnel since the collection and testing of samples is time- and labor-intensive. It is therefore unlikely that developing countries will adopt this strategy extensively. Biotech companies can contribute to monitoring and compliance programs by being required to develop and implement resistance-monitoring protocols as a condition of registration. In the USA, the development of a resistance-management and -monitoring plan has been a requirement for registration of transgenic Bt crops.

## Appropriate policy implementation

An appropriate policy for transgenic plant deployment will have broad applicability and be tailored to available resources. However, if it does not take into account the resistance-management concerns described above, then the usefulness of such a policy will be lost.

Researchers emphasize that decisions to deploy Bt transgenic crops should be made on a case-by-case basis within each crop production system (Hokkanen and Wearing 1994; Kennedy and Whalon 1995). A case-by-case (country-by-country) assessment process requires identification and understanding of both the environmental and agricultural conditions as well as the biology and host plant interactions of the target species. It may begin by assessing data on three main elements: (1) features of the crop that could impact on selection for resistance, (2) the target pest, its host range, and its propensity to develop resistance to Bt, and (3) applicability and utility of available transgenic technology in light of the crop and pest data. This process will determine if the available transgenic technology is suitable to the crop-pest complex under consideration. If so, it will help determine which deployment strategies are appropriate to maintain susceptibility in the pest population. In the absence of this approach, regulatory policymakers can choose to take an "educated guess" and implement a strategy that has a high probability of success with minimum investment (Wearing and Hokkanen 1995).

# Regulatory options

Pest resistance can be managed through various regulatory approaches such as licensing, central control of seed, and labeling. The choice depends on a consideration of several components, including target(s) of the regulatory method, enforcement method(s), cost/resources required for implementation, and complexity or ease of implementation.

Regulatory methods may target the originator of the transgenic Bt seed (a biotech company), seed distributors, or end users. The success of a regulatory policy will depend on the number of entities or people regulated, as well as the particular crop and production system involved. Success for any resistant-pest transgenic regulatory policy will be based on its ability to delay the development of resistance in the target pest(s). This will require monitoring and assessment of the progress of resistance in the target pest population(s). The likelihood of success of the policy also depends on the policy's level of complexity.

## Licensing

Licensing of any type requires an efficient and stable bureaucratic apparatus for its implementation and enforcement. While licensing is common in the developed world, countries with less evolved institutions may find it more difficult to implement. Licensing targets could include seed companies, seed distributors, sales representatives, or end users. The enforcement apparatus would vary based on the particular target. For example, a national government agency could enforce licensing agreements at all target levels, although it would probably be easier for such an agency to implement licensing of seed companies than of individual farmers. Likewise, seed companies could enforce licensing at the level of seed distribution and sales. For example, Monsanto (USA) has used licensing to achieve compliance on use, production, and sales of the company's transgenic Bt potato, cotton, and corn crops. Further, a local growers' organization may act as licenser and enforcer at the growers' level.

In countries that lack strong institutions/agencies for enforcement, local seed distributors or growers' organizations may provide the most effective means for regulation. In addition, the establishment of a penalty (fine, forfeiture of seed, etc.) for license infringement or, conversely, an incentive for compliance will be necessary in most cases to insure effective enforcement. In most instances, an understanding of resistance management strategies is not sufficient incentive for compliance except at the industry or sales level (Kennedy and Whalon 1995).

A licensing program increases in complexity as the regulatory apparatus becomes more removed from the target. In other words, the licensing of growers by local growers' associations is relatively simple compared with the complexity of grower licensing by a national agricultural agency. In all cases, the establishment of a licensing policy requires the input of capital, personnel, and facilities, and educational campaigns will be necessary to train individuals in applying and using the seed properly. Extension workers would be well-suited to train the end users, but the extensionists themselves will need initial training from national-agency or seed-company personnel. The burden of these costs will most likely fall on the national government, but in some cases countries may be able to establish partnerships with the seed companies that require a commitment of resources, at least in education and training, as a condition for importing and selling the seed or transgenic product.

## Central control of seed

Central control of seed is a regulatory option that would appeal to governments with "command and control" structures in their national agriculture programs. In this scenario, the regulatory targets would be limited to the seed companies and/or seed distributors. Governments with strong national agencies could then directly regulate sales (imports/exports) and distribution via the seed companies. While implementation and enforcement would require a new agency branch and administrative staff, some of the education costs associated with licensing may be eliminated. However, the complexity of the program could increase as governments attempt to regulate national distribution of seed. In addition, the government may need to implement strong penalties or incentives for compliance at both the seed company/distributor level and the grower level to prevent the misuse of seed once it has been distributed. Alternatively, the government could pass along to the seed companies some of the burden of educating the end users and then hold the seed companies accountable for subsequent misuse of the seed. However, this too would increase the complexity and cost of enforcement. Ultimately, the success of a central control of seed policy depends on the same three factors: monitoring, assessment, and feedback.

## Labeling

Package labeling is an obvious starting point for countries considering deployment of transgenic products, because it is straightforward, low-cost, and easy to enforce. Additionally, it can serve as a foundation for formulating more extensive regulation. Since many biotech companies already provide "proper use" labels on most products, industry cooperation and compliance with this type of regulation can be very high. Label requirements for Bt seed should include:

1. identification of transgenic status
2. recommended planting ratio (% transgenic vs nontransgenic seed) or recommended refugia for resistance management
3. a warning against misuse or over planting of the transgenic seed
4. an agency or industry contact in the event that resistance to the transgenic crop becomes evident in the pest population. Should such an event occur, the proper authorities should be alerted and seed use terminated in the next growing season. If possible, the resistant-pest population should be destroyed in the field by other means (mechanical, physical, or population suppression with insecticides).

Many countries have never used some of these options in pesticide-related policy. The current regulatory apparatus, the particular transgenic crop under consideration, and the existing seed-handling system will all be important factors in choosing the most effective regulatory options. Since most countries already have in effect a unique form of pesticide regulation, national policy experts will need to evaluate the options and choose an appropriate policy. In all cases, transgenic crop deployment should be done in conjunction with a monitoring program to ensure proper use of seed, regulatory compliance, and control of resistance development. Some countries may also choose to implement an integrated "package" of policies designed to give maximum benefit, e.g., labeling and seed licensing.

## Conclusions

Resistance management strategies provide a means to manage both the use of transgenic plants and the development of insect resistance to these plants. Resistance management strategies include: (1) diversification of key pest mortality sources, (2) reduction of selection pressure on key pests and maintenance of a susceptible population via refugia or immigration, (3) selection of gene strategies and tactics, (4) prediction and monitoring of resistance development, and (5) appropriate policies to implement these strategies.

Since the technology for producing transgenic plants is still expanding, current and future developments may provide new approaches for implementing resistance management strategies. These evolving technologies may include: modifying the number and type of genes expressed in the plants, fine-tuning the level of toxin expression (low, moderate, high), and controlling the timing or location of expression in specific plant tissues. Field tactics such as seed mixes, rotations, and use of refuges (areas where key pests escape control) are also important in the proper development and implementation of a resistance management plan. Probably the best long-term policy that a country could adopt to reduce the likelihood of resistance development would be to require a high dose Bt expression coupled with adjacent refugia of at least 20% non-Bt plants within an IPM program.

The social, economic, and technical infrastructure of developing countries may be quite different than in countries where many transgenic plants originate. Therefore, the decision to deploy transgenic crop plants should be based on an assessment of indigenous ecological, environmental, socioeconomic, and agricultural conditions. Criteria to consider include the availability of refugia to counteract resistance development, economic importance of the target pests, and the level of cooperation among growers and industry in the management of transgenic resources. The assessment should include input from scientists, policymakers, agricultural specialists, public and private institutions, and local farmers.

The implementation of an appropriate regulatory policy requires consideration of several components, including the regulatory target group, the enforcement method, the cost/resources required for implementation, and the complexity of implementation. Final policy will vary on a country-by-country basis, but regulatory options could include licensing, central control of seed, regulation of seed distribution, product labeling, and monitoring of both seed use and resistance development.

Licensing could be implemented at various levels, but complexity and cost increase as the size of the target group increases. Some developing countries may find it easier to implement licensing through local growers' associations rather than direct governmental licensing of individual growers.

Central control of seed may be appropriate in countries with "command and control" structures in their national agriculture programs. Governments with strong national agencies could then directly regulate sales and distribution via the transgenic seed companies. Such a program would require a system of incentives to ensure proper use of the seed once it has been distributed.

Labeling of transgenic products is the most optimal starting point for initiating transgenic regulatory policy in any country, but labeling will not likely be effective if used alone. As industry cooperation could be high under this policy, implementation and enforcement costs would be low. Labels should include "proper use" information, including details of appropriate resistance management and resistance monitoring strategies. All countries that choose to deploy transgenic plants should do so only in conjunction with a strong resistance management program.

Ultimately, the effectiveness of any transgenic regulatory policy will require continual monitoring, assessment, and feedback to ensure compliance and control of resistance development. Should incidences of resistance in target pests occur, the current policy should be modified immediately and deployment of transgenic crops should be halted until the pest population has been destroyed by other means.

## References

Agne, S., H. Waibel, F. Jungbluth, G. Fleischer. 1995. Guidelines for Pesticide Policy Studies: A Framework for Analyzing Economic and Political Factors of Pesticide Use in Developing Countries. Publication Series 1. Hannover: The Pesticide Policy Project.

Alstad, D.N. and D. A. Andow. 1996. Implementing management of insect resistance to transgenic crops. *AgBiotech News & Information* 8:177-81.

Georghiou, G. P. and A. Lagunes-Tejeda. 1991. The Occurrence of Resistance to Pesticides in Arthropods. Rome: Food and Agriculture Organization of the United Nations.

Gould, F. 1988. Evolutionary biology and genetically engineered crops. *BioScience* 38(1): 26-33.

Gould, F. 1991. The evolutionary potential of crop pests. *American Scientist* 79:496-507.

Gould, F., A. Martinez-Ramirez, A. Anderson, J. Ferre, F.J. Silva, and W.J. Moar. 1992. Broad-spectrum resistance to *Bacillus thuringiensis* toxins in *Heliothis virescens*. *Proceedings of the National Academy of Science* 88: 7986-7990.

Grafius, E. 1996. Economic impact of insecticide resistance in the Colorado potato beetle (Coleoptera: Chrysomelidae) on the Michigan potato industry. *Journal of Economic Entomology* 89.

Green, N.G., H.M. LeBaron, and W.K. Moberg. 1990. Managing resistance to agrochemicals: From fundamental research to practical strategies. Washington, D.C.: American Chemical Society.

Hokkanen, H.M.T. and C.H. Wearing. 1994. The safe and rational deployment of *Bacillus thuringiensis* genes in crop plants: Conclusions and recommendations of OECD workshop on ecological implications of transgenic crops containing Bt toxin genes. *Biocontrol Science and Technology* 4: 399-403.

Hoy, M.A. 1992. Proactive management of pesticide resistance in agricultural pests. *Phytoparasitica* 20(2): 93-97.

Kennedy, G.G. and M.E. Whalon. 1995. Managing pest resistance to *Bacillus thuringiensis* endotoxins: Constraints and incentives to implementation. *Journal of Economic Entomology* 88(3): 454-460.

McGaughey, W.H. and D.E. Johnson. 1994. Influence of crystal protein composition of *Bacillus thuringiensis* strains on cross-resistance in Indianmeal moths (Lepidoptera: Pyralidae). *Journal of Economic Entomology* 87: 535-540.

McGaughey, W.H. and M.E. Whalon. 1992. Managing insect resistance to *Bacillus thuringiensis* toxins. *Science* 258: 1451-1455.

Natural Resources Defense Council. 1996. Unpublished EPA data released as Congress contemplates weakening pesticide laws, allowing more chemicals in food, environment. Press release, May 28, 1996.

Riggin-Bucci, T. M. and F. Gould. 1997. Impact of intraplot mixtures of toxic and nontoxic plants on population dynamics of diamondback moth and its natural enemies. *Journal of Economic Entomology* 90:

Roush, R.T. 1994. Managing pests and their resistance to *Bacillus thuringiensis*: Can transgenic crops be better than sprays? *Biocontrol Science and Technology* 4: 501-516.

Tabashnik, B.E. 1994. Delaying insect adaptation to transgenic plants: Seed mixtures and refugia reconsidered. *Proceeding of the Royal Society of London* 255: 7-12.

Tabashnik, B.E., N.L. Cushing, N. Finson, and M.W. Johnson. 1990. Field development of resistance to *Bacillus thuringiensis* in diamondback moth (Lepidoptera: Plutellidae). *Journal of Economic Entomology* 83(5): 1671-1676.

Wearing, C.H. and H.M.T. Hokkanen. 1995. Pest resistance to *Bacillus thuringiensis*: Ecological crop assessment for Bt gene incorporation and strategies of management. *Biological Control: Benefits and Risks* (Series: Plant and Microbial Biotechnology Research Series 4): pp: 236-252.

Zadoks, J. C. 1992. The costs of change in plant protection. *Journal of Plant Protection in the Tropics* 9(2): 151-159.

# SECTION V

# Managing IPR, Proprietary Science, and Technology Transfer

Agricultural development has benefited from a long history of public-sector/public-good investment, at the core of which has been the wide availability of plant genetic resources. However, public-good investments in agriculture face an uncertain future, with increased emphasis on market mechanisms forcing publicly funded organizations to respond to broader economic opportunities, tendencies to restrict the availability of germplasm to those working for national agricultural research, and changes occurring in the management of intellectual property.

The primary purpose of intellectual property ownership by national agricultural research organizations (NAROs) is to promote the fundamental research mission of the institute, keeping in mind the needs of clients and production of public goods. Clarification of ownership of assets has an important role to play in this regard. Assets include research inputs, including patent rights for a gene sequence or for a laboratory or industrial process, and outputs. Copyrights and trade secrets may govern access to and use of experimental techniques and laboratory notes. Patents for research outputs may be sought for novel processes and products, while plant variety protection is sought for new crop varieties. Management needs to ensure that (1) ownership of intellectual property used by a research organization is respected by all who use the property and (2) organizations are in a position to identify, secure, manage, and exploit the intellectual property that they generate.

This section begins with an examination of the implications of intellectual property rights (IPR) in a biotechnological context (chapter 18, "Intellectual Property Rights and Agricultural Biotechnology"). The differences between principal categories of IPR are explained, including plant variety rights, patents, trademarks, confidential information, trade secrets, and copyright. It also discusses material transfer agreements, farmers' rights and developments as related to *sui generis* systems. The national, regional, and international dimensions of these forms of protection and agreements are reviewed, with details provided of the patenting process, estimates for filing patents, and claims for plant variety rights.

The next two chapters explore the establishment of intellectual property statutes and infrastructure to manage intellectual property in a research institute. Chapter 19 ("Agricultural Research and the Management of Intellectual Property") shows how the intellectual property resources of a NARO can be managed in conjunction with a research liaison office or by using the intellectual property services offered by an equivalent office within an associated institution. The research liaison officer improves understanding of legal rights given for protecting creative effort and helps to further the institution's research mission. For a NARO, the research liaison office is responsible for protecting and maintaining the intellectual property assets of the organization and assisting in developing an awareness and appreciation of the use of patent documents and registered plant variety data as research resources.

Chapter 20 ("Managing Intellectual Property in Embrapa: A Question of Policy and a Change of Heart") highlights the management challenges encountered by the Brazilian agricultural research corporation Embrapa as it develops and implements a new internal policy regarding intellectual property protection. This policy was designed to maintain research quality, enhance competitiveness, and prepare Embrapa for the demands and challenges of modern agriculture. The chapter analyzes the design and implementation of this new policy, the legal instruments developed and dialogue being held with clients and society. It explores why Embrapa used a centralized office for providing IPR information and decision making while working for and with geographically dispersed research entities. This is a situation shared in common with the CGIAR institutions and other national agricultural research systems.

Chapter 21 ("Managing Proprietary Science and Institutional Inventories for Agricultural Biotechnology") confronts major decisions needed to clarify the status of proprietary assets in international agricultural research. It presents insights from inventories of proprietary biotechnology inputs used by national and international agricultural research organizations. It reviews the most important findings derived from these inventories. It documents the difficult and often confusing situation that institutions face regarding the use and dissemination of products resulting from proprietary science where others hold rights.

Addressing these issues prepares institutes for dealing with the agricultural biotechnology-intellectual property interface. The lack of an understanding and appreciation of the implications of intellectual property issues may impact on the commercialization of research products and affect certain international collaborations, particularly those involving the private sector. Chapter 22 ("International Collaboration: Intellectual Property Management and Partner-Country Perspectives") draws on experiences gained through a project for international collaboration in biotechnology that includes commercial research. It highlights the need for national policies and effective enforcement to transfer available technologies to developing countries, and it identifies specific challenges for the project's management. These include raising awareness and understanding of the importance of intellectual property; developing a system to access, use, and manage intellectual properties; and building functional national and institutional intellectual property policies.

The remaining two chapters in this section present insight into use of IPR from the perspective of commercial and joint-venture opportunities in biotechnology. Chapter 23 ("Industrial Research and Business Development: Experiences from the Singapore Institute of Molecular Agrobiology") presents approaches for industrial research, using experiences from the author as well as from the Institute of Molecular Agrobiology. It analyzes strategies applied for establishing industrial research and development. Emphasis is placed on issues important to industrial R&D management, technology transfer, and business development. Chapter 24 ("Introducing Transgenic Crops in India: A Joint Venture Approach") describes a joint venture company, Proagro-PGS in India, established to advance conventional and biotechnological breeding. It describes the process that led to forming the joint venture. Critical factors for its success included the driving force and foresight of individuals participating in its formation, time taken to prepare an exhaustive memorandum of agreement, and a spirit of partnership extending beyond the written agreements.

## Recommendations

This section focuses on understanding the current environment for IPR as it affects biotechnology research in national and international research organizations. Recommendations for addressing the challenges of intellectual property management are the following:

- Ensure participation in intellectual property and biotechnology policy formulation at the national level, thereby ensuring that the biotechnology research viewpoint is heard at various international fora.
- Formulate an institutional intellectual property policy that reflects stakeholder's expectations, is consistent with the mission of the institute and provides clarity on all matters of institutional intellectual property, meeting both internal and external needs.
- Formulate legal and intellectual property mechanisms that facilitate collaboration with the commercial sector, allowing for development of business opportunities (when appropriate) and participation in joint ventures.
- Develop intellectual property management capacity by ensuring that such capacity building is part of institutional human resource development planning. This can be done through workshops; placing researchers as interns at appropriate local and foreign institutions; facilitating the use of intellectual property literature; and identifying projects which may generate protectable assets and to determine the most suitable intellectual property protection regulation to undertake.
- Establish an intellectual property coordinator or research liaison office with responsibility for identifying, securing, managing, and transferring appropriate technologies. Consideration should be given to building centralized services that can provide expertise to decentralized research institutes.
- Build regional cooperation in intellectual property and technology transfer, particularly through linkages with regional trade and intellectual property organizations, subscriptions to data bases, and participation in collaborative biotechnology research networks.
- Conduct institutional inventories of intellectual property and proprietary technologies as a first approximation of institutional use and intellectual property assets. This should include time for a thorough explanation of the process and results with staff, leading towards more formal intellectual property audits if needed to clarify right to use and disseminate future products.

# 18 Intellectual Property Rights and Agricultural Biotechnology

*Michael Blakeney, Joel I. Cohen, and Stephen Crespi*

## Abstract

*Scientists, research directors, and policymakers face complex questions and decisions when managing intellectual property rights (IPR) for agricultural research. This chapter discusses the impact of IPR in a biotechnological context. It explains differences between principal categories of IPR relevant to agricultural research: plant variety rights, patents, trademarks, confidential information, trade secrets, and copyright. Material transfer agreements and farmers' rights are also discussed. National, regional, and international dimensions of these various forms of protection and agreements are reviewed, as well as the interrelationship between intellectual property and biodiversity rights.*

## Introduction

The decision to adopt an intellectual property rights (IPR) system for agricultural research hinges on several factors affecting the use of genetic material and germplasm. Research programs in developing and developed countries therefore seek clarification on the rights of and access to research innovations and genetic resources. The Agreement on Trade-Related Aspects of Intellectual Property Rights (TRIPs) and the Convention on Biological Diversity (CBD) are meant to provide such clarification. Countries pursuing commitments under both the TRIPs agreement and the CBD must also address the possible conflict between provisions regarding IPR in the two agreements. In addition, decision makers want to improve their understanding of innovation in farming communities. In some countries, they are examining these alongside the innovations produced by scientists.

This chapter demonstrates that IPR protection in agricultural research is growing, offering public research a "defensive" strategy. It cautions that there are difficulties in documenting any significant gains from using IPR protection as a strategy for generating new, external funds for research. The chapter underlines the fact that patents simply protect innovation and secure the potential rights for future development. The chance for national programs to earn financial benefits from research comes mainly from working with the private sector and by providing for technology transfer.

© CAB *International*. 1999. *Managing Agricultural Biotechnology—Addressing Research Program Needs and Policy Implications* (ed. J.I. Cohen)

The most important mechanisms for legally protecting agricultural innovations are plant variety rights (PVR)[1] and patents (extended to cover plants, animals, and microorganisms) (see table 18.1 for a comparison between PVR and patents). Other forms of protection can be provided through trademarks, trade secrets, and copyright. Alternatives to these include material transfer agreements (MTAs) of a private contractual nature. If no form of protection is taken, then research results are generally placed in the public domain, mostly in the form of publications, making results available to all without restrictions on use.

## PVR

Patent law was originally considered unsuitable for protecting new plant varieties developed by traditional breeding methods. Some countries therefore introduced special national laws for PVR in the 1960s, as did the International Union for the Protection of New Varieties of Plants (UPOV), established in 1961. These rights are granted by the state to plant breeders to exclude others from producing or commercializing material of a specific plant variety for a minimum of 15 to 20 years.

To be eligible for PVR, the variety must be novel, distinct from existing varieties, and uniform and stable in its essential characteristics. At first, this form of legal protection was limited to commercializing reproductive or vegetatively propagated material taken from a new variety. It was implied or specified that certain exemptions were allowed to farmers

Table 18.1. The Main Provisions of Plant Variety Rights and Patent Law

| Provisions | Plant variety rights (under UPOV 1991) | Patent law |
|---|---|---|
| Protection coverage | Plant varieties of all genera and species | Inventions |
| Requirements | Novelty, distinctness, uniformity, stability | Novelty, inventiveness, non-obviousness |
| Protection term | Minimum 20 years | 17–20 years (OECD) |
| Protection scope | Commercial use of all material of the variety | Commercial use of protected matter |
| Breeder's exemption (research only) | Yes | Variable |
| Breeder's exemption (commercial use) | Not for essentially derived varieties | No |
| Farmer's privilege | Up to national laws | limited form in prospect in EU countries |

UPOV: International Union for the Protection of New Varieties of Plants
Revised from Van Wijk et al. 1993.

---

[1] The terms "plant variety rights" and "plant breeders' rights" are synonymous. For purposes of clarity, only the term "plant variety rights" is used here, because this corresponds most closely to the nature of the legal protection that is obtained.

and researchers (breeders). Such exemptions under PVR systems are termed farmer's privilege and breeder's privilege (or research exemption).

## PVR under UPOV

The original 1961 version of UPOV was revised in 1972, 1978, and 1991. The 1991 version has come into force in some countries and preparations for ratification are underway in most others. Originally, the scope of PVR concerned "the production for purposes of commercial marketing, the offering for sale, the marketing, of the reproductive or vegetative propagating material, of the variety" (UPOV 1978). This has been broadened under UPOV 1991 (see later).

UPOV 1978 specified that any member state could provide *either* patent protection *or* PVR protection for the same botanical species or genus. This prohibition of double protection is not present in UPOV 1991. Researchers using biotechnology techniques alongside traditional breeding methods will be able to obtain both types of protection as appropriate. The status of PVRs in UPOV member states is shown in box 18.1.

---

**Box 18.1.** The 43 Member States of UPOV as of May 23, 1999

| | | |
|---|---|---|
| Argentina (2) | France (2,4) | Portugal (2,4) |
| Australia (2,5) | Germany (3,4) | Republic of Moldova (3) |
| Austria (2,4) | Hungary (2) | Russian Federation (3) |
| Belgium (1,4) | Ireland (2,4,5) | Slovakia (2,5) |
| Bolivia (2) | Israel (3) | South Africa (2,5) |
| Brazil (2) | Italy (2,4,5) | Spain (1,4) |
| Bulgaria (3) | Japan (3) | Sweden (3,4) |
| Canada (2) | Kenya (2) | Switzerland (2) |
| Chile (2) | Mexico (2) | Trinidad and Tobago (2) |
| China (2) | Netherlands (3,4) | Ukraine (2) |
| Colombia (2) | New Zealand (2) | United Kingdom (3,4) |
| Czech Republic (2) | Norway (2) | USA (3) |
| Denmark (3,4) | Panama (2) | Uruguay (2) |
| Ecuador (2) | Paraguay (2) | |
| Finland (2,4,5) | Poland (2,5) | |

(1) 1961 Act as amended by the Additional Act of 1972 is the latest act by which the state is bound (2 states).
(2) 1978 Act is the latest Act by which the state is bound (30 states).
(3) 1991Act is the latest Act by which the state is bound (11 states).
(4) Member of the European Union, which has introduced a (supranational) plant variety rights system based upon the 1991 Act.
(5) Has already amended its law to conform to the 1991 Act; most other states are in the process of doing so.

Belarus, Costa Rica, Croatia, Estonia, Georgia, Kyrgyzstan, Morocco, Nicaragua, Romania, Slovenia, Venezuela, and Zimbabwe, as well as the European Union are applying with the UPOV Council to become members. Many other nonmember states currently have laws to protect plant varieties or proposals for laws before their legislatures.

Information courtesy of C. Rovere (1999).

## Procedures, fees, and scope of protection

Under UPOV 1991, a plant breeder is conferred the exclusive right to do or to license the following acts in relation to propagating material of the variety:

- produce or reproduce the material
- condition the material for the purpose of propagation
- offer the material for sale
- sell the material
- import the material
- export the material
- stock the material for the purposes described above

The general duration of PVR is 25 years in the case of trees and vines and 20 years for any other variety.

### Registrable plant varieties

As with patents, PVR are established after a registration process. A plant variety is considered to be registrable if it has a breeder; if it is distinct, uniform, and stable; and if it has not been or only recently been exploited. A plant variety is considered *distinct* if it is clearly distinguishable from any other variety whose existence is a matter of common knowledge. It is *uniform* if, subject to the variation that may be expected from the particular features of its propagation, it is uniform in its relevant characteristics on propagation. A plant variety is *stable* if its relevant characteristics remain unchanged after repeated propagation. A plant variety is taken *not* to have been *exploited* if it or propagating material has not been sold to another person by or with the consent of the breeder. The test of no commercial exploitation is easier to satisfy than the test for novelty under patent law, and the choice between the two forms of IPR is a matter to be considered by the agricultural research institute.

The most important procedure in getting PVR is examining the biological material itself. Extensive field trials are necessary to determine whether the variety meets the legal requirements of distinctiveness, uniformity, and stability. The breeder must also supply an objective description of the new variety and list its characteristics in a qualitative or quantitative way so that it can be clearly distinguished from already known varieties. Table 18.2 presents the fee structure for obtaining and maintaining PVR in the UK for 10 groups of crop species.

Each country must have a means of registering and certifying material selected for PVR to guarantee that seed or planting material distributed to growers remains "true to type" (retains the qualities originally stated on the application). To maintain confidence in the PVR system, there must be agreement among breeders and growers on the validity and usefulness of the system, with the benefits of compliance fully understood. The system must ensure that a variety for which protection is sought meets the requirements of distinctiveness, uniformity, and stability. Apart from the financial constraints in maintaining a PVR system, finding technically qualified personnel to staff a PVR office may present a major difficulty.

Table 18.2. UK Plant Variety Rights Fees (in UK £), Regulations 1999

| Species groups* | Fees per test application/annual renewal fees | | | | | | | | | |
|---|---|---|---|---|---|---|---|---|---|---|
| | 1* | 2 | 3 | 4 | 5 | 6 | 7 | 8 | 9 | 10 |
| Application | 275 | 275 | 275 | 275 | 275 | 275 | 75 | 275 | 135 | 275 |
| Tests or examination | 745 | 455 | 695 | 645 | 555 | 455 | 120 | 185 | 230 | 185 |
| Granting of PVR | 145 | 145 | 145 | 145 | 145 | 145 | 60 | 145 | 120 | 145 |
| Continued exercise of rights each year | 435 | 435 | 435 | 170 | 435 | 435 | 70 | 330 | 175 | 330 |

\* 1. Cereals (excluding maize); 2. Maize; 3. Potato; 4. Beetroot, brussel sprout, cabbage, celery, fenugreek, leek, turnip, carrot, curly kale, onion, radish, fodder crops; 5. Field pea, vegetable pea, field bean, broad bean; 6. Herbage, oil and fiber (including oilseed rape); 7. Rose; 8. Chrysanthemum; 9. Other decoratives; 10. Fruit.

*Source:* Plant Variety Rights Office and Seeds Division, Ministry of Agriculture, Fisheries and Food, UK.

## Exemptions under UPOV

Until recently UPOV has provided exemptions to breeders' rights to allow farmers and breeders to use protected material. The breeder's exemption, also called research exemption, is only for research purposes. The farmer's exemption, known as farmer's privilege, arose as a consequence of the limited scope of the variety right. According to Article 5 of UPOV 1978, the rights of the breeder of a protected plant variety were restricted to commercial dealing in the *reproductive material* through (1) production for purposes of commercial marketing, (2) offering for sale, and (3) marketing. This meant that a farmer growing a first crop from purchased seed of the protected variety was legally free to save seed from the first harvest, which could be used for sowing a second and subsequent crops on his own farm.

PVR exemptions allowing farmer's privilege are especially important to those producers who do not rely on purchased inputs but tend to save their own seed and exchange some of it among themselves every year. In India, any attempts to weaken farmer's privilege and therefore rural communities have led to strong protests by farmers (Dhar et al. 1995).

The research exemption, provided in UPOV 1978, stated that the breeder's authorization would "not be required either for the utilization of the variety as an initial source of variation for the purpose of creating other varieties or for the marketing of such varieties."

The two exceptions to the breeders' rights have been modified in the latest amendment to the UPOV convention (1991). Article 14(1) extends the breeders' rights to all acts pertaining to production and reproduction of seeds and other planting material. Thus, unlike the provisions of UPOV 1978, there is no longer an implicit right of farmers to save and reuse seed from protected varieties without the breeder's authorization.

Article 15(2) does, however, provide some leeway for farmers: "each Contracting Party (to UPOV 1991) may, within reasonable limits and subject to the safeguarding of the legitimate interests of the breeder, restrict the breeders' right in relation to any variety in order to permit farmers to use for propagating purposes, on their own holdings, the product of the harvest which they have obtained by planting, on their own holdings, the protected variety." Here, "legitimate interests" refers to the royalties that should be paid to the breeder for reuse of seed. Thus, new provisions allow farmers to reuse the protected material, but only if they pay.

Another provision of UPOV 1991, article 14(5), strengthens the breeders' rights by extending protection to "essentially derived varieties and certain other varieties" of the protected varieties. However, article 15(1) reaffirms the free availability of protected varieties as a source of germplasm for the introduction of further variation. This provision says that the breeders' right shall not extend to "acts done for the purpose of breeding other varieties." The freedom of research is therefore safeguarded. The extension of breeders' rights to cover essentially derived varieties may be limited to varieties that take over virtually the whole of the genome of the protected variety.

## Patent protection

The international standard for the domestic protection of inventions is prescribed in article 27 of the TRIPs agreement, which is binding for all members of the World Trade Organization (WTO). Article 27(1) provides that "patents shall be available for any inventions, whether products or processes, in all fields of technology, provided that they are new, involve an inventive step and are capable of industrial application". Article 27(2) envisages that inventions may be excluded from patentability to protect "*ordre public* or morality, including to protect human, animal or plant life or health or to avoid serious prejudice to the environment". Article 27(3) permits signatories to exclude from patentability "plants and animals other than microorganisms, and essentially biological processes for the production of plants or animals other than non-biological or microbiological processes". Additionally, Article 27(3) requires that "members shall provide for the protection of plant varieties either by patents or by an effective *sui generis* system or by any combination thereof." The TRIPs agreement will be reviewed in late 1999, and there are already indications that amendments to Article 27(2) will be sought.[2] National patent statutes, in line with Article 27(1) of TRIPs, grant patent protection to inventions that are novel and industrially applicable and that involve an inventive step.

### *Inventiveness*

The first requirement that a biotechnological innovation has to satisfy, if it is to be patented, is that it constitutes an invention. In this regard a distinction is often drawn between an invention and a discovery, which is considered not patentable. The development of a number of new biotechnologies is based on the discoveries of researchers, which initially provided a problem for potential patentees. Thus in 1948 the US Supreme Court in *Funk Brothers Seed Company v Kalo Innoculant* held that a scientist's combination of bacterial strains found separately in nature did not constitute "an invention." According to the Court, "a product must be more than new and useful to be patented; it must also satisfy the requirements of invention."

A shift by the Supreme Court in favor of biotechnological patenting occurred in its 1980 decision in *Diamond v Chakrabarty*. By a 5-4 majority the court held that a

---

[2] The US government, for example, intends to revisit this in late 1999.

hybridized bacterium, used in the treatment of oil pollution could be patented because it was "the result of human ingenuity and research." The Supreme Court in *Chakrabarty* drew a distinction between naturally occurring bacteria, which *Funk* had held not to be patentable, and genetically manipulated strains that were patentable because of the alteration involved.

In the USA this narrow majority in *Chakrabarty* laid the foundation for granting intellectual property protection for products of modern biotechnology. Relying on this decision, the United States Patent Office was prepared to grant broad patents for hybridized and genetically engineered organisms, such as the grant to Agracetus in 1992 of a patent in relation to the insertion of the Bt gene into cotton plants. This practice has been criticized on the grounds that the Patent Office failed to distinguish between the manipulation of a genome and its design or invention. The breadth of the Agracetus patent and its application to plants other than cotton resulted in a successful challenge by the United States Department of Agriculture in 1998, and it is expected that such patents will now be more closely scrutinized in the USA.

### Novelty

Novelty means that an invention has not been previously published or used in the market. Protection is not provided for inventions that are obvious. The test of whether an innovation is novel is assessed by reference to a person having ordinary skill in the art to which the subject matter of the invention pertains. The US Patent Office considered this question, arising in the context of the patenting of plant varieties, in a 1992 determination concerning a patent application in relation to disease resistance that was bred into soybeans. The claimed, novel soybean plant differed from prior soybeans by having *Phythopora* root rot resistance. The Patent Office ruled against the applicant because it was well known that resistance to root rot could be bred into a soybean line by crossing it with one that possessed the desired phenotype.

### Industrial applicability

The next requirement for patentability is that an invention is capable of an industrial application, i.e., that products can be produced or that industrially useful results can be achieved through the application of a process. Agricultural patents are considered to satisfy this requirement. To ascertain whether an invention is industrially applicable, most laws require that the patent application describe the invention in sufficient detail to enable others in the field to make the invention. The patent specifications must be able to explain how to make the invention and describe the means for carrying it out. However, describing the best way to make a biological organism may be difficult, not enabling others in the field to reproduce the organism.

## The registration process

In Australia, applications for patents are filed in the appropriate national or regional patent office. The application is accompanied by a written specification that fully describes the invention. The relevant national or international law prescribes the application procedure. The following procedure is presented as an example.

The period of protection begins with the date of filing a complete specification. Following the filing process the application and specification are examined to see whether they comply with the formality requirements of the law and whether the legal requirements are satisfied. The examiner may object to the application. The applicant is provided with an opportunity to contest objections or amend the specification. All objections must be addressed within 21 months from the date of the first official report. Once objections have been overcome, the application will be accepted and the acceptance advertised. Then follows a three-month period during which parties can oppose the acceptance. If the patent is opposed, both parties are given the opportunity to file evidence and an opposition hearing may be made before a delegate of the Commissioner of Patents, with an opportunity of an appeal from this decision into the court system. If there is no opposition, or the opposition is decided in favor of the applicant, the patent will be granted.

Where a country is a signatory to the Patent Cooperation Treaty 1970 (PCT), a single application is made in a signatory country, designating other signatory countries with regard to which protection is sought. Where an application is filed under the PCT an application search is carried out, which will be made available to all of the national offices in the countries that have been designated in the application. This avoids the necessity for searches of technology in each country in which protection is sought, and it has the advantage of deferring costs and providing an applicant with an indication of the likelihood of success of multiple applications.

## Disclosure of inventions

The cornerstone of patent protection and the protection of new plant varieties is the obligation of disclosure of the invention by the applicant in return for intellectual property protection. An inevitable tension exists between the desire of the applicant to retain as much information as possible to preserve the competitive advantages that the first person in a field will have and the desire of the granting authority to have as much technological information placed in the public domain as possible. Patent laws generally require that an applicant shall disclose the invention in a manner sufficiently clear and complete for the invention to be carried out by a person skilled in the art. This may require the applicant to indicate the best mode for carrying out the invention known to the inventor at the filing date or, where priority is claimed, at the priority date of the application. An exception to the fullness of disclosure will exist with regard to inventions concerning microorganisms, where deposit of a culture with an international deposit authority under the Budapest Treaty will be equivalent to a full description of the characteristics of the microorganism.

## Differences between patenting and PVR

To avoid the confusion of the double protection of PVRs under both patent and PVR laws, the patent laws of many countries exclude the protection of plant varieties. For example, Article 53b of the European Patent Convention excludes patents for "plant and animal varieties" and "essentially biological processes" for the production of plants and animals. Since the mid-1980s, a normal US "utility" patent can cover innovations in the production of new plant varieties or specific genes and their corresponding traits.

The process of obtaining patent protection depends heavily on an examination of the written word. In the case of microorganisms and other living matter, it is usually necessary to deposit a culture of a new organism in an official culture collection. The written specification contains the claims that define the protected technology. Claims almost always cover a range of products or processes extending beyond the specific application of the innovation by the inventor.

Since 1989, about 250 patents have been awarded per year (Joly and de Looze 1996), which is quite high for agricultural biotechnology, as this research is still an emerging field with great legal uncertainty. Some of the uncertainty is being clarified as companies involved in genetic engineering have recently consolidated their positions and are licensing or selling technologies. This patenting is occurring mainly in anticipation of technological breakthroughs, rather than based on its proven economic importance to agricultural research (Joly and de Looze 1996).

Of most immediate interest to research scientists are patents covering genes and transformed plants that use those genes. This type of patent can cover a number of claims, such as isolated proteins, nucleic acid sequences coding for a protein, and plasmids containing that particular genetic sequence. The actual claim protects the patent holder against use of the gene by other scientists but still leaves anyone else free to use and breed with organisms containing the gene as it occurs naturally (Barton 1998).

Another patent category protects basic processes and inventions. There are already many important patents covering transformation processes, plant growth promoters, and virus coat proteins that confer particular forms of resistance. The variety and scope of these are so broad that it is likely to be very difficult to develop transgenic plants without infringing one or another of these patents (Barton 1998).

PVR have been highly successful. However, the use of patent law is increasingly viewed as better suited for the protection of recombinant methods for producing transgenic plants and the resulting products (Suwantaradon 1995). PVR are highly specific to the variety and their scope is limited by reference to the physical (propagating) material itself, combined with the description of the variety given in the documentary grant of the rights. The freedom to undertake research is safeguarded under both patent and PVR law (see table 18.1 and Van Wijk et al. 1993). The freedom to commercialize the resulting products of research, however, depends on whether they infringe on patent claims or are "essentially derived" under PVR legislation.

## Decision making and procedures for patents

Exercising judgment is more of a challenge when inventors are pursuing patent protection than when they are looking for variety protection. The legal and technical complexities, plus the time and money involved in navigating through the patent application process required for full international protection, are considerable.

As soon as the invention has been clearly described, it is time to consider filing a patent application. The first consideration is whether the gene, plant, process, or product is truly new (in patent law terms) rather than an obvious development of what is already known. It must have potential industrial or other utility. In its official examination of the application, the patent office will carefully apply these criteria of patentability.

An application for patent protection is normally first made in the country of residence or place of business of the applicant (see table 18.3). This establishes a priority date that will be recognized in most of the other countries of the world under the provisions of an international agreement known as the Paris Convention. In practice, this means that the

**Table 18.3.** The Patenting Process

| Moment | Action by inventor | Action by patent office |
|---|---|---|
| Before first application | • Invention and preliminary appraisal of patentability | |
| First application | • First patent application filed in home country<br>• Establishing priority date | |
| Within 12 months after first application | • Further development of invention and technical/commercial assessment by internal staff, consultants, industrial, and government contacts<br>• Decision taken to proceed or abandon, and costs estimated<br>• Patenting route selected (national, European, international) | • Official prior art search (novelty search) |
| 12 months after first application | • Home filing consolidated<br>• Foreign applications filed based on priority application | • Official examination starts, precise moment depends on backlog |
| 18 months after first application | | • Official publication of application (in some countries) |
| At a later moment variable in time | • Further prosecution of patent application by applicant and attorney | • Patent granted or refused |

major expense of a foreign patenting program can be postponed until one year after the initial filing date in the home country.

The value of this one-year interim period, both to industry and to other organizations that have to assess the potential industrial importance of new research results, is considerable. Moreover, under the Paris Convention the patent applicant can publish details of the invention at any time after the priority date without detriment to patent prospects. The only provisos are that the invention be clearly defined and well supported by data in the first application and that the foreign applications be filed no later than one year after the first application.

Applicants who file priority applications in their home country normally use the following year as a "breathing space" to consolidate their position. In most countries, the patent office will not fully examine the first filing immediately. The first filing will remain secret until the formal publication stage is reached, which gives the applicant a limited period of effective trade secrecy in which various important matters can be further considered.

## Technical development and assessment

Any further technical development of the invention during the year following the first filing can be made the subject of further patenting efforts. These can either stand alone or may be merged into a final overall application filed near the end of the first year. A final application thus contains the totality of the inventor's results over the periods preceding and following the first filing. It is usually the basis of the foreign patent applications, which can be filed abroad at this time and which claim one or more of the various priority dates given by the earlier national filings.

## Commercial assessment and cost factors

Apart from the technical assessment, marketing and other commercial factors need to be considered before the end of the first year after filing. Whether to proceed and, if so, on what territorial scale are essentially decisions about market potential and the corresponding financial expenditure on patents that can be justified.

Filing the first application in one's home country protects the home market and establishes a base from which to secure wider coverage. Although costs begin to mount if international coverage is required, a large proportion of these can, fortunately, be distributed over time. At each stage of patenting, separate estimates of financial costs and benefits can be made in light of the prevailing commercial climate. A proper judgment about costs must take market size into account. For example, a cost of US $20,000 or more to secure a US patent may not be considered excessive in relation to the potential US market.

These calculations or estimates of costs and benefits are far more difficult to make in developing countries than in industrialized countries, because market opportunities are normally harder to identify. Scientists in national research programs and universities of developing countries also have far fewer opportunities to enter into commercial or strategic alliances that may help produce, market, or advertise a particular invention. Thus, the costs of filing at home and securing international protection may far outweigh the expected economic returns.

Patent costs arise at various stages over an extended period but can be terminated if the value of the protection diminishes over time. They consist of official fees charged by patent offices and professional fees charged by patent attorneys. Professional charges are based on standard fee scales and on the time spent. Time-based charges will vary from one patent application to another and according to professional rates in effect in the country of filing. Initial filing to set up a patent base may cost as little as $2,000, but it could also be 10 times that amount if the legal and technical complexities of the case are demanding. One year later, at the foreign filing stage, additional expenditures are incurred, as services of foreign agents are needed.

The Patent Cooperation Treaty (covering international or "PCT" applications) covers a wide range of countries and offers a system of initial and deferred costs. In selecting countries in which to file for protection, it is important to distinguish between those in which the product will or could be manufactured and those that are simply markets. Various cost structures for patents are shown in table 18.4, which summarizes initial filing fees under the European Patent Convention and the Patent Cooperation Treaty. The initial cost of a Patent Cooperation Treaty application covering the major European countries, the USA, Japan, and many others can be up to $10,000 and even up to $20,000 in legally and technically complex applications.

**Table 18.4.** Initial Filing Fees

| Type of fee | Fee in US $ |
|---|---|
| European Patent Convention[1] | |
|   Filing fee | 150 |
|   Search fee | 950 |
|   Designation fee (per country) | 80 |
|   Examination fee | 1,500 |
|   Grant fee (first 35 pages) | 750 |
| Patent Cooperation Treaty[2] | |
|   Transmittal fee | Up to 75 |
|   Basic fee | 450[3] |
|   Designation fee, per country or regional patent | 105[3] |
|   Search fee | 500-1,200 |
|   Examination fee (handling fee plus exam fee) | 1,000-1,500[3] |

[1] EPC, March 1999.
[2] PCT, March 1999.
[3] A 75% reduction in the basic fee, the designation and confirmation fees, and the handling fee is available to applicants from most developing countries.

Actual fees vary by country, examination provider, and filing location.

## Other protective mechanisms

### Trademarks and registrable marks

A trademark is a sign used to indicate the origin of goods or services. Legal protection is provided for trademarks through a system of registration. To be registered as a trademark, a sign must be represented in a visible form. In some countries, audible or olfactory "signs" may be registered if they can be recorded in an appropriate manner. Visible signs typically include names, invented or existing words, letters, numbers, pictures, and symbols, or combinations of these signs. The TRIPs agreement envisages that the shape of goods or their containers, as well as colors and smells may be registered as marks. Excepted from registration in most countries are marks that are not distinctive or that are deceptively similar to existing marks and marks that violate public order or morality.

### Registration

The application process usually includes an examination by the granting office to ensure that (1) the application complies with the formal registration requirements, (2) the mark meets the substantive requirement of distinctiveness, (3) the mark is not in conflict with prior rights. After an application is published, most countries provide for an opposition process whereby interested third parties may protest the registration of a mark. Common objections include prior rights or deceptive similarity with another mark.

Upon acceptance by the registering office, registration is conferred for a term of between 10 and 20 years, with a possibility for renewal. A registered mark will expire if a renewal is not sought. Cancellation of a mark may also be sought where its use becomes deceptive or where the mark becomes generic of goods or services. Vaseline and gramophone are two examples of marks that became generic descriptions of the type of goods to which they were appended.

A requirement of some trade mark laws is that registration is contingent on its use with the classes of goods or services for which it is registered. A similar requirement provides for the removal of the registration after a prescribed period of non-use. Protection without registration may be extended to "well-known marks," i.e., those with a significant reputation in a country. Such marks invariably have a substantial international reputation through advertising and use.

## Trademark rights

Registration of a mark confers protection against emulation by traders using identical or substantially similar marks. Most systems of registration permit assignment or licensure. A system of registered user may be provided to record trademark licenses. In the event of infringement of a registered mark, a trade mark proprietor may seek relief in the form of injunction, compensation orders, and seizure of infringing goods.

A trademark serves as a warranty of the quality of the goods or services supplied under that mark. The name or acronym of an agricultural research institute is often a warranty of the quality of the services or products supplied by that institute. An example is the IRRI prefix for rice types developed by the International Rice Research Institute. Its designation is worthy of protection, particularly because unauthorized traders may falsely represent an affiliation with the institute.

## Geographic marks and appellations of origin

A specialized form of trademark, which has been identified as the subject of a separate system of protection, are those marks that identify that a product or service originates in a country, region, or particular place. The false or deceptive indication of source is actionable. An appellation of origin is a mark that indicates that in addition to the geographic source of goods, the place of origin decisively influences the character or quality of the goods. For example, the soil and climatic influences in a wine-producing district, such as Burgundy or Champagne, can be demonstrated to produce a wine of such a particular quality that it would be deceptive to permit other wine producers to use those appellations of origin.

## Confidential information and trade secrets

IPR cannot be effectively used without the deployment of substantial know-how. However, there is no system of registration for know-how. It can be protected only if it is regarded as confidential or if a restriction is placed on its unauthorized communication. In the case of contracts with employees or researchers, most common law systems imply a contractual term obliging employees not to divulge information that is considered to be an employer's property. This contractual term is enforceable only against an employee. In determining whether information is protectable as a trade secret, the courts have regard to (1) the extent to which the information is known by employees and by persons outside the relevant business, (2) the extent of measures made to guard its secrecy, (3) the value of the information to the business and to competitors, (4) the amount of effort or money expended in developing the information, and (5) the ease or difficulty with which the information could be properly acquired or duplicated by others. The TRIPs agreement, article 39, permits restrictions on the disclosure of confidential information.

## Copyright

Copyright law is concerned with the protection and exploitation of the expression of ideas in a tangible form. The law developed in response to the invention of the printing press, which made possible the mass production of print works. The technological context in which the law developed explains the way in which copyright concepts have been formulated. For example, the law focuses on the rights of authors and those who claim through the author; it protects "original works" and only if those works are fixed in a material form. The central right that the law confers is to prevent unauthorized persons from copying a work. To be protected as copyright, ideas have to be expressed in an original way; i.e., they must have their origin in the labor of the creator. Works are protected irrespective of their quality.

### Subject matter of protection

Originally, the subject matter of copyright protection was printed literary artistic and literary works. Research notes and reports will be protectable as copyright works. More recently, copyright protection has been extended to computer programs and databases.

### Rights comprised in copyright

The owner of a copyrighted work may exclude others from using it without authorization. Authorization of the copyright owner is usually required for copying or reproducing the work, performing the work in public, making a sound recording of the work, making a motion picture of the work, broadcasting a work through the electromagnetic spectrum or through cable diffusion, and translating or adapting the work.

In addition to these rights, the Berne Convention for the Protection of Literary and Artistic Works also recognizes certain "moral rights." These include the right to claim authorship of a work and the right to object to any distortion, mutilation, or other modification of, or other derogatory action in relation to a work that would be prejudicial to an author's honor or reputation. These moral rights usually remain with an author, even after the transfer of the various economic rights mentioned above.

## Material transfer agreements

Contractual in nature, MTAs offer a form of intellectual property protection that can cover material not generally protected by patents (Barton and Siebeck 1994). Most international agricultural research centers use such agreements for the genetic resources they hold in trust for the world community. They are also widely used among public-sector research organizations in industrialized countries. MTAs provide interim protection for material sent to collaborating organizations for advanced research. They can thus be used until more formal IPR is sought.

MTAs are becoming especially important in the exchange and use of plant genetic resources, particularly since open access to such material is essential for the development of food and agriculture. In effect, these agreements can help clear the way for research and breeding by setting out the conditions that govern each germplasm exchange.

## Farmers' rights

Although not formally a part of the Convention on Biological Diversity, farmers' rights[3] were seen as a related concept in public debate. First formulated in Resolution 5/89 of a 1989 conference of the Food and Agiculture Organization of the United Nations (FAO), "'farmers' rights' means rights arising from the past, present and future contributions of farmers in conserving, improving, and making available plant genetic resources, particularly those in the centers of origin/diversity..." (FAO 1989).

The original intent of farmers' rights was to recognize farmers and members of indigenous rural or traditional communities for their role in creating, domesticating, and building sources of agricultural varieties and diversity for food and agriculture. However, it is not clear how farmers' rights are to be given practical legal expression. It may prove difficult to graft this kind of right to traditional intellectual property law, in which case it will almost certainly be necessary to create a specialized new legal framework. Focusing on developing countries in particular, article 27(3) of the TRIPs agreement of April 1994 envisaged special types of legal systems for plant and animal varieties, although these have not yet been defined or detailed.

## The continuing debate on IPR

Agricultural development, including the release of improved planting materials through formal breeding and production, has benefited from a long history of public-sector/public-good investment. At the core of this system has been the wide availability of plant genetic resources.

But such public-good investments face an uncertain future. First, an increased emphasis on market mechanisms has forced publicly funded organizations to respond to broader economic and market opportunities and to position themselves to be part of the future global agricultural research system. Second, there is a tendency to restrict the free availability of germplasm to breeders working in publicly funded national agricultural research programs.

While many of those representing the formal and informal sectors oppose the use of patents on agricultural improvements, public institutions are increasingly being encouraged to protect their intellectual property (Baenziger et al. 1993). But many developing countries are being cautious about extending intellectual property protection to agricultural crops (Rai 1994). For example, current thinking in India on the country's IPR framework attempts to take into account the interests of those using planting material. The preference is to continue to leave research results in the public domain.

These problems and issues arise as IPR generally and patents in particular are adapted to cover living organisms, genes, and biological processes related to agriculture. But even early on, many countries judged patent systems to be inappropriate for protecting living organisms because they imposed practical restrictions (ODI 1993). The increased use of IPR protection in agricultural research does not seem to account fully either for the

---

[3] Farmers' rights should not be confused with farmer's privilege, mentioned earlier in the chapter.

long-standing tradition of public-sector investment or for the innovations contributed by international agricultural research and by informal or indigenous communities. It is feared that such protection destroys the public-good nature of agriculture, especially as it relates to the needs of the rural poor. MTAs (if carefully prepared to ensure agreed-on use) and research exemptions could allay fears regarding access to material protected by patents.

The WTO agreement on TRIPs provided impetus for domestic intellectual property legislation to comply with international norms. Applicants for membership of the WTO have to sign this agreement. The TRIPs agreement promulgated minimum standards for intellectual property laws and enforcement regimes (Blakeney 1996). Industrialized countries were obliged to give effect to the agreement's provisions within one year of its commencement in 1995. Developing countries had an additional four years and least-developed countries a total of 10 years within which to bring their laws into line with the TRIPs norms. The TRIPs norms also form the basis for the intellectual property rules for a number of regional commercial unions such as the Association of South-East Asian Nations (ASEAN), the European Union, Mercado Común del Sur (Mercosur), and the North American Free Trade Agreement (NAFTA) (see Blakeney 1998).

## Sui generis systems

There has been a vigorous debate on the sorts of *sui generis* systems that might comply with Art.27.3(b) of the TRIPs agreement. The TRIPs provision makes no reference to UPOV, which is considered to provide some leeway in the formulation of *sui generis* systems. Furthermore, key elements for the shaping of *sui generis* systems are either unclear or not defined. First, there could be several ways of defining the term plant variety. For granting protection under the traditional PVR system, plant varieties must meet the criteria of being distinct, uniform, and stable. It has been suggested that "uniformity" and "stability" could be replaced by the criterion of identifiability, allowing the inclusion of plant populations that are more heterogeneous, thus taking into account the interests of local communities. The scope of protection could be limited to cover only the reproductive parts of plants, or could be extended to include also harvested plant materials.

Second, the TRIPs agreement does not prohibit the development of additional protection systems, nor does it prohibit the protection of additional subject matter to safeguard local knowledge systems and informal innovations as well as to prevent their illegal appropriation. Several elements could be added, such as community gene funds and the establishment of mediation procedures (public defender) for the protection of local interests or local registers.

Darrell A. Posey and Graham Dutfield have conceived of traditional resource rights as an approach to *sui generis* protection. Traditional resource rights are posited as "a process and framework to develop multiple, locally appropriate systems and 'solutions' that reflect the diversity of contexts where *sui generis* systems are required" (Posey 1996). "Traditional resources" include tangible and intangible assets and attributes deemed to indigenous and local communities. Traditional resource rights are described as "an integrated rights concept that recognizes the inextricable link between cultural and biological diversity" delineating a constellation of "overlapping and mutually supporting bundles of rights" that "can be used for protection, compensation and conservation" (Posey and Dutfield 1996).

## Costs and benefits of IPR protection: The lack of empirical social and economic analysis

In evaluating options for IPR protection, we must recognize that virtually no empirical analyses, either sociological or economic, have been done on the impact of IPR on food and agriculture, especially in developing countries. In industrialized countries there is a clear correlation between plant variety protection and the willingness of companies to produce varieties. Without strong protection, there would be few new varieties available for the public benefit (Price and Lamola 1994). In addition, when national legislation allows public institutions to retain property rights, the number of patents increases since these facilitate licensing agreements.

While the up-front costs of obtaining IPR protection and building national competence in this area of expertise are fairly clear, it is harder to predict what the benefits to developing countries would be. Where domestic research is not internationally competitive or where IPR laws are ignored and protected material is reproduced illegally, it is especially difficult to expect any substantial payoff.

A recent review has found that there is still insufficient evidence to generalize about the benefits of establishing property rights for plant material. However, it is clear that private incentives for research on crops and the amount of plant variety protection sought increases with the value of the crop (Butler 1996). Expected benefits from IPR protection would be very low for plants or farming systems depending on low-value or open-pollinated crops.

If a good portion of national research is targeted for international or global application, then investments in IPR protection may be justified. But it is not appropriate to shift all national program research efforts toward products that can be patented or protected by IPR. In this regard, the introduction of IPR protection to agriculture is often cited as a means to counter the decline in funding to national agricultural research. It is argued that IPR protection mobilizes additional money from the private sector (because it creates an expectation of return on investments) and that it gives scientists access to protected material (Smith 1996).

While an evaluation of the utility of a patent system for plants, including cost-benefit analysis, is recommended, in practice this is very difficult to do. In addition, the international agreements mentioned above do not allow a lot of time for such analysis, nor do the needs of those wishing to protect innovations and seeking financing for their development within the developing countries themselves (Butler 1996).

## Conclusions

Selecting among the types of protection to be applied to innovations arising from agricultural research is a complex management decision. In many public organizations, research liaison officers or offices of intellectual property help with these decisions (see chapter 19 of this volume). These individuals will help assess the accountability requirements and public expectations regarding innovations produced with public funds. Such offices can also help the national research program anticipate means to license, develop, and move its innovations into production. Since such technology transfer agreements are especially important for new products serving farmers who rely on purchased inputs and make capital expenditures, technology transfer will usually include the licensing of some proprietary right.

In deciding on which forms of IPR protection to adopt, it is important to consider whether an innovation will have only national application or perhaps wider, even global, relevance. Applying innovations to the needs of farming communities that do not traditionally rely on purchased inputs requires no IPR protection. In fact, the costs of such protection would far outweigh any commercial benefits. However, if that same innovation has global implications, then some form of protection may well be advised.

Decisions to extend IPR protection to agriculture will also depend on the assessment of the impact such decisions may have on farmers who use farmer-saved seed for planting in subsequent years. In most developing countries, small- and medium-scale farmers and those operating in a resource-limited environment form the core of the agricultural system. Any system of IPR protection must take into account the needs of this community as well as the services provided to the commercial or highly productive sector, including limitations that may be imposed on a farmer's ability and rights to replant saved seed.

It should be noted that the application of IPR to agricultural products is a very recent phenomenon. There is little record of the overall utility or success of patenting innovations. Revenues gained from IPR protection may help pay the costs of maintaining the structures necessary for providing researchers with advice on IPR, documenting innovations, and preparing applications, but not necessarily much more.

## References

Baenziger, S. P., R. A. Kleese, and R. F. Barnes (eds). 1993. Intellectual Property Rights: Protection of Plant Materials. Executive Summary and Work Group Reports. CSSA Special Publication No. 21. Madison: Crop Science Society of America.

Barton, J.H. 1998. The impact of contemporary patent law on plant biotechnology research. *In* Global Genetic Resources: Access and Property Rights. CSSA Special Publication. Madison: Crop Science Society of America.

Barton, J. H. and W. E. Siebeck. 1994. Material Transfer Agreements in Genetic Resources Exchange: The Case of the International Agricultural Research Centers. Issues in Genetic Resources No. 1. Rome: International Plant Genetic Resources Institute.

Blakeney, M. 1996. Trade Related Aspects of Intellectual Property Rights: A Concise Guide to the TRIPs Agreement. London: Sweet & Maxwell.

Blakeney, M. 1998. The Role of Intellectual Property Law in Regional Commercial Unions. *Journal of World Intellectual Property* 1: 691-710.

Butler, L. J. 1996. Plant Breeders' Rights in the U.S.: Update of a 1983 Study. *In* Intellectual Property Rights and Agriculture in Developing Countries, edited by J. van Wijk and W. Jaffe. Amsterdam: University of Amsterdam.

Crespi, R.S. 1995. Intellectual Property in Agricultural Biotechnology: Issues for Developing Countries. *In* Turning Priorities into Feasible Programs: Proceedings of a Regional Seminar on Planning, Priorities and Policies for Agricultural Biotechnology in Southeast Asia. Singapore, 25-29 September 1994. The Hague / Singapore: Intermediary Biotechnology Service / Nanyang Technological University.

Dhar, B., B. Pandey, and S. Chaturvedi. 1995. Plant Variety Protection in India: The Prospects. New Delhi: Research and Information System for the Non-Aligned and Other Developing Countries.

FAO. 1989. Interpretation of the International Undertaking on Plant Genetic Resources, 89/24. Rome: Food and Agriculture Organization of the United Nations.

Joly, P. B. and M. A. de Looze. 1996. An Analysis of Innovation Strategies and Industrial Differentiation through Patent Applications: The Case of Plant Biotechnology. *Research Policy* 25:1027-1046.

Mudur, G. 1995. New Rules Push Researchers Closer to Biotech Industry. *Science* 269:297-298.
ODI. 1993. Patenting Plants: The Implications for Developing Countries. ODI Briefing Paper, November 1993. London: Overseas Development Institute.
Posey, D. 1996. Traditional Resource Rights. International Instruments for Protection and Compensation for Indigenous Peoples and Local Communities. Gland, Switzerland: World Conservation Union.
Posey, D.A. and Dutfield, G. 1996. Beyond Intellectual Property: Towards Traditional Resource Rights for Indigenous Peoples and Local Communities. Ottawa: International Development Research Center.
Price, S. C. and L. M. Lamola. 1994. Decision Points for Transferring Plant Intellectual Property to the Private Sector. *In* Conservation of Plant Genes II: Utilization of Ancient and Modern DNA. Monograph in Systematic Botany 48, edited by R.P. Adams, J.S. Miller, E.M. Golenberg and J.E. Adams. St. Louis: Scientific Publications.
Rai, M. 1994. The Current Situation of the Plant Breeding and Seed Industries in India: The Policy on the Protection of New Varieties. Seminar on the Nature of and Rationale for the Protection of Plant Varieties under the UPOV Convention, organized by UPOV, Beijing, China.
Rovere, C. 1999. International Union for the Protection of New Varieties of Plants. What it is, What it Does. UPOV Publication No. 437 (E). Geneva: International Union for the Protection of New Varieties of Plants.
Rzucidlo, E. C. and D. R. Auth. 1996. Will the Real Inventor Please Stand Up? *Nature Biotechnology* 14:358-359.
Smith, S. 1996. Farmers' Privilege, Breeders' Exemption and the Essentially Derived Varieties Concept: Status Report on Current Developments. *In* Intellectual Property Rights and Agriculture in Developing Countries, edited by J. van Wijk and W. Jaffe. Amsterdam: University of Amsterdam.
Suwantaradon, K. 1995. A Private Company's Perspective on National Policies Affecting Biotechnology Research in Thailand. *In* Turning Priorities into Feasible Programs: Proceedings of a Regional Seminar on Planning, Priorities and Policies for Agricultural Biotechnology in Southeast Asia. Singapore, 25-29 September 1994. The Hague / Singapore: Intermediary Biotechnology Service / Nanyang Technological University.
UPOV (International Union for the Protection of New Varieties of Plants). 1961. 1968. 1972. 1978. 1991. International conventions for the protection of new varieties of plants. UPOV, Geneva.
Van Wijk, J., J.I. Cohen, and J. Komen. 1993. Intellectual Property Rights for Agricultural Biotechnology: Options and Implications for Developing Countries. ISNAR Research Report No. 3. The Hague: International Service for National Agricultural Research.

# 19 Agricultural Research and the Management of Intellectual Property

*Michael Blakeney*

## Abstract

*Intellectual property rights (IPR) are of increasing significance to the research conducted at agricultural research institutes. First, they are part of the exploitable capital of those institutes. Second, research may depend on securing permission to use the IPR of others. Third, where there is a research team, there will be visitors and institute members, and there will be a difficult question of identifying ownership of any intellectual property that is generated. Using IPR presents management challenges, including ways of disclosure of intellectual property, questions of ownership, remuneration, forms of intellectual property exploitation, and whether the institute should manage the IPR rights itself or through a research liaison office.*

## Ownership of intellectual property by agricultural research organizations

Intellectual property is a broad term for the various rights that the law gives for the protection of economic investment in creative effort. The principal categories of intellectual property that are going to be relevant to agricultural research organizations are patents, plant variety rights and trademarks (see chapter 18 of this volume; van Wijk et al. 1993). Because of investments of public moneys to establish and run national agricultural research organizations (NAROs), governments and funding bodies increasingly insist that NAROs are accountable for the intellectual property rights (IPR) that they generate and that they initiate new discipline for the management of their intellectual property. A prerequisite for implementing an intellectual property management regime is securing ownership of the intellectual property that a NARO generates.

### Accountability

Much of the funding for NAROs is obtained from public sources, such as governments, international organizations, and charitable foundations. Consequently, institutional accountability is needed for the intellectual property emerging from the institution's research. NAROs may take the view that their research results are freely available to all. However, an unauthorized person may seek to register a patent arising from that research, thereby preventing its use as a public good. This can be pre-empted by the institute seeking

---

© CAB *International*. 1999. *Managing Agricultural Biotechnology—Addressing Research Program Needs and Policy Implications* (ed. J.I. Cohen)

its own intellectual property protection. Patent or plant variety rights, thus obtained, can be licensed to users on a nonexclusive basis.

As owner of the intellectual property, the institute can apply quality assurance procedures to its development, evaluation, protection, and exploitation through administrative procedures dedicated to this task. It can also assist innovators in commercializing the intellectual property by concentrating expertise in an area available to all research teams.

The purpose of intellectual property ownership by a NARO should be to promote the fundamental research mission of the institute. Intellectual property ownership helps a NARO in the following ways:

- It protects its integrity and welfare, providing a resource for industry.
- It obtains appropriate return for the use of facilities, resources, and services provided by the institution.
- It encourages the growth, progress, and success of the research institute through ventures with industry.
- It seeks commercial returns to provide fair and reasonable reward (and incentive) to staff and students who apply their intellectual activity.
- It increases the acumen and accountability for management and use of public funds and fosters the identity of the research institute and an *esprit de corps*.

## Enhancing public-private cooperation

Specific examples exist where public national agricultural research institutes have conducted biotechnology research in collaboration with industry. Cohen and Komen (1995), for example, mention a collaborative project for maize transformation between ICI Seeds (USA) and the Central Research Institute for Food Crops (Indonesia), a project to develop transformation technology for the development of virus-resistant sweet potato between Monsanto (USA) and the Kenya Agricultural Research Institute, the development of ELISA kits for local maize viruses between Pioneer Hi-Bred (USA) and the National Research Center for Maize and Sorghum (Brazil), and the development of insect-resistant genes for potatoes between Agricultural Genetics (UK) and the International Potato Center (Peru).

Industry operates in an environment that uses IPR to protect the value of its products and the money expended on research and development. As most corporations will not invest in, or transfer technology to, countries that do not have an intellectual property infrastructure, corporations are increasingly beginning to insist that research institutes have an equivalent infrastructure and an intellectual property regime. Where collaborative research is conducted, the private collaborator will insist that the products of research, representing significant investments, are not appropriated by unauthorized third parties.

Enhancing public and private cooperation also depends on the ability of all parties to clearly understand each other's position with regard to eligibility for protection of intellectual property. The public status of NAROs facilitates access to the genetic material held "in trust" by genebanks within the research centers of the Consultative Group on International Agricultural Research (CGIAR). Problems have arisen when, through collaborative projects with industry, plant breeders' rights have been sought for varieties developed from this material. Blakeney (1998), for example, describes how a number of agricultural research institutes in Western Australia had difficulties in seeking plant breeders' rights in relation to material obtained from the International Crops Research Institute for the Semi-Arid Tropics (ICRISAT) and the International Center for Agricultural Research in the Dry Areas (ICARDA). The problems arose in part because it was not clear whether IPR could be sought in relation to any improvements of the material. NAROs and other organizations need a clear statement from CGIAR centers as to how they can use material obtained from the centers. Such information and conditions pertaining to IPR are to be provided in the material transfer agreements (MTAs) used by the centers and in their intellectual property policy statements, a copy of which should be obtainable from each center.

### Securing ownership of intellectual property

As a general rule a research institute will own all the intellectual property generated by its staff (and research students) if the intellectual property has been created in the course of employment and if the research institute's resources and services have contributed to the intellectual property. These resources include pre-existing intellectual property belonging to or licensed by the research institute. Typically, the research institute will waive its rights in relation to conventional scholarly output, such as books, book chapters, journal articles, book reviews, and conference papers.

A research institute can affirm its ownership of IPR in the research conducted by its staff by inserting a term along the following lines into the conditions of employment of its staff:

> "Although the institute has adopted the general principle that the results of its research and development should be made generally available, the institute is prepared to consider statutory protection (e.g., patenting) for work arising in the course of institute duties. The ownership of intellectual property created in the course of employment by the institute, and hence the sole right to use such intellectual property, belongs to the institute. The originator(s) of such intellectual property are entitled to an equitable proportion with the institute of any net income that may arise from the exploitation of the property. Staff members may be required to assign their interest in such intellectual property. Where intellectual property is developed by members of institute staff as a result of work not related to its institute duties, it will be for the staff member concerned to decide whether to seek statutory protection (e.g., a patent, trade mark, or design), but this should be reported on a confidential basis to the director of the institute."

The question of intellectual property ownership is also clarified by adopting an intellectual property statute along the lines of the attached precedent by the Universiti Putra Malaysia. Providing for the insertion of an intellectual property clause in the contract of employment, it is an example of an effective way of establishing an appropriate intellectual property infrastructure for agricultural research institutes. It enacts an intellectual property statute or set of regulations. Annex 19.1 contains the intellectual property statute recently adopted by Universiti Putra Malaysia because of problems arising in the research departments.

A delicate question that can arise regarding the intellectual property generated by a research institute is the question of ownership of that property when a visiting researcher has contributed to the research. This issue arose recently in Malaysia. A visitor who was spending his sabbatical leave at research institute X began to work as part of a research team to investigate the elimination of the toxicity to sheep of stock feed made from palm kernel cake. He continued this research in his home institute Y with the palm kernel cake sent to him from the institute. During the research at his own institute the researcher solved the toxicity problem, but a dispute arose between the institutes over patent rights. Institute X has since adopted an intellectual property policy that requires visitors to agree to assign to the institute any intellectual property that is developed by visitors with the use of the institute's resources and facilities.

## Managing the intellectual property of a research institute

Providing a central point of expertise to manage intellectual property is one way to help NAROs confront the complexities of the legal protection of IPR and commercial arrangements needed to acquire and exploit those rights. Additionally, some research may depend on the acquisition of a third party's IPR. For example, a private company may wish to contribute a patented gene. This may be on the basis that it is exclusively for the use of the institute or on condition that any improvements or applications are jointly patented. This will also be more conveniently managed centrally.

### Choosing between a research liaison office and an intellectual property coordinator

To manage its IPR the research institute can choose between contracting a dedicated research liaison office (RLO) and assigning an official within the institute. The decision usually depends on funds available to pay salaries and to house relevant staff. Another critical factor is the amount of revenues expected from IPR generated by the research institute. When the research institute is associated with a university, it can often use the services of a central university institution.

*Examples of intellectual property managed by universities*

The University Business Centre of Universiti Putra Malaysia, established in 1995, provides intellectual property coordination for the agricultural research institutes that are located on the university's campus. Similarly, the Intellectual Property Institute at Chulalongkorn University, Thailand, established in the same year, provides advice to the university's agricultural research institutes. In Nagpur, India, a Patent Documentation and Information Centre, established in 1989, offers intellectual property management services to agricultural research institutes in that area. At Murdoch University in Perth, Australia, the Rhizobium Research Institute and the State Agricultural Research Centre, established in 1996, have resort to a university company for the management of their IPR. The commercial exploitation of those rights is undertaken by a technology management company that was established by a consortium of three Western Australian universities in 1993. Agricultural research institutes that are unable to use the facilities offered by associated universities or government bodies are advised to appoint one of their officers to perform the intellectual property coordination function.

## Functions of the RLO or intellectual property coordinator

The RLO is responsible for securing professional assistance under appropriate conditions of secrecy. The RLO may decide not to exploit ownership of certain intellectual property, in which case the rights will be released to the innovator(s), usually subject to a nonexclusive license to use the intellectual property for research in the research institute. If the RLO forms the view that the research institute should be involved in the exploitation of the intellectual property, it will work with the innovator(s) in evaluating the market and in finding commercial collaborators to exploit their intellectual property. As each case is different, an appropriate strategy must be devised in conjunction with the research staff involved.

The ultimate vehicle for commercialization may be the research institute itself, the RLO, or the institute's commercial arm, if one exists. Some research institutes have established a specific company to sell goods or services to the public, such as library or advisory services, as well as the services of researchers as expert advisors. Maintaining a balance between commercial and research activities for public organizations is an obvious problem that requires close attention. The RLO will ensure the coordination of commercialization activities and prevent conflicts of interest.

## Disclosure and evaluation of inventions

To enable the RLO to evaluate the commercial potential of inventions and discoveries, researchers should be obliged to divulge their research activities in a form that captures all preliminary information required for initial evaluation. Initial evaluation will involve assessing the invention's patentability (or other intellectual property protection) and commercial potential and fulfilling any obligations to research sponsors. For example, according to the research agreement, companies supporting specific research projects may have the first right to negotiate licenses for any inventions resulting from that research. Nonprofit research sponsors may also have certain rights under previously negotiated research agreements. These rights and obligations must be fully defined and resolved before the invention can be made available for licensing.

The time needed to complete initial evaluation varies greatly for each invention. For example, efforts to determine commercial interest may need to await receipt of further research data or resolution of outstanding sponsor obligations.

All contracts relating to research grants, collaboration, and intellectual property should be vetted by the RLO for conformity to NARO requirements and appropriateness. The RLO should also ensure that the ownership of any intellectual property is in the hands of the research institute and that any grant of intellectual property does not conflict with previous contracts or with the terms of any MTAs that may have been entered into.

## Intellectual property registration

Due to the high cost of prosecuting and then maintaining patents, the RLO will usually have to delay filing of patent applications until strong commercial potential interest in technologies has been confirmed. In many cases initial filing is authorized only when companies commit to inventions through letter or option agreements. In the course of the patenting and licensing process, multiple patent applications may be filed to maximize patent coverage for a given technology and to cover developments that are made during the preparation period. Applications also may well be filed in foreign countries. Again, the commercial potential of the invention must be considered when making subsequent decisions to file or maintain patents.

### Negotiation

Once it has been decided to commercialize the IPR, the RLO will approach the most appropriate firms from a list prepared in conjunction with the innovators. The RLO will prepare executive summaries of inventions and submit these to relevant publications and databases. Marketing packages can be targeted to potential licensees identified in discussions with inventors and in various internal and external resource lists. The RLO will negotiate agreements in conjunction with the innovators to ensure conformity with the role and mission of the research institute.

Once potential licensees have been identified, negotiations proceed in a variety of ways. Short-term letter agreements may be used to confirm company interest in technologies and establish their commitment to pay patent costs during an evaluation and negotiation period. Option agreements define each partyies' rights and responsibilities and grant licensees the right to use the institute's intellectual property.

Depending on the nature of individual technologies and the estimated resources necessary for development and commercialization, the research institute may grant licensees exclusive rights, rights exclusive for a particular field of use, or nonexclusive rights. For example, most inventions in the health care field require very large investments of time and resources to develop and commercialize. Companies will often only make the necessary investment within the context of an exclusive license agreement. In contrast, plant varieties will have already been tested and are ready for commercialization, and are therefore licensed (typically) nonexclusively to multiple companies.

### Incentives to researchers

Incentives to researchers to engage in commercializing their research results usually include the disbursement of a fair share of royalties, based on a sliding scale that also provides income to the research institute for further research. These incentives can be disclosed in the context of an intellectual property statute, or as part of the operating procedure of an institute. Some institutes provide for the division of remuneration to be negotiated on a case-by-case basis. For Australian and US universities, a conventional method is to divide the revenues, after development costs are deducted, equally between the researcher, the research center, and the university. The university's share is used to cover the administrative costs of the RLO or coordinator.

## Summary and conclusions

The principal objective of a NARO's intellectual property management policy is to further the research mission of the institution. This is achieved by creating a fruitful climate for innovation and invention and facilitating the establishment of strategic linkages with researchers in other research institutes and with the private sector. The management of a NARO's intellectual property resources may be through a specially established research liaison office (RLO) or by using the intellectual property services that may be offered by an equivalent office within an associated institution, such as a university. An RLO provides researchers with a better understanding of the rights that the law gives for the protection of creative effort, and funding will be secured through collaborative research and development agreements. The RLO will be responsible for protecting and maintaining the intellectual property assets of the NARO and assist in developing an awareness and appreciation of the use of patent documents and registered plant variety data as research resources.

An RLO also assists the NARO in discharging its obligations to ensure the preservation of IPR in materials that may be obtained from gene banks of CGIAR centers and the private sector. The RLO supervises the publication of the results of research undertaken by the NARO, consistent with the preservation of any IPR. It may also be obliged to provide for the equitable remuneration of originators of intellectual property and to account to the providers of its research funds for the use of institute resources and facilities in the generation of that intellectual property. Finally, a NARO may be obliged to take steps to secure IPR in its research as a means of making it available to farmers and to other researchers.

Establishing an RLO to provide cousel on matters of intellectual property is one means of addressing institutional obligations for accountability (as imposed by funders of the NAROs) for the intellectual property which they generate, the obligation of responsibility that the law imposes upon NAROs in relation to the intellectual property of third parties that they handle, and their research agreements with CGIAR research centers and other advanced research institutes. The effective operation of an RLO can convert these obligations from a burden into an asset, with the RLO working in direct support of the research mission of the NARO.

## References

Blakeney, M. 1996a. Trade Related Aspects of Intellectual Property Rights: A Concise Guide to the TRIPs Agreement. London: Sweet & Maxwell.
Blakeney, M. 1996b. The Impact of the TRIPs Agreement in the Asia Pacific Region. *European Intellectual Property Review* 18: 544-554.
Blakeney, M. 1998. Intellectual Property Rights in the Genetic Resources of International Agricultural Research Institutes- Some Recent Problems. *Bio-Science Law Review*.
Cohen, J.I. and J. Komen. 1995. Research Collaboration, Management and Technology Transfer: Meeting the Needs of Developing Countries. *In* Plant Biotechnology Transfer to Developing Countries, edited by D.W. Altman and K. Watanabe. San Diego: Academic Press.
Kjeldgaard, R.H. and D.R.Marsh. 1997. Recent Developments in the Patent Protection of Plant-based Technology in the United States. 19: 16-20.
Roberts, T. 1996. Patenting Plants Around the World. *European Intellectual Property Review* 18: 531-536.
Van Wijk, J., J.I. Cohen, and J. Komen. 1993. Intellectual Property Rights for Agricultural Biotechnology: Options and Implications for Developing Countries. ISNAR Research Report No.3. The Hague: International Service for National Agricultural Research.

# Annex 19.1. Universiti Putra Malaysia Intellectual Property Statute 1997

In exercise of the powers conferred by section 26 of the Constitution of Universiti Putra Malaysia, the Chancellor of Universiti Putra Malaysia makes in accordance with the provisions of the said section the following Statute:

## Citation

1. This Statute may be cited as the Universiti Putra Malaysia (Intellectual Property) Statute 1997.

## Interpretation

2. In this Statute, unless the context otherwise requires-

"academic publication" means any book, journal, periodical, thesis, magazine or like publication, or any part thereof, which contains material or articles or text written by members of educational or research bodies on areas of educational or scholastic learning, research or debate;

"intellectual property" means information, ideas, inventions, innovations, art work, designs, literary text and any other matter or thing whatsoever as may be capable of legal protection or the subject of legal rights in any of the ways set out in this Section and includes the following rights recognized by Malaysian and/or foreign law:

(1) patents;
(2) information which is of a kind and which has been communicated in such a way as to give rise to a duty of confidentiality;
(3) trade secrets and information which is subject to an employee's duty of fidelity to the employer;
(4) copyright vesting in literary works (including computer programs), dramatic works, musical works, artistic works, films, sound recordings, broadcasts, published editions and certain types of performances;
(5) registered trademarks;
(6) unregistered trade marks used or intended for use in business;
(7) registered designs and designs capable of being registered;
(8) new plant varieties and the rights of breeders of such varieties;
(9) layout designs of integrated circuits; and
(10) other rights resulting from intellectual activity in the industrial, commercial, scientific, literary and artistic fields.

"commercially exploit" means the application, publication, development, use, assignment, licensing, sub-licensing, franchising, exploitation or other utilization of intellectual property for the purpose of generating financial or other commercial gains, and "commercial exploitation", "commercially exploited" and "commercially exploitable" have corresponding meanings;

"computer program" means a computer program as defined by the Copyright Act of 1987 as amended or replaced from time to time;

"confidential information" means information of any kind which, because of its confidential character, is capable of protection by contractual or equitable means, and includes information of a valuable commercial or technical character;

"copyright work" means any "artistic work", "literary work", "dramatic work", musical work", "sound recording", "cinematograph film", "television broadcast", "sound broadcast", "published edition of work" and "photograph", as those terms are defined by the Copyright Act of 1987 as amended or replaced from time to time;

"create" means produce, invent, develop, generate, discover, make, originate or otherwise bring into existence, and "creation", "creating" and "created" have corresponding meanings;

"design" means a design as defined by the Copyright Act of 1987 as amended or replaced from time to time;

"invention" means an invention (whether or not qualifying for registration) under the Patents Act of 1983 as amended or replaced from time to time;

"literary work" means a literary work under the Copyright Act of 1987 as amended from time to time;

"originator" means any staff or student who creates or contributes to the creation of any intellectual property.

"patent" means a patent within the meaning of the Patents Act of 1983 as amended or replaced from time to time;

"plant variety" means a plant variety within the meaning of the International Convention for the Protection of New Varieties of Plants of 2 December 1960 (UPOV) as revised from time to time;

"specific contribution" means funding (excluding scholarships), resources, facilities or apparatus of the University used for the purpose of creating intellectual property;

"the committee" means the Intellectual Property Committee established under sub-section 5(1) hereof;

"trade mark" means a trade mark as defined by the Trade Marks Act of 1976, as amended or replaced from time to time, whether or not registered under that Act;

"staff member" refers to and includes any member of the full or part-time academic or general staff of Universiti Putra Malaysia whether engaged in or holding a permanent post or not.

"student" means a graduate student, an undergraduate student and/or any other person designated or defined as a student in the Universiti Putra Malaysia Constitution. Any reference to a student in this Statute shall also apply to a visiting academic, scholar or other person in so far as that person undertakes studies, scholarship or research with or at the University and in so doing uses University resources, works as part of a University research team responsible for developing the intellectual property, or develops intellectual property through the use of facilities or confidential information belonging to the University.

## *Ownership of Intellectual Property*

3. (1) Subject to the further provisions of this Statute, the University is the owner of all intellectual property to which this Statute applies.

   (2) This Statute applies to:
      (a) intellectual property created by a staff member in the course of his or her employment by the University;
      (b) intellectual property created by a student in the course of her or his studies, scholarship or research with or at the University; and
      (c) intellectual property, the creation of which has been substantially contributed to by the University (or by any third person either on behalf of the University or by virtue of an agreement with the University) by the provision of resources, facilities, apparatus, supervision, salary or other funding.

   (3) This Statute applies to intellectual property which may come into existence prior to the date of this Statute, provided it is created by the persons referred to in sub-section 3(2) during the course of their employment, studentship or research with the University.

(4) Notwithstanding anything contained in sub-sections 3(1), 3(2) and 3(3) hereto:
   (a) A staff member is the owner (and the University hereby formally waives any claim it would otherwise have to ownership) of the copyright in his or her conventional scholarly output. Likewise, a student and a visitor to the university, is the owner of the copyright in all original work produced in the course of, or for the purposes of his or her studies, scholarship or research with or at the University.
   (b) A staff member and a student may deal with and exploit such material freely without any requirement of explicit approval by the University;
   (c) If requested to do so by a staff member or a student, the University must, without delay and at no cost to that person, execute a formal written assignment in favor of that person of such copyright as it may enjoy in respect of any written work to which sub-section 3(3) applies.
   (d) Subject to the preceding provisions of this section, an originator (whether a staff member or a student) of intellectual property which belongs to the University shall not make a public disclosure concerning that intellectual property without the written consent of the Vice-Chancellor. An application for consent to public disclosure must be made in writing and addressed to the Vice-Chancellor.
   (e) Intellectual property which is not the property of the University, may be dealt with and exploited freely by the originator of that intellectual property (whether or not he or she is a staff member or a student of the University).

## Disclosure of Intellectual Property

4. (1) To enable the University to assess the ownership of intellectual property generated by staff members or students, no staff member or student may take steps to disclose, protect or to commercially exploit intellectual property originated by her or him, other than copyright in material for publication, which falls within subsection 3.4(a), without first having made a written report to the Vice-Chancellor setting out the nature of the intellectual property concerned and the circumstances in which it was created, devised or originated.

   (2) The written report referred to in sub-clause 4(1) above shall include all relevant details about a work including:
   (a) the date upon which the work was created;
   (b) the identity of any person or persons who contributed to the creation of the work;
   (c) the details of any pre-existing intellectual property which was used in creating the work;
   (d) the details of any person other than the originator who claims any entitlement or interest in the intellectual property in the work;
   (e) the details of any University facilities or resources used to create the work; and
   (f) the details of any known existing or potential use or commercial exploitation of the work.

## Intellectual Property Committee

5. (1) There shall be an Intellectual Property Committee that shall be a Committee of the Senate (hereinafter called "the Committee").

   (2) The Committee shall comprise –
   the Deputy Vice-Chancellor (Academic Affairs) who shall be Chair;
   the Director of the University Business Centre/Company;
   two persons appointed from the Senate; and two persons possessing expertise in the relevant field, to be nominated by the Vice-Chancellor.

   (3) The Vice-Chancellor shall seek the advice of the Intellectual Property Committee on matters relating to whether the University has any intellectual property rights and in what circumstances the University should seek to manage such intellectual property by any of the following acts:
   (a) lodging a provisional specification for a patent in its own name;
   (b) commencing the process leading to the registration of a design;

(d) commencing investigations leading to the registration of a trade mark;
(e) providing notification of its ownership of copyright; or
(f) commencing negotiations leading to licensing or confidentiality agreements concerning the University's intellectual property; or
(g) relinquishing ownership of the intellectual property in favor of a member of staff or student.

(4) The Vice-Chancellor shall, as soon as practicable, notify the originator of the advice received from the Intellectual Property Committee under sub-section 5(3) hereof.

## Procedures for the Protection of Intellectual Property Belonging to the University

6. (1) Where statutory intellectual property rights are or may be available in respect of particular intellectual property belonging to the University, and it is decided that an application for statutory intellectual property right should be made, then the Faculty in which the originator(s) is or are located shall bear the initial costs of filing such an application.

(2) If the University decides to proceed through subsequent stages of the application process, all further costs and expenses associated with that application shall be met by the University.

(3) Each originator shall, at the request of the University and in a timely fashion, execute all such documents and do all such other acts or things as may be necessary or desirable in order to enable the University to properly and efficiently protect and commercially exploit its intellectual property and to give full effect to the provisions of this Statute.

(4) If the University decides not to initiate or proceed with an application for statutory intellectual property rights, the originator(s) may apply on their own behalf and own expense. Any such originator may, to the extent of that person's interest therein, by written notice to the Vice-Chancellor require an assignment of all of the University's rights in or over the intellectual property. Within 28 days of receipt of such notice, the University shall notify that originator of the terms upon which it would be prepared to assign the rights. Any offer to assign made by the University must take into account and seek to recoup (as far as reasonably practicable) all costs and expenses incurred or likely to be incurred by the University and its agents up to the date of the assignment. In this regard, provision may be made for the University to recover any such costs, which are not paid on assignment, from income generated by the intellectual property.

### Distribution of Benefits Derived from Intellectual Property

7. The University shall enter into an agreement with the originator(s) of intellectual property, including both staff and students regarding the division of revenue between the University and the originators where any intellectual property owned by the University is to be assigned, licensed, or otherwise commercially exploited by the University.

### Application of this Statute as a Contractual Term

8. From the date of this Statute, the preceding provisions of this section shall be implied into and/or be deemed to be terms of:
   (1) the contract of employment between the University and each present and future staff member;
   (2) the contract between the University and each student pursuant to which each student is admitted to the University.

## Inconsistency

9. In any case where the ownership, licensing or exploitation of any intellectual property to which this Statute applies is governed by a written agreement between the University and a student, a staff member and/or any other person, the provisions of that agreement prevail, to the extent of any inconsistency between that agreement and this Statute.

Made the         day of                1997

# 20 Managing Intellectual Property in Embrapa: A Question of Policy and a Change of Heart

*Maria José Amstalden Sampaio and Elza A.B. Brito da Cunha*

## Abstract

*As a result of changing national and international requirements with respect to intellectual property, the Brazilian Agricultural Research Corporation (Embrapa) has developed and begun to implement a new, internal policy for intellectual property protection. The purpose of the policy is to maintain research quality, enhance competitiveness, and prepare Embrapa for the demands and challenges of modern agriculture, including new business opportunities. Different operating structures to support the policy have been tested. This chapter analyzes the design and implementation of the new policy and considers related managerial challenges. The chapter also reviews the essential legal instruments (statutes, national legislation) that have been developed, as they highlight the importance of capacity building and the acquisition of new management skills.*

## Introduction

Embrapa (Empresa Brasileira de Pesquisa Agropecuaria; Brazilian Agricultural Research Corporation) is a public research institution associated with the Ministry of Agriculture and Food Supply. Established in 1973, it has a strong social component to its research. Its programs cover the entire range of agricultural research, such as genetic resources, commodities and social crops, livestock, forestry, soil management, integrated pest control, and conservation of the environment. Embrapa has recently been tasked by the federal government to cover 30% of its total costs from ventures with private initiatives or alternative funding sources.

Brazil, as one of the founding members of the 1883 Paris Convention, has a historical association with intellectual property. The treaty established certain minimum agreed-upon standards and procedures for dealing with industrial property. It formed a unique basis for the treatment of agriculture-related intellectual property rights (IPR) that were considered. However, as the treaty left considerable freedom to individual members to tailor their laws according to their developmental and technological requirements (Watal 1998), in 1994, member countries moved to approve the Agreement on Trade-Related Aspect of Intellectual Property Rights (TRIPs) (WIPO 1996). Brazil adopted the TRIPs agreement in 1995, which set the scene for important changes in the application of IPR. External

© CAB *International*. 1999. *Managing Agricultural Biotechnology—Addressing Research Program Needs and Policy Implications* (ed. J.I. Cohen)

economic changes that were taking place at the same time called for rapid adjustments in some long-standing policies and practices in agriculture.

Embrapa has had the opportunity to interact with policymakers in the government and with those responsible for making national law. However, Embrapa researchers and administrators rarely interacted on such matters until the institute became involved in the preparations for the United Nations Conference on the Environment and Development (also known as the Earth Summit), which took place in Brazil in 1992. Since then, the legislative and executive branches of the government have worked hard to design, discuss, and approve the legal backbone that was to support the implementation of agreements and treaties such as TRIPs, the Convention on Biological Diversity (CBD)[1], and other related agreements of interest to the country (see table 20.1) (Sampaio 1998).

## IPR at Embrapa: Preparing for the future

Based on its strong participation in national affairs, Embrapa gives high priority to its responsibility for protecting intellectual innovations so that they may become institutional assets. The institute began to discuss necessary internal changes in 1995 by considering the economic and social consequences of forthcoming policy changes. It identified the following challenges regarding intellectual property protection:

Table 20.1. IPR-Related Legislation Approved by Congress and Laws Currently under Discussion

| Date | Event |
| --- | --- |
| 1993 | Convention on Biological Diversity ratified |
| Dec. 1994 | GATT/TRIPs agreement signed |
| Jan. 1995 | TRIPs agreement implemented |
| Jan. 1995 | Law no. 8.974 (Biosafety) approved by Congress |
| Dec. 1995 | Biosafety law, Regulatory Decree no. 1.752 published |
| | Senate proposal no. 306 to regulate access to biological resources submitted to Congress (under discussion) |
| May 1996 | Law no. 9.279 (Industrial Protection—new "patent" law) approved by Congress. |
| 1997–1998 | Several regulatory decrees published by the National Industrial Property Institute |
| May 1997 | Law no. 9.279 fully active after the end of the one-year pipeline period |
| April 1997 | Law no. 9.456 (Cultivar Protection) approved by Congress |
| Nov. 1997 | Regulatory Decree no. 2.366 published, establishing the National Service for Cultivar Protection |
| April 1998 | Cultivar Protection Law and Decree approved by the International Union for the Protection of New Varieties of Plants (UPOV), according to the 1978 UPOV version |
| Jun. 1998 | Proposal for a new seed law, better adapted to the Cultivar Protection legislation, submitted to Congress (under discussion) |
| Aug. 1998 | New proposal for a law to regulate the access to biological resources submitted to Congress (under discussion) |
| May 1999 | Brazil acceded to the 1978 Act of the UPOV Convention |

---

[1] For background, see http://www.biodiv.org

1. **Implement an internal intellectual property policy.** Embrapa will implement an internal policy, in conjunction with Congress approving the necessary legal framework.
2. **Raise awareness of intellectual property.** The institute should launch an internal awareness-raising campaign through lectures, courses, and workshops to promote and diffuse the new intellectual property policy. This campaign would also help researchers understand that they need to have their research results prescreened for possible intellectual protection before publication.
3. **Create assets from intellectual property.** Embrapa should protect all assets coming from its research programs (including processes and products worthy of patents, cultivars, trade secrets, copyrights, trademarks, and software). Thus, revenues can be obtained through licensing, or the institute can allow a third (resource-poor) party to use an asset for free.
4. **Establish regulatory infrastructure.** Embrapa should hire and train personnel to manage the implementation of its policies and intellectual property laws. It has to take into account that this includes a learning curve for preparing and filing patents and negotiating and licensing a protected technology.
5. **Modify licensing system.** Embrapa should modify its cultivar licensing system and its basic seed production program to suit the IPR legislation and the growing presence of a much stronger and competitive private seed industry in the country.

To address these challenges Embrapa conducted several internal meetings during 1995–96. In 1996 it published "Deliberation no. 22," (IB[BCA]n. 30, 1996), the institutional policy for managing intellectual property. The basic guiding principles of the policy can be summarized as follows:

- Embrapa has to maximize its capacity to use IPR to facilitate the transfer or the licensing of technology, processes, and products (such as cultivars, improved germplasm, isolated genes and molecules, software, CDs, books, and periodicals) without sacrificing its social mission.
- Embrapa has to seek legal protection for the technologies, processes, and products derived from its research program, giving credit to employees as inventors.[2]
- Embrapa may authorize the use of its protected assets through a royalty-free license only when doing so contributes to its social objectives and only after approval from its Intellectual Property Committee (see "Implementing a new policy: An intellectual property secretariat" below).
- Embrapa research centers cannot release a new cultivar or disclose any process or product without previous analyses by the designated committee of the possibility, convenience, and opportunity for protection.

To complement its institutional policy for the management of intellectual property and to address challenge no. 5 (modify licensing system), Embrapa published a set of regulations to adapt the operation of its breeding program and seed business to the cultivar protection environment (IB no.34, 1998). The regulations establish (1) the creation of a special

---

[2] As part of this requirement, each research unit has established an intellectual property committee to review applications from researchers and to determine those that can be exempted from the obligation to seek protection. Scientists approach this committee for advice regarding disclosure and dissemination of information.

germplasm bank for the maintenance of all cultivars (to be immediately protected or not), coordinated by the National Research Center for Genetic Resources and Biotechnology (CENARGEN), located in the city of Brasilia and already responsible for the National Germplasm Storage System, (2) rules for the denomination of new and essentially derived cultivars in accordance with the national law and the regulations of the International Union for the Protection of New Varieties of Plants (UPOV), (3) rules for the conditional negotiation of third-party-owned biotechnological assets before beginning a new project, and (4) rules for establishing partnerships for the development of improved genetic material that may give rise to new cultivars. These rules include a set of requirements, which, if complied with during the development of a cultivar, process, or product, may permit co-ownership with Embrapa.

## Implementing a new policy: An intellectual property secretariat

To become effective, a policy requires a meticulous implementation phase that ensures that relevant changes are absorbed into the institution's internal culture. Embrapa established its Intellectual Property Committee in 1997 to help implement the intellectual property policy (challenge no. 1). Composed of Embrapa researchers and reporting to the head of the Intellectual Property Secretariat, this committee meets twice a year to deliberate on internal policies and other intellectual property issues associated with processes, products, and technologies arising from the research pipeline. A secretary coordinates the committee's correspondence and forwards consultations via e-mail to the members. In 1997, the committee prepared guidelines for the functioning of laboratories, and it decided on the need for confidentiality in research projects and grant applications. It also prepared documents that stipulate the responsibilities of and confidentiality from visitors to Embrapa, such as ad hoc consultants, grantees, international consultants, and undergraduate and graduate students who develop joint research projects.

In order to complement the work of the committee and to help disseminate the policy, smaller intellectual property committees were created in 1997 in each of Embrapa's 39 research centers (designated as "local" intellectual property committees). The diverse nature of Embrapa's research program makes it very difficult to train staff with the expertise that is needed to respond to the complex issues involved in implementing the policy at each research center. Based on the experiences gained during 1997–98 and to better respond to challenges no. 3 (create assets from intellectual property) and 4 (establish regulatory infrastructure), Embrapa created a centralized unit, the Intellectual Property Secretariat, in December 1998. Located at Embrapa's headquarters in Brasilia, the secretariat has a direct link with Embrapa's president. Its location at headquarters enables the secretariat to work with each local intellectual property committee (see table 20.2).

During the past year, the rationale for having an institutional Intellectual Property Committee has diminished somewhat as the intellectual property rules are becoming firmly internalized and the Intellectual Property Secretariat is taking on more responsibilities. The secretariat now coordinates the technology acquisition and technology transfer associated with processes and products that have Embrapa- or third-party-owned intellectual property. The secretariat is not involved in activities that have no protectable assets. This was clarified in a special statute that prevents the secretariat from interfering with Embrapa's normal technology transfer activities, as carried out during the last 25 years.

In addition, the secretariat serves as the policy-making body that adapts, adjusts, and updates Embrapa's IPR policy and its implementation in accordance with global developments.

Table 20.2. IPR at Embrapa: Time Frame from Planning to Implementation

| Date | Event |
| --- | --- |
| **Phase A: Intellectual property infrastructure and initial awareness** | |
| June 1995 | To design the terms of an internal intellectual property policy, working groups were formed with one board director as coordinator and five to six researchers from various disciplines. |
| March 1996 | Preliminary results presented by working groups to a selected group of researchers. |
| April 1996 | Final results presented to the Board of Directors. |
| | Deliberation no. 22 (IPR management and policy in Embrapa) published |
| May 1996 | Embrapa's IPR committee nominated with 10 members and two ad hoc members (multidisciplinary composition). The committee meets twice per year. |
| | Local, four-member intellectual property committees formed in the 39 research centers (1996–97). |
| | A special IPR coordination unit composed of two junior lawyers and two support staff formed to link between the institutional committee and the local committees and to take care of patents and cultivars documents to be deposited at the national offices. |
| June 1996 | Two short-term international consultants (from Stanford University and Cornell University) invited (1996–97) (cost: $ 9,000). |
| September 1997 | Local committees in operation. |
| October 1997 | First workshop to debate the implementation of the intellectual property policy within Embrapa for the chairpersons of the local committees. |
| February 1998 | 20 members of the local committees engage in a special event organized by Embrapa and the University of Viçosa to discuss IPR-related issues. |
| May 1998 | Need identified for committee members to receive training in negotiation, protection, and licensing of products and processes and access to world patent data banks. |
| **Phase B: Review of the breeding program in view of the implementation of PVP legislation** | |
| March 1997 | Working groups called to meet. Two board directors and four researchers met and called on experts for the different plant species (core: four people – part time). |
| April 1997 | First workshop to draw up major guidelines in multidisciplinary approach |
| December 1997 | Breeders' workshop (40 breeders) to discuss and decide on major policies. |
| March 1998 | Results presented and approved by the Board of Directors. |
| April–May 1998 | New guidelines prepared and distributed to research centers for review. |
| July–August 1998 | First version of guidelines published. |
| | Ongoing seed production and licensing contracts and new contracts to be negotiated under the new guidelines. |
| December 1998 | Guidelines modified for clarification and republished. |

The running costs for meetings in phase A were estimated at $92,300 and for phase B at $45,200, excluding staff salaries.

## Implementation challenges

As the implementation of the policy progressed, a new set of challenges confronted those managing IPR at Embrapa. The outstanding issues deal directly with breeding (cultivar protection and licensing) and biotechnology programs (patenting, licensing), because these activities demand immediate action and the Intellectual Property Secretariat allows for a coordinated negotiation strategy to be developed and implemented.

The more recent challenges include the following:

1. **Licensing of biotech assets used by Embrapa's research teams.** Many biotech assets are not bound to proper material transfer agreements (Sampaio 1998), and the renegotiations of these agreements only allow for research use, with no provisional clauses for a possible future commercial agreement.
2. **Developing model agreements.** Models are needed to deal with issues of confidentiality and to standardize contracts, including transfer agreements. Special attention must be given to projects involving the joint collection, research, and development of biological samples. This includes issues regarding the conservation of the environment and the respect for indigenous knowledge, and intellectual property issues.
3. **Developing research and commercial contracts.** Commercial contracts are needed to promote interest in and from private companies, whereby the third party can own genes of immediate relevance to Brazilian agriculture and Embrapa can own the "elite" cultivars.
4. **Developing best practices for patenting or cultivar protection.** In practical terms, developing these best practices means having an operating interface between the National Industrial Property Institute (INPI) (for patent protection) and the National Service for Cultivar Protection (for cultivar protection).
5. **Encouraging researchers towards intellectual property.** Researchers should be encouraged to produce sufficient data to allow for the protection of the invention. The Brazilian legislation does not accept a preliminary application, as do some other countries. The application must therefore be in a final format before submitted to INPI.
6. **Motivating managers.** Research and program managers should be given priority to develop biotechnological assets such as genes and processes that would strengthen Embrapa's capacity to negotiate in the international market.

## Addressing past and future management challenges

As shown in this chapter, designing and implementing an intellectual property policy has been a sizeable task for Embrapa, and it will continue to be an exercise with many challenges for at least the coming five years. The new legal background requires dramatic changes in the management of Embrapa's human resources at the research and administrative levels. These changes present opportunities to stimulate scientific production by distributing royalties derived from proprietary technology. However, Embrapa will have to make further administrative adjustments to regulate the division of revenues among inventors, breeders of new cultivars, research groups and other components. Several models available from intellectual property and technology transfer offices linked to US universities are under discussion.

Embrapa is considering a new policy to change researchers' values and practices. For example, visitors may have to be kept away from sensitive areas in biotechnology laboratories, and some information and data will have to be treated confidentially. This means that researchers have seen some quite dramatic changes in the daily routine at Embrapa, and some of these changes have not been easy to implement. Few researchers have immediately adopted the new mode of operation; most of them will need some time to adjust. The ongoing capacity building program has to be improved to raise awareness among researchers, laboratory support personnel, and administrative staff. This is a top priority of the Intellectual Property Secretariat. Research directors of each center have been reminded repeatedly of their responsibility to supervise the implementation of the policy and to monitor the research programs to detect any new opportunities or problems related to intellectual protection.

The Intellectual Property Secretariat has created a demand for specialized personnel to develop contracts and licensing agreements. Until hiring such personnel becomes possible, Embrapa uses its own staff, including three lawyers and two support personnel, to staff the secretariat. At the time of writing, six months after the secretariat became operational, it is clear that at least one additional position needs to be filled by a person with a biological/agronomic background and a sound understanding of the international environment. This is critical for establishing a dialogue with third parties. Implementation of Embrapa's new intellectual property strategy has been supported by placing some researchers[3] abroad to work on technology assessment and related intellectual property aspects.

Due to the peculiarities of the Brazilian legal framework on intellectual property, Embrapa also anticipates future studies on the interrelationship between the patenting of genes and the protection of transgenic cultivars. These studies will help answer questions raised by researchers, research managers, and potential business partners regarding these forms of intellectual property protection. The studies will examine how cultivars produced and protected abroad by third parties enter the national seed market and how umbrella patents or "pipeline"[4] patents of importance for the agricultural field are assessed by INPI as to whether they will be patented in Brazil. Many of Embrapa's negotiations with private firms may be affected by the answers to these questions.

One of the most important challenges ahead is shifting Embrapa's priorities in biotechnology programs to give more importance to in-house development of genes and processes useful for tropical agriculture. This is a research area where Embrapa has a natural competitive advantage. To discuss these new priorities, Embrapa organized a workshop with national and international scientists in August 1997. Important strategic decisions where taken following a meeting by Embrapa's directors, who decided to strengthen a few centers of excellence to cover the major areas of biotechnology research. To support these centers, increased financing for the programs has to be negotiated, which will allow basic scientific developments to be made in parallel with practical approaches that address agricultural demands.

---

[3] The project, entitled Labex, began in November 1998 and is implemented by Embrapa with support from the World Bank through an agreement with USDA/Agricultural Research Service.

[4] Pipeline patents are patents deposited in Brazil during 1996–97 during the 12-month period stipulated in law no 9.279. Patents deposited during this period are directly accepted by INPI if they fully comply with the law.

## Implications regarding trade, farmers, and public acceptance

While adapting to the new legal environment, Embrapa has been urged to form new partnerships with private multinational companies interested in negotiating the introduction of proprietary genes in elite commodity cultivars (soybean, cotton, rice, corn, and beans) owned by Embrapa. Farmers who have access to the latest inputs from new technologies have urged Embrapa to develop cultivars containing new transgenic traits so that they can compete with other farmers in the Mercosur[5] region that already have access to transgenic soybean and corn seeds. To develop such materials, Embrapa must comply with national and international biosafety regulations and support the National Technical Biosafety Committee in its task to control and implement the safe use of genetically engineered plants, animals, and microorganisms in the country.

The recently published intellectual property regulations are Embrapa's guide to clarifying to all partners, potential clients, and society how it will treat the improved genetic plant material and other proprietary processes and products resulting from its research programs. It is expected that Embrapa and its potential partners will negotiate the use of proprietary assets, i.e., protected cultivars and genes, as they will no longer be freely available.

Initial reactions from multinational businesses with long-standing presence in Brazil seem to indicate that they welcome Embrapa's new policy, mainly because it shows that Embrapa is making an effort to internalize revenues to give continuity to its research programs. Inevitably, some national groups are dissatisfied with the change, because they are not properly acquainted with the new legal framework or because previous appropriation was not bound to rules and products from research were released free of charge. Some international and national nongovernmental organizations have tried to prevent changes in defense of small-scale farmers. However, under the Brazilian Cultivar Protection Law, small farmers have already been awarded a special article that allows them to freely multiply and exchange seeds of protected materials, provided that they participate in a governmental program for small-farm agriculture (Law no. 9.456, Art.10 – IV, 1997).

The fact that many of Embrapa's cultivars exhibit superior agronomic quality should facilitate negotiations, under the new directives, with multinational firms for the use of proprietary genes. During this process, Embrapa will face crucial points when discussing royalty shares and licensing agreements, as it currently lacks experience in this area and will seek to gain experience while negotiating and implementing such partnerships. Initial contacts have shown that mutual confidence and patience are necessary, and that business contracts materialize only when a balanced agreement can be reached.

Some factors, however, may preclude a balanced contractual agreement. Embrapa has an obligation to provide improved agricultural inputs to a diversity of farmers. However, four or five multinational companies alone currently own most genes expressing high agronomic value. They use these genes and other biotechnological assets as one means of penetrating seed markets, thus forming one part of a process of agricultural business verticalization. This leads to a completely new panorama established by market forces and the need to gain public acceptance of genetically engineered food over the next few years. The existence of approved intellectual property laws in the country has enormously

---

[5] The Mercosur (Mercado Común del Sur; the Southern Common Market) region comprises Argentina, Brazil, Paraguay, and Uruguay, representing a total population of 190 million.

facilitated dialogues[6] with such important private-sector companies, facilitating Embrapa's entrance into biotechnology for the coming century.

## Concluding remarks

Since Embrapa began building its capacity in intellectual property, Brazil has been preparing for the TRIPs review taking place in 1999. Industrialized countries, playing a leading role in agricultural trade and export, support present developments as related to TRIPs. However, countries importing agricultural products are expressing the need for a conceptual review, including many developing countries that wish to examine articles specifically related to intellectual property protection.

Therefore, changes may occur regarding present trends on IPR and trade-related issues. In addition to Embrapa examining these matters, it must also keep up with growing concerns among the general public regarding biotechnology-derived agricultural products and the life science industry's movement towards vertical integration of the global seed market. This is also taking place among the world's leading agriculture- exporting countries. For Embrapa, it is essential to be aware of these global changes and be prepared to adapt to them so that Brazilian agriculture and agribusiness retain an edge with regard to global competitiveness.

## References

Embrapa. 1996. Internal Bulletin (BCA) no.30. Brasilia: Embrapa Press.
Embrapa. 1998. Internal Bulletin (BCA) no. 34. Brasilia: Embrapa Press.
Government of Brazil. Law no. 9.456. April, 1997. Diário Oficial da União de 28/04/97.
Sampaio, M.J.A.M. 1998. Intellectual Property Rights: An important issue for the Brazilian agriculture R&D and related agribusiness. Paper presented at the workshop on IPR in Agriculture: World Bank's Possible Future Role in Assisting its Borrowers/Member Countries. Washington, D.C., June 1998.
Watal, J. 1998. Intellectual property rights in agriculture: The position in India. Paper presented at the workshop on IPR in Agriculture: World Bank's Possible Future Role in Assisting Its Borrowers/Member Countries. Washington, D.C., June 1998.
WIPO. 1996. Publication no. 223(E). Geneva: World Intellectual Property Organization.

---

[6] As this chapter was revised and prepared for publication, Embrapa modified its cultivar licensing agreements to comply with the IP policy. The market response will be analyzed for the first time during the 1999–2000 planting season. It is therefore too early to speculate on the socioeconomic impact.

# 21 Managing Proprietary Science and Institutional Inventories for Agricultural Biotechnology

*Joel I. Cohen, Cesar Falconi, John Komen, Silvia Salazar, and Michael Blakeney*

## Abstract

*New inputs derived from biotechnology, especially those coming from the private sector, are widely used in agriculture. Intellectual property rights protect most of these inputs. Protected materials are used and developed by public, national, and international agricultural research organizations working for and with developing countries. This chapter presents insights from inventories of proprietary biotechnology inputs used by national and international agricultural research organizations. It reviews the most important findings derived from these inventories. It documents the difficult and often confusing situation that institutions face regarding the use and dissemination of products resulting from proprietary science where others hold rights. The chapter also looks at means for conducting inventories of intellectual property.*

## Introduction

Developing-country national agricultural research organizations (NAROs), research centers of the Consultative Group on International Agricultural Research (CGIAR), and advanced research institutes are all affected by changes in proprietary rights related to agricultural research. Two developments in particular have affected them: the increasing importance of biotechnology in research at CGIAR centers and NAROs and the growing position of the private sector in international agricultural research. These developments have led the International Service for National Agricultural Research (ISNAR) to study the magnitude of and procedures for using inputs that are protected by intellectual property rights (IPR) at selected CGIAR centers and NAROs.

Proprietary technologies and materials are those that are privately owned, managed, or protected through some sort of IPR. Such materials and technologies may have restrictions placed on their use during the research stage or in a later stage, when products derived from the protected materials are ready for wide dissemination. A growing number of research inputs are protected as intellectual property. This chapter focuses on the use of such protected or proprietary materials and technologies in seven CGIAR centers, and it studies preliminary information from five countries in Latin America.

© CAB *International*. 1999. *Managing Agricultural Biotechnology—Addressing Research Program Needs and Policy Implications* (ed. J.I. Cohen)

Since the early 1980s, the centers of the CGIAR have invested in strengthening their infrastructure and human resources for biotechnology research. Most of them now have specialized units or divisions to do research on molecular biology and other techniques covered by the term biotechnology. In 1996, the CGIAR adopted some guiding principles on genetic resources and intellectual property. These principles reaffirm that the resources maintained in the centers' genebanks should be freely available and that the centers should not seek legal protection for their innovations unless it is absolutely necessary to ensure that developing countries have access to new technologies ("defensive patenting"). The centers should not seek intellectual property protection for income-generating purposes and will not view potential returns from intellectual property protection as a source of operating funds. The 1996 document also states that any IPR acquired by a center should be exercised without compromising in any manner the fundamental position of the CGIAR regarding free access by developing countries to knowledge, technology, materials, and genetic resources.

At the annual CGIAR Mid-Term Meeting in May 1997, an expert panel was established to focus on issues of proprietary science and technology. The panel explored legal issues and their ramifications regarding proprietary science and the complex partnerships arising in agricultural research (TAC 1998). The panel felt that gaining an understanding of the current technologies and practices employed by the various centers would be an important first step in this process. The study of the seven CGIAR centers was done on behalf of the panel. It was the first step in a long-term project focusing on the access, research management, and dissemination of products resulting from proprietary biotechnology inputs by developing countries.

In a second step, a similar study focused on the application of proprietary research inputs, and prospects for generating proprietary products from these inputs in selected NAROs of Brazil, Chile, Colombia, Costa Rica, and Mexico. These countries have been developing their agricultural research biotechnology capacity for the last 10 years.

## Purpose and methodology of the studies

### Purpose

The studies aimed to (1) assess the extent to which proprietary applications of biotechnology (technologies and materials) are being used in NAROs and CGIAR centers, (2) present potential legal implications regarding use of the identified proprietary technologies and materials, and (3) synthesize findings and recommendations to stimulate further discussion.

### Methodology

In December 1997, ISNAR conducted a survey among CGIAR centers on the application of proprietary research inputs and prospects for generating proprietary products from these inputs. Assisted by several technical experts, it constructed a list of the most relevant proprietary technologies and materials. It grouped the technologies and materials into eight categories: (1) transformation systems, (2) promoter genes, (3) insect-resistance genes, (4) disease-resistance genes, (5) selectable marker genes, (6) genetic markers, (7) diagnostic probes, (8) others.

A survey was designed to determine which institutes use which proprietary technologies or materials from the above categories and for what purposes. Respondents were asked to provide information on specific applications, the means by which intellectual property protection is provided (patents, plant breeders' rights, or other means), and how the center obtained permission for research (e.g., material transfer agreement (MTA) or license). In addition, information was requested on the research products to be derived from the technologies or materials identified, the dissemination of results from this research, and if any intellectual-property protection was to be sought by the CGIAR center or NARO.

The Latin American study found that some organizations are just beginning transformation, a few institutes are quite advanced, and that some academic institutions are using transformation. Using these distinctions, a number of organizations were selected in each of the five countries. The results showed that NAROs with a reasonably extensive capacity for transformation had the best understanding of IPR and how it affects their work. However, a great deal of uncertainty regarding IPR implications existed among all respondents. Consequently, significant time was taken to provide preliminary explanations of the topics covered by the survey. After the survey was completed, more time was taken to address specific results. During these discussions, it was found that researchers felt that concerns about use of third-party intellectual property were yet another barrier to research, another problem being added to the already long list confronting researchers in Latin America. Providing adequate time for this explanatory and diagnostic process was not possible for the CGIAR study due to the deadlines imposed by the CGIAR panel. This reflects a major difference between the two studies, highlighting the importance given to the extensive follow-up needed to clarify identified problems and concerns.

As confidentiality was essential for conducting the surveys, the study gives no specific information collected from the NAROs and CGIAR centers. It focuses on general trends and gives recommendations on the use of proprietary science in the CGIAR centers, NAROs, and their partners.

## Results from the CGIAR center survey

All the CGIAR centers that responded to the survey currently use proprietary inputs for biotechnology research. ISNAR recorded 166 applications of proprietary research inputs. In total, 46 different technologies and materials were reported over the eight technology categories introduced above. Most centers apply these technologies and materials in research on several mandated commodities. Of the technology categories surveyed, three had the broadest utility across centers: selectable marker genes, promoters, and transformation systems (table 21.1). This clearly demonstrates the important role that proprietary technologies and materials have assumed in research in the CGIAR, as is true for advanced research centers globally.

### *Utility across research categories*

Based on a list of possible research targets for the CGIAR centers, the main categories were defined as cereals, noncereals, and other.[1] Survey results reveal an almost equal use of

---

[1] Cereals: maize, rice, wheat, sorghum, finger millet, pearl millet. Noncereals: beans, cassava, tropical forages, potato, sweet potato, chickpea, cowpea, pigeonpea, groundnut, lentils. Other: diagnostics, livestock health, microbial systems, general technology development.

**Table 21.1.** Applications of Proprietary Technologies and Materials by Research Category as Identified in CGIAR Survey

| Technology Category | Number of applications per research category | | | |
|---|---|---|---|---|
| | Cereals | Noncereals | Other | Total |
| Selectable markers | 17 | 25 | 2 | 44 |
| Promoters | 18 | 14 | 3 | 35 |
| Transformation systems | 12 | 14 | 3 | 29 |
| Insect-resistance genes | 8 | 11 | 0 | 19 |
| Disease-resistance genes | 6 | 5 | 0 | 11 |
| Genetic markers | 4 | 4 | 2 | 10 |
| Diagnostic probes | 0 | 0 | 3 | 3 |
| Others | 7 | 6 | 2 | 15 |
| **Total** | **72** | **79** | **15** | **166** |

inputs among centers working on cereal and noncereal crops, with fewer applications recorded at the noncrop centers and programs. The distinction reflects the emphasis placed on crop research in the CGIAR.

## Means of protection and permission for use

Figure 21.1 describes the means by which the technologies and materials are protected. Not all proprietary inputs pose difficulties related to intellectual property or the dissemination and use of resulting products. Results from the survey help centers explore applications where potential difficulties could be foreseen. For many applications of proprietary technology, centers were not able to obtain clear knowledge or information regarding the type of IPR provided for a particular proprietary tool. The rapid consolidation among biotechnology providers (Williams 1998; chapter 2 of this volume) seeking ownership of technologies complicates obtaining such information. The extensive litigation occurring among commercial, public, and farmers regarding details of license and use agreements is another complicating factor (Barton 1998).

Depending on the means of protection involved, institutes could inadvertently infringe upon legal conditions regarding the dissemination of future products derived from these inputs. In addition, written agreements on use are often lacking when collaborating scientists exchange technologies or materials that are needed to address a particular research objective. This leaves centers in an unclear position regarding legal obligations, both to the owner of the technology and to other research partners. CGIAR centers are now studying situations where there is a lack of information about conditions for using these inputs.

Figure 21.2 shows that when permission for use was obtained, it was generally achieved through MTAs, licenses, or sublicenses. However, for each technology category, except diagnostic probes, the centers are using a number of technologies and materials when written agreements are lacking, or for which permission status is unknown. Obtaining permission is the general rule, but exceptions exist. For example, in some transformation systems, the permission to use specific systems is unclear. Similar problems exist for the use of selectable markers, genetic markers, and various insect- and

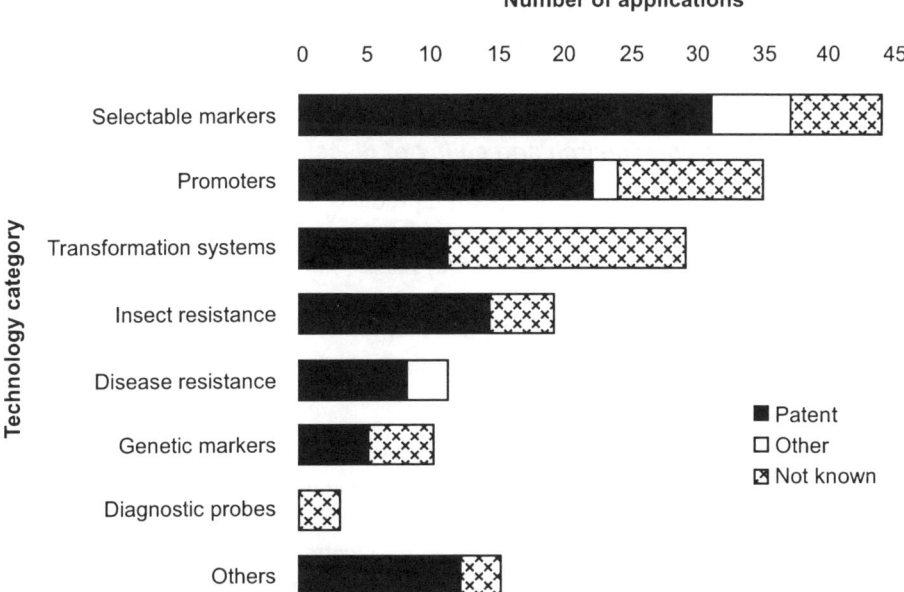

**Figure 21.1.** Proprietary technologies and their means of protection as identified through CGIAR survey

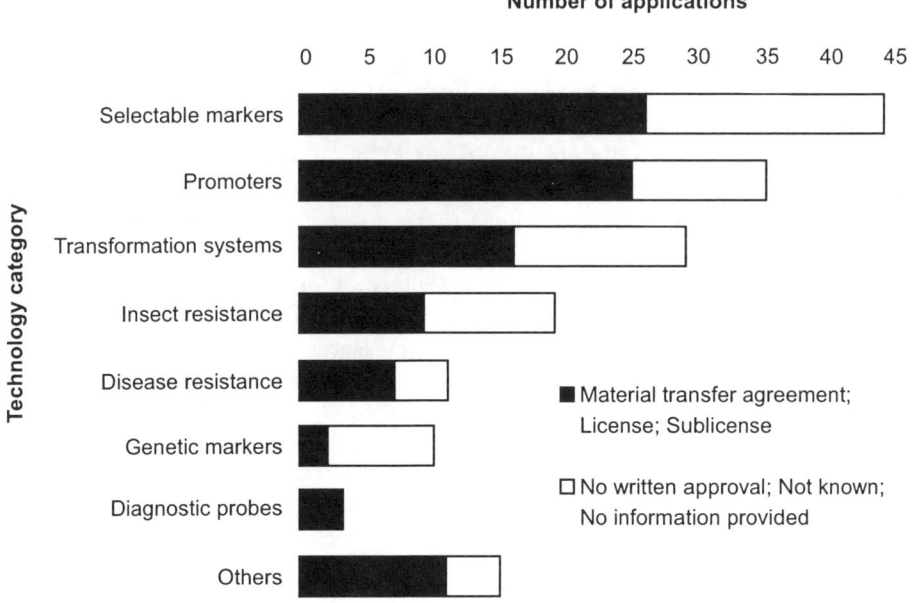

**Figure 21.2.** Agreements and potential difficulties in permission to use proprietary technologies and materials as identified through CGIAR survey

disease-resistance genes. However, it should also be noted that while MTAs may secure permission for research, they do not generally confer permission for subsequent dissemination.

## Expected products from research and ability of dissemination

The survey also sought information on research products that were expected to encounter difficulties in their dissemination. As inputs are proprietary, restrictions may exist for their use, dissemination, or further production. Survey results showed that the centers expected a total of 58 outputs or products (table 21.2).

More than 40% of the responses indicated that centers often lack critical information or knowledge needed to anticipate difficulties in post-research use and the dissemination of outputs that are generated from proprietary technologies. Consequently, some outputs developed with proprietary technologies are likely to encounter problems in their use and dissemination, especially when they are exported to countries where the technologies are protected. Many CGIAR centers have yet to take into account extensions of protection and permission requirements for such technologies.

However, centers have considered the implications of legal agreements for disseminating research outputs, with 14% of the responses foreseeing some limitations. For example, a contractual arrangement between a CGIAR center and a private multinational as owner of the input technology specifies that outputs can be distributed only in certain countries. In this case, there is *a priori* understanding of the restrictions and limitations for dissemination that should be studied further by the CGIAR. Such studies have already begun on a center-by-center basis.

## CGIAR-center patents

Along with other institutions producing international public goods, the CGIAR is considering options to ensure that developments from its research programs reach their intended beneficiaries. According to the End of Meeting Report of the CGIAR's 1997 International Centers Week, "the CGIAR stands for free flows of germplasm and it has no profit motive. However, it may have to think of defensive patenting in order to stake out a claim and ensure access."

The study assessed the degree to which centers are planning to patent or otherwise protect inventions. Only three outputs were identified that may be patented (see table 21.3). Centers anticipated the use of other protective measures for 11 outputs. This limited amount of intellectual property protection being sought by centers can be attributed to

Table 21.2. Products Expected from the Application of Proprietary Tools at the CGIAR Centers

| Product category | No. of products expected | Examples |
|---|---|---|
| Improved crops | 36 | Improved cereal and noncereal varieties with enhanced insect, fungal, and virus resistance |
| Diagnostics | 11 | Diagnostic tests for tropical livestock diseases |
| Vaccines | 1 | Vaccine for East Coast fever |
| Others | 10 | Transformation protocols, genetic markers |

**Table 21.3.** Expectation of Centers to Obtain Intellectual Property Protection for New Products (in Number of Products)

| Product category | Expectation for protection or registration | | |
| --- | --- | --- | --- |
| | No | Yes (patent or other means) | Maybe |
| Improved mandate crops | 19 | 1 (patent) | 16 |
| Diagnostics | — | 11 (other) | — |
| Vaccines | — | 1 (patent) | — |
| Others | 4 | 1 (patent) | 5 |

many factors, including lack of familiarity with IPR issues, the fact that suitable IPR options are not yet developed and approved, and the tradition that goods and services are developed as international public goods. Further, many bilateral donor and civil society organizations are opposed to applying IPR protection to products of CGIAR research.

## Results from Latin American NAROs

Between July and September 1998, ISNAR conducted a similar survey among NAROs in Brazil, Chile, Colombia, Costa Rica, and Mexico. At the time of the survey, none of the institutions that were surveyed had suitable institutional or legal frameworks for related IPR topics. With the exception of two research centers, none of the institutions had an office or person responsible for assisting the researchers in issues of intellectual property, access to adapted technologies, technology transfer, or ways to protect their own inventions. There was little coordination between institutions in the same country and even between researchers in the same institution. For example, within the same institution, there were cases where the same vector was requested from different sources, all under different conditions. There were also cases where one scientist has an agreement for the transfer of biological material for a vector, while a neighboring scientist used the vector without permission. The researchers were functioning without the institutional support needed to address these issues for their research (for further analysis of this issue, see chapter 19 of this volume).

Figure 21.3 summarizes the range of proprietary technologies and materials used by the research organizations and the number of applications reported in each category. In total, 34 different technologies and materials as well as 386 specific applications of proprietary research inputs were reported. Of the eight technology categories surveyed, selectable marker genes (GUS, kanamycin), promoters (CaMV/35s), and transformation systems (Agrobacterium) show the broadest utility across the institutes involved in the survey.

### Means of protection and permission for use

Figures 21.4 and 21.5 show what means of protection have been given to the technologies and materials analyzed in the survey and how permission for use was obtained. Again, while not all proprietary inputs pose difficulties regarding intellectual property or for the dissemination and use of resulting products, this study has helped research organizations explore areas where potential difficulties may occur.

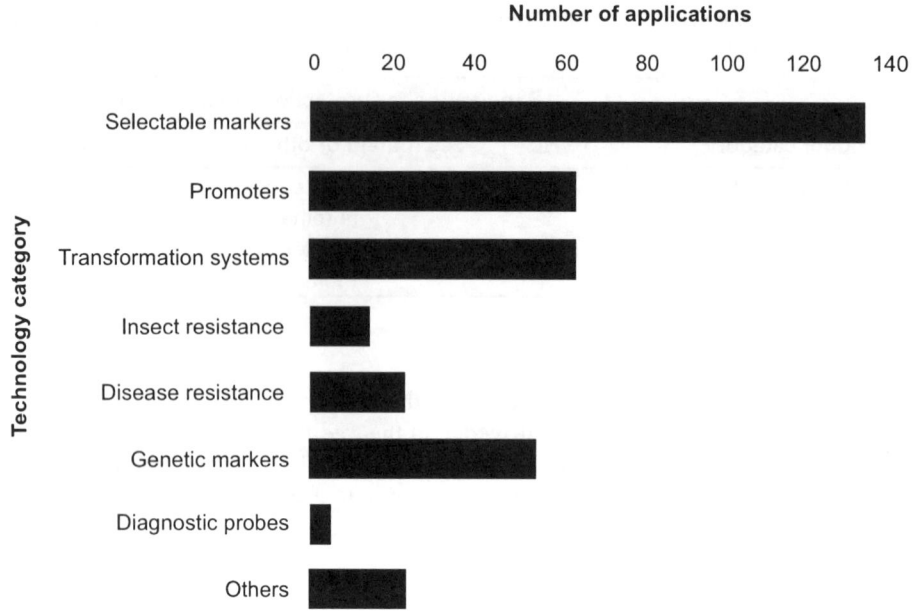

**Figure 21.3.** Proprietary technologies and materials covered in LAC survey

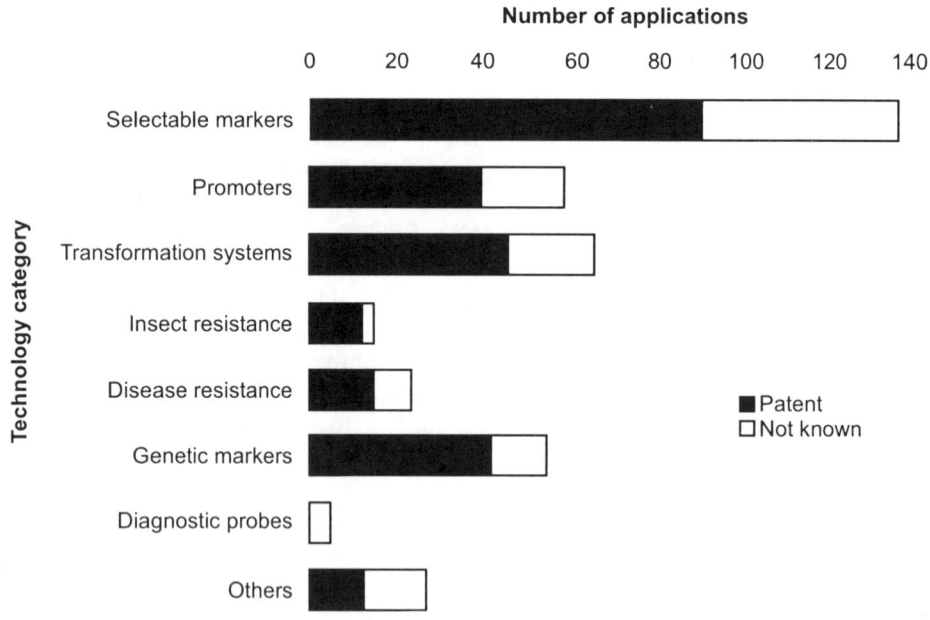

**Figure 21.4.** Applications of proprietary technologies and their means of protection as identified in LAC survey.

*Managing Proprietary Science and Institutional Inventories* 257

**Figure 21.5.** Proprietary technologies and their permission for use as identified in LAC survey

As indicated in figure 21.4, in more than 30% of the applications of proprietary technology NARO respondents indicated that they lack clear knowledge or information on the type of means of protection provided for a particular proprietary tool. These cases need to be further examined to avoid potential infringement of legal conditions on the use of these proprietary technologies and materials. Figure 21.5 indicates that MTAs are the most common means of acquiring technologies, as they are in CGIAR centers. However, it also highlights the importance of international collaboration as an important mechanism for acquisition, and it indicates that some of the NAROs purchase proprietary technologies for their own use. Many applications that are being used lack written agreement, leaving the institute wondering about its legal responsibilities and how the use of the proprietary tool has been achieved. In addition, the use of license, as a legal alternative for technology transfer, is very limited.

### NARO patents

The results show that the Latin American NAROs have high expectations of obtaining intellectual property protection for their new products. They expect that 74% of the 50 products that are expected from the application of proprietary technologies will be protected either by patents or by plant variety protection (see table 21.4).

**Table 21.4.** Expectations of Protecting Products as Identified by the LAC NAROs

|  | Brazil | Chile | Colombia | Costa Rica | Mexico | Total |
|---|---|---|---|---|---|---|
| Protection | 11 | 9 | 0 | 0 | 17 | 37 |
| Nonprotection | 0 | 3 | 6 | 0 | 0 | 9 |
| Not known | 0 | 0 | 2 | 2 | 0 | 4 |
| Total | 11 | 12 | 8 | 2 | 17 | 50 |

## Conclusions and management lessons

The ISNAR survey findings indicate a pressing need for establishing competent legal expertise in issues of proprietary rights for the CGIAR centers and their partners, including NAROs and other agricultural research institutes. Based on these inventories, a number of specific managerial lessons and recommendations can be made.

### First approximation of proprietary inputs

The inventories provide a first approximation of the use of third-party innovations that are protected through patents or other forms of intellectual property. The CGIAR-center study shows the extent to which centers are using and integrating biotechnology into their ongoing research programs. The centers now use biotechnologies and materials to develop a new generation of inputs, many of which come from transformation. Similar findings occurred for the NAROs, showing wide use of selectable markers, promoters, and transformation systems.

Once reviewed, these first approximations provide a source of common information regarding the use of proprietary technologies and materials in research programs. When several centers or organizations are found to use the same proprietary tools, it may be advantageous to collaborate in the acquisition of such technologies and to seek common understanding for the particular IPR involved. These inventories provide one means for CGIAR centers and NAROs to confirm the use of proprietary technologies and materials in their research programs and follow up their first assessments of such use with more formal IPR audits.

### IPR and proprietary science: A proactive or a "wait-and-see" approach?

As noted earlier, international and national organizations using biotechnology for agricultural development are caught in a complex environment, reflecting a transition from earlier periods where products and processes for research resided in the public domain. The increasing acquisition of proprietary technologies, their use in research serving the public good, and the vast array of developing countries where such use occurs raises questions regarding appropriate IPR arrangements. However, for many scientists and institutions, such concerns are overwhelming. Yet their work continues, trusting that as final products are developed, no legal instruments will block the dissemination of improved materials to their clients.

To adopt a more proactive strategy requires significant time and investment in taking steps to find institutional mechanisms for addressing these complex challenges. Such advancements are being made by both the CGIAR centers and by larger national research organizations, such as Embrapa (see chapter 20 of this volume). However, national and international institutions are exploring both of these options. They realize that no one clear position exists regarding the use of third-party IP for or with developing countries, as given by commercial biotechnology providers owning much of the IPR for applications identified. For this reason, it is recommended to raise IPR awareness by developing systemwide expertise for managing intellectual property, providing an economy of scale for CGIAR centers and for NAROs.

## Extensive use of MTAs

The most common legal arrangement by which CGIAR centers obtain permission for using proprietary inputs is through MTAs, followed by licensing. The extensive use of agreements and licenses is a new fact of life for publicly funded agricultural research organizations. Proprietary technologies and materials that are developed by or that are under the protection of private-sector research organizations have made important contributions to CGIAR-center and NARO research. The use of such materials means that as centers transform mandate crops, develop vaccines, do diagnostic probes, and provide for marker-assisted breeding, their dependence on licenses, MTAs, and other agreements with the private sector increases. The experiences with and conditions of MTAs should therefore be analyzed and exchanged among CGIAR centers, NAROs, and other agricultural research institutes to explore possibilities for standard formats and to determine legal implications regarding possible restrictions on use and dissemination. Where CGIAR centers or NAROs in one country are using the same category of proprietary tools, it may be advantageous to cooperate in acquiring such technologies. Precedents for such coordination have been set for research collaborations supported by international biotechnology research programs.

## Enforcement of obligations

Where MTAs impose legal obligations, the receiving institute must be in a position to honor its obligations under the agreement. For example, where the proprietary technology involves the supply of a trade secret and imposes confidentiality obligations, the institute must be able to police the handling of the material supplied. This may require the establishment of a secure system of operation and placing of researchers and visitors under confidentiality obligations. The increasing legal complexity involved in the supply of proprietary technologies may raise matters of contract law, intellectual property law, biodiversity and biosafety law, technology transfer, and competition law (where restrictive provisions are imposed). A second point is whether research institutes anticipate any dissemination constraints due to their current legal arrangements. Some centers were not clear if such constraints exist. This information confirms the need for CGIAR centers and NAROs to (1) develop IPR expertise to analyze potential limitations and (2) designate an officer to administer legal obligations.

These increasing legalities indicate the need for research institutes to have a primary legal administrator. The administrator would ensure the compliance of staff with obligations generated by MTAs and with the terms of intellectual property licenses. Because MTAs and licenses are among the most common ways to acquire proprietary technologies, it should be reiterated that MTAs, licenses, and sublicenses impose direct legal obligations upon the research institutes.

## IPR sought by CGIAR centers and NAROs

All of the centers surveyed indicated that they use applications of biotechnology that are protected by others. While the centers receive numerous technologies, the number of products that they expect to be patented or otherwise protected are very few. However, the NAROs that were surveyed expected a far greater number of products for patent filing, which requires an interface such as a research liaison office (see chapter 19 of this volume). The IPR expertise needed by the CGIAR centers for their potential patents needs to be further evaluated as well. Each CGIAR center is currently considering these matters through a recently established Central Advisory Service for IPR, located at ISNAR.

### *Partners in a challenging environment*

CGIAR centers, research institutes, and NAROs have formed many partnerships to undertake joint or contractual biotechnology research where access to proprietary materials has been achieved. The technology or licensing agreements held by the partners may cover the use of proprietary inputs by the centers. In other cases, partner organizations may hold their own rights to inputs. In that case, the centers expect that they are undertaking research consistent with the mission of the CGIAR and to produce international public goods. However, such partnerships may only cover the actual research. The use and dissemination of products are a matter for further negotiation; this will often require legally binding understandings or license agreements prior to extensive dissemination (see chapter 22 for examples of agreements constructed in the context of international development).

## References

Barton, J.H. 1998. The impact of contemporary patent law on plant biotechnology research. *In* Global Genetic Resources: Access and Property Rights. CSSA Special Publication. Madison: Crop Science Society of America.

Cohen, J.I., C. Falconi, J. Komen and M. Blakeney. 1998. Proprietary biotechnology inputs and international agricultural research. ISNAR Briefing Paper 39. The Hague: International Service for National Agricultural Research.

Salazar, S. 1999. The use of proprietary biotechnology research inputs at selected Latin American NAROs. Draft project report. June 1999. The Hague: International Service for National Agricultural Research.

TAC. 1998. Report of the CGIAR Panel on Proprietary Science and Technology. SDR/TAC:IAR/98/7.1. Rome: Technical Advisory Committee of the CGIAR.

Williams, A. 1998. Consolidation. *Journal of Commercial Biotechnology* 5(2): 106-112.

## 22 International Collaboration: Intellectual Property Management and Partner-Country Perspectives

*Catherine L. Ives, Karim M. Maredia, and Frederic H. Erbisch*

### Abstract

*Managing intellectual property in the context of international development presents many complex decisions regarding research. This chapter discusses three key intellectual property management challenges arising from collaborative research projects implemented through the Agricultural Biotechnology for Sustainable Productivity project, supported by the United States Agency for International Development. The project managers and collaborating scientists are faced with specific challenges regarding intellectual property. For each of these challenges, details are provided on how the collaborating institutions developed an approach to managing intellectual property within the legal framework of the USA and in conjunction with a range of scientific partnerships.*

### The Agricultural Biotechnology for Sustainable Productivity project

As we approach the 21st century, technology transfer in the agricultural community is changing from the informal and free exchange of materials to a more formal and legal process. International treaties, as well as the privatization and consolidation of the agricultural biotechnology sector are some of the forces behind this shift. Developing countries are now faced with the need to acquire new skills to access, use, and manage biotechnology. To help address this need, development agencies such as the United States Agency for International Development (USAID) and their clients needed answers to questions, such as:

- Can researchers in developing countries access new technology to address local and regional constraints if this technology is to be found within the private sector (or held as proprietary information by the public sector) in developed countries?
- How can vastly different cultures be encouraged to join together to address critical problems for farmers and consumers in the developing world while protecting the environment, human health, and biodiversity?

---

The authors acknowledge the help of Colm Lawler, Office of Intellectual Property, Michigan State University, in the writing and editing of this document.

© CAB *International*. 1999. *Managing Agricultural Biotechnology—Addressing Research Program Needs and Policy Implications* (ed. J.I. Cohen)

This chapter focuses on particular management challenges faced by a USAID-supported international collaborative project in agricultural biotechnology, Agricultural Biotechnology for Sustainable Productivity (ABSP). The project takes an integrated approach to generating, transferring (exchanging), and managing biotechnology (see also chapter 11 of this volume). The ABSP project was designed and implemented in a consortium of public- and private-sector institutions in the USA and abroad, with Michigan State University (MSU) as the lead institution

One of the main goals of the project is to help build capacity around the world for the safe and legal exchange of intellectual property. To do so, the project developed a flexible process for managing research and technology transfer involving intellectual property. Components of this process include (1) developing research/technology transfer agreements, (2) developing capacity in accessing and managing intellectual property, and (3) communicating clear responsibilities and lines of authority for each project partner.

Established in 1991, the project sought an integrated approach by combining applied research, product development, and policy development in the areas of biosafety and intellectual property rights (IPR). The project assists developing countries not only in accessing and generating technology, but also in using that technology in an environmentally and legally responsible manner in order to increase the productivity of the agricultural sector.

The project goals have been implemented in two, partly overlapping phases (technology generation and technology transfer), and individual research projects are currently at different phases. Technology-generation activities include developing technology as well as human capacity building and policy assistance. Technology transfer includes all activities that assist in transferring or exchanging materials to partner countries and commercializing in-country research programs. This includes support to research laboratories, legal and biosafety consultations, collaborative field tests, and developing commercial partnerships. ABSP technologies include genes and gene constructs, protocols, germplasm, transgenic plants, and biocontainment technologies.

The project objectives included production of genetically engineered crops and field-testing them in the USA and collaborating countries. The project also aimed to develop innovative micropropagation methods for producing large quantities of uniform, disease-free planting material of a number of crops, including banana and pineapple, and to carry out genetic stability tests of these cultures. (The crops and constraints addressed by the ABSP project are outlined in table 22.1.) Another priority was to train scientists, administrators, and policymakers in applying biosafety procedures and IPR in biotechnology.

The project has three different modes of partnership: public-public, public-private, and private-private (see table 22.2).

The innovative nature of biotechnology requires managers to adopt novel approaches to managing research projects. Those responsible for the project face the following main management challenges:

1. raising general awareness and understanding of the importance of intellectual property
2. developing a system, both within the project and within partner countries and institutions, for accessing, using, and managing intellectual property
3. building functional national and institutional policies and their enforcement

**Table 22.1.** Crops and Constraints Addressed by the ABSP Project

| Crop | Insect/Virus Pest | Technical Approach |
| --- | --- | --- |
| Potato | • potato tuber moth | • Bt toxin genes/transgenic plants |
| Sweet potato | • feathery mottle virus<br>• sweet potato weevil | • coat protein/transgenic plants<br>• proteinase inhibitor gene/<br>• transgenic plants |
| Maize | • corn borers | • Bt toxin genes/transgenic plants |
| Cucurbits<br>(melon, squash) | • zucchini yellow mosaic virus<br>• watermelon mosaic virus<br>• papaya ringspot virus, watermelon strain | • coat protein/transgenic plants<br>and traditional breeding of<br>varieties |
| Tomato | • tomato yellow leaf curl virus | • replicase and other genes/<br>• transgenic plants |
| Banana | | • micropropagation/<br>• bioreactor technology |
| Pineapple | | • micropropagation/<br>• bioreactor technology |

**Table 22.2.** Primary Institutional Linkages of Research Groups within the ABSP Project

| US institution | Sector | Developing-country institution | Sector | Commodity |
| --- | --- | --- | --- | --- |
| MSU | public<br>public | CRIFC<br>AGERI | public<br>public | potato<br>potato |
| MSU/Cornell | public | AGERI | public | cucurbits |
| Scripps | public | AGERI | public | tomato |
| Advanta (Garst) | private | CRIFC | public | maize |
| Monsanto | private | KARI | public | sweet potato |
| Pioneer Hi-Bred Intl. | private | AGERI | public | maize |
| DNAP | private<br>private | Fitotek Unggul, Indonesia<br>Agribiotecnología de Costa Rica | private<br>private | pineapple<br>pineapple, banana, coffee |

MSU: Michigan State University, USA
CRIFC: Central Research Institute for Food Crops, Indonesia
AGERI: Agricultural Genetic Engineering Research Institute, Egypt
DNAP: DNA Plant Technologies, USA
KARI: Kenya Agricultural Research Institute

## Raising awareness and understanding of intellectual property

Already in its initial design the ABSP project included support for raising awareness of intellectual property and recognized the importance of intellectual property in developing collaborations and accessing agricultural biotechnologies. USAID provided broad guidelines, regarding how intellectual property is to be managed, as defined by the Bayh-Dole Act[1]. The US government passed this act to become Public Law 96-517 to facilitate the transfer of technology from US universities to private industry, either through licensing to established companies or through the creation of start-up companies. This law allows US universities and other public institutions to own patents arising from federally sponsored research and provides the legal framework for technology transfer from US public institutions and universities as well as other institutions that receive US economic assistance. Under this law, universities are allowed to obtain royalties from licensing of technologies developed by university inventors (Youngers 1994).

Certain conditions for commercializing technologies have been described since the introduction of this legislation. These conditions include the formal notification of the funding source of any new intellectual property derived from the assisted research, including agriculture. For projects such as ABSP, it is necessary to inform all parties of the new paradigm for agricultural research and explain how to develop collaborations and access technology under certain restrictive conditions.

Within the ABSP project, general capacity building efforts have helped developing-country collaborators formulate favorable intellectual property policies for their particular situation at both the national and institutional levels. One of the unique features of the ABSP project was the recognition that not only policymakers, but also scientists and administrators need to receive information and training in this area. The project attempts to bring individuals from the legal, scientific, and administrative spheres together so that disparate views can be gained and policies crafted that provide mutual benefits to all parties involved.

The ABSP project has used a number of ways to raise awareness of the importance of intellectual property in agricultural biotechnology. The following are two examples: (1) law school professionals provided legal consultations to assist policymakers in assessing their intellectual property needs and designing model legislation and institutional policies. Led by Stanford University, ABSP developed an IPR internship program at Stanford Law School to educate both the policymakers and scientists in various biotechnology-related intellectual property issues. (2) ABSP sponsored in-country workshops on general intellectual property issues in Egypt, Indonesia, Morocco, and the USA. In Cairo, Egypt, the ABSP project organized an IPR workshop in January 1994 for over 150 scientists, administrators, and policymakers from various public- and private-sector institutions (Bedford and Maredia 1994).

Complications in IPR of course inhibit the formation of partnerships, but these complications are not always due to an absence of clear policies or procedures for managing intellectual property. The lack of clear ownership for a molecular tool can equally inhibit a partnership. This was evidenced by negotiations between a research institute in Egypt and a US-based company that wanted to develop virus-resistant Cucurbits. The US company's lack of clear title to a promoter complicated negotiations in

---

[1] 35 USCA sections 200-212; 37 CFR section 401. See http://www.ucop.edu/ott/bayh.html for an on-line version of the law from the Council of Government Relations.

forming the research collaboration. (The collaboration was subsequently postponed when the company was sold to a large multinational vegetable seed company.)

## Developing a system to access, use, and manage intellectual property

### *Developing a system within the ABSP project*

Many of the technologies and innovations developed using agricultural biotechnology are proprietary and must be managed in a different manner than the traditional nonproprietary technologies. The ABSP project was initially managed by MSU's Department of Crops and Soil Sciences. However, it quickly became clear that the level of integration of this project required additional management skills than those commonly found in a single university department. The project was therefore transferred to MSU's Institute of International Agriculture, where additional resources were available. At its inception, ABSP had no intellectual property or technology-transfer specialist assigned to the project. This resulted in confusion surrounding the use of some proprietary research tools. For example, one researcher at MSU was designing vectors for transformation using genes for which no permission had been received to transfer to third parties.

When the project was faced with the possibility of designing and developing transgenic material that might never be used outside the university's laboratory, it became clear that the project needed assistance in intellectual property management. With that requirement in mind, MSU's Office of Intellectual Property agreed to assist the project in ascertaining the ownership of research tools used in the project and in formulating intellectual property policies and technology transfer protocols.

These are not straightforward issues, as ownership within the agricultural biotechnology community is in constant flux, and there are numerous legal actions in progress to determine outright ownership of certain genes and processes. This is an ongoing area of concern for the project, as researchers continuously combine technology components. For example, a researcher may use a promoter, a marker gene, and a target gene all from different sources, which have different restrictions on their transfer to third parties. However, an initial assessment of research tools for clear ownership and/or negotiated use and licensing might have avoided the problem described above. So in 1995, a technology transfer coordinator was appointed to assist the project in managing intellectual property generated within the ABSP project. This person also serves as an important link and information intermediary between the ABSP Management Office, MSU's Office of Intellectual Property, scientists, and administrators within the project. Together with the director of the Office of Intellectual Property, a new framework was designed for handling and managing intellectual property, as described below.

The main elements of the ABSP intellectual property agreements cover the handling of new inventions/materials resulting from joint ventures; the use, exchange, ownership, and protection of new inventions; the notification to the funding agency (USAID); the sharing of royalties; the publication of research results; and the handling of other parties' confidential information.

All subcontractors are now required to sign subagreements that contain a research or joint-venture agreement outlining intellectual property responsibilities before any funds are allocated. Although this was not an initial component of the project, it was necessary because problems were discovered when attempting to move research from the laboratory to the field and from the US to a partner country. During subsequent negotiations with subcontractors, (which can include universities, companies, and developing-country institutions), a research agreement is negotiated to clearly define ownership of current and future technologies.

The agreements developed through the ABSP project define the conditions under which both parties collaborate on intellectual property. The agreements benefit all parties: ABSP project managers, MSU, collaborators (scientists), collaborators' institutions, and funding agencies (USAID). The agreements provide ABSP with a management tool and MSU with a basis to manage intellectual property beyond the very broad guidelines of the US government. Since many scientists are not skilled in legal issues, the agreements help collaborators avoid possible mistakes and confusion. The project helps them handle and manage intellectual property.

These agreements are a crucial component of any biotechnology project whose ultimate goal is to develop and disseminate products. However, there are drawbacks to these agreements, all centering on the time that is often required to negotiate them. This may disturb the funding agency, which wants to see the project get underway, as well as the researchers, who understandably want to get started. However, if the goal is to transfer technology in a safe and legal manner, then these negotiations are critical and should be completed before research begins. Problems arising in negotiations, such as an inability to agree on use of materials and location of practice, may signal difficulties in implementation. The project managers could then decide early on in the process if a collaboration is worth pursuing. Additionally, it forces researchers to look "downstream" at how their research outputs will be used and who their clients may be.

One example of this process is the negotiation that took place between MSU, Egypt's Agricultural Genetic Engineering Research Institute, and Pioneer Hi-Bred International, Inc., which had discussed plans for a collaborative research project to develop insect-resistant maize in varieties/hybrids of importance to the Middle East. Negotiations for the research collaboration required input from MSU's Office of Intellectual Property, patent attorneys, and technology acquisition specialists at Pioneer Hi-Bred. Requiring a number of face-to-face meetings, the negotiation process took about 10 months to complete. Scientists now are doing collaborative work, with some Egyptian scientists based at Pioneer Hi-Bred. A smooth transition from laboratory work to field-testing and product development in Egypt is anticipated.

However, providing general information on intellectual property and proprietary issues in ABSP research did not fully address specific legal issues arising from the project's collaborative research. The project therefore subcontracted the director of MSU's Office of Intellectual Property to work with the ABSP technology transfer coordinator to develop a process for transferring materials developed through the project to host countries. This provided constant attention to and expertise in the legalities of technology transfer.

### *Process of technology transfer within ABSP: Disclosure agreements*

As mentioned above, USAID is the main supporter of the ABSP project, and as such, ABSP is subject to the Bayh-Dole Act. One of the primary responsibilities of the ABSP scientific and management team, in keeping with the intent of the act, is the formal disclosure of any new invention that may arise from the project. Disclosure occurs in two stages: (1) scientists inform their host institution, which in turn informs MSU, and (2) MSU reports to the funding agency USAID within 60 days of notification of the invention. This responsibility is part of MSU's contract for coordinating the ABSP project.

An invention disclosure normally takes the form of a brief description that highlights the novel aspects of a technology and clearly states who the inventors are. Once this disclosure is received, it provides a starting point for the handling of the technology, including determining the intellectual property status or commercial potential. The purpose of the invention disclosure process is (1) to inform the ABSP management team that a proprietary technology may emerge and (2) to initiate the technology transfer process while at the same time fulfilling any sponsor obligations. It is essential to stress that ownership of any invention depends on the intellectual input of the inventors, not on the funding source. Therefore, within the ABSP project, intellectual property may be jointly or solely owned by any of the participants and can impact on the technology transfer process.

This process is used when researchers disclose new inventions that result from a collaborative project to MSU and ABSP management. A simple form, especially developed for this purpose, outlines the technologies involved and indicates whether technologies are proprietary. The director of MSU's Office of Intellectual Property then meets with the ABSP Technology Transfer Coordinator to review the disclosure and to decide on the next steps. In some cases, they meet with the inventor to get additional information, such as any material transfer agreements signed with other parties. This notifies project management of potential complications, such as the need to obtain permission from or notify other parties before sharing their technologies. The invention disclosure thus serves as a basis for how to handle the inventions whether in their initial protection, current, or future transfer phases, or the ultimate marketing of technologies.

The disclosure form allows new inventions to be recorded in a uniform manner. It alerts the ABSP management team that proprietary materials may be coming out of ABSP research efforts and allows the team to notify those involved in the ABSP technology transfer process that a new technology might soon be available. This information is useful for evaluating the discovery. Patent attorneys can use in their initial review of the technology. The completed form is sent to USAID and/or other research sponsors as a formal notification of a discovery. It should be noted that failure to report an invention to USAID or failure to specify that USAID provided funding on a patent application is a violation of the Bayh-Dole Act. This can result in disallowance of project expenditures and may ultimately lead to a patent being declared invalid.

### *Technology transfer within ABSP*

Technology transfer agreements are negotiated for two kinds of technologies: those coming to the ABSP project and those going from the ABSP project to collaborating countries. For either kind of technology transfer, the ownership of intellectual property has to be determined and respected. As already stated, the use of invention disclosures helps clarify the intellectual property status and ownership of any new technology. Inventorship is determined by the intellectual input that went into developing the invention. As the ABSP project has participants in a number of countries, this may result in a situation where intellectual property is shared among a number of institutions in different countries. The technology transfer process has the best chance of success when an integrated system is in place with institutional and intellectual property policies along with sufficient human and financial resources.

In the ABSP project, this integrated system currently exists only in MSU, which means that MSU has taken the lead in the ABSP technology transfer process. As part of the process to further the developing-country objectives within the ABSP project, the ABSP management team consults with developing-country managers. However, since one of the core objectives of the ABSP project is to strengthen the partners' intellectual property and technology transfer management capacities, it is anticipated that developing-country managers will eventually lead this process for the technologies in which they are intellectual property stakeholders. The overall aim therefore is to allow the commercialization of ABSP inventions with particular potential for the developing-country partners.

Within the ABSP project, before products can be transferred, appropriate individuals review and sign the agreements that partners have proposed during the research. These agreements include MTAs for components needed to conduct the research and similar agreements that may restrict any usage of the resulting ABSP research technology/discovery. These agreements ensure that the condition for using any such materials is included with the technology profile. (The profile is developed prior to the approval.) The technology is then transferred in different ways, resulting in either temporary or permanent transfer.

*MTAs*

MTAs for biologically active materials allow the transfer of materials and stipulates their handling, use, commercialization, and liabilities. The agreement allows the recipient to work with the materials under the terms agreed by both parties. It aims to provide a short-term arrangement for conducting additional research with very little or no revenue required through the agreement. Commercialization usually requires a license agreement.

Some years ago, MSU's Office of Intellectual Property and the ABSP project, in collaboration with the Agricultural Genetic Engineering Research Institute in Egypt, successfully executed an MTA to transfer a *Bacillus thuringiensis* gene from Egypt to MSU. The ABSP project has also executed an MTA with Advanta/Garst Company for the use of a CryV gene for research purposes.

*License agreements*

With a license agreement, an institute can commercialize the technology. It lays out terms and conditions for the supplier and the recipient for the commercialization process. A license agreement is usually a long-term arrangement that requires sharing of commercialization receipts.

While MSU has executed many license agreements, the ABSP project has not yet executed any. Projects usually take a number of years before generating a technology requiring intellectual property protection. However, the process is in place, and the ABSP project is currently educating scientists and policymakers and disseminating information about the process.

*Option to license agreement*

The option to license agreement helps reserve rights to use the technology for a limited period of time while exploring commercialization. It gives the recipient time to explore commercialization of the technology or discovery without having to enter into a permanent licensing agreement, and reserving the technology for the recipient. This process is generally inexpensive as it does not require any extensive long-term payments for the technology or discovery. No actual option to license agreements have been executed under the ABSP project, but the process has been developed and has been communicated to collaborating partners.

All partners need to be kept informed of the products that may be restricted and those that may be available for wider distribution. Partner countries are informed about the available technologies within the ABSP project in national technology-transfer workshops and individual visits. For example, in 1995, a technology transfer workshop was conducted in Indonesia to set the stage for technology-transfer activities in the country.

The importance of communicating with all parties involved (researchers, administrators, policymakers from both developed- and developing-country institutions) cannot be overemphasized. As the technology-transfer "business" heavily depends on the people involved, it needs individuals who are technically competent, commercially savvy, and culturally sensitive. They are in great demand within the university and industry.

The development of intellectual property policies is an ongoing process that calls for constant education and communication. It is often seen as an issue between researchers on the one hand and technology-transfer specialists on the other. In much the same way that cultural differences exist between countries, cultural differences exist between research scientists and technology transfer/commercialization specialists. The language, skills, and interests differ, and both groups must be willing to learn from and support each other. It became clear during the ABSP project that focal points for these issues were needed, both within the ABSP management office and within our host-country collaborators.

### Development of technology-transfer systems within partner countries

ABSP's educational efforts described above have helped a number of countries develop technology-transfer systems. This is critical, because partner countries must eventually have the sole capacity to enter into negotiations and develop collaborations outside the ABSP project. But many developing countries currently lack facilities and people trained in intellectual property management. ABSP therefore saw a need to build capacity in the day-to-day handling of intellectual property and technology transfer management.

In collaboration with MSU's Office of Intellectual Property, ABSP organized a two-week internship program in 1996, which was repeated in 1997. The purpose of the program was to provide hands-on experience with intellectual property and transferring technologies from the public to the private sector. The program emphasized the ideas, concepts, and processes used in handling, transferring, and managing intellectual property in various US institutions. Seventeen participants from Africa, Asia, and Latin America attended the program.

Two examples illustrate the spin-off that such exchanges can have. In 1996, an attorney from Costa Rica visited MSU under ABSP sponsorship. This led to an interest in sharing the knowledge acquired with countries in Central America. With support from USAID, a two-day seminar was organized in Costa Rica in 1998 for Central American countries to raise awareness and educate more than 125 scientists and policymakers in various technology-transfer and biodiversity-related IPR issues.

Another internship at MSU in 1997 was sponsored by a fellowship from the Brazilian Ministry of Science. This fellowship enabled a chemical engineer from the Division of

Technology Management of the Oswaldo Cruz Foundation (an agency of the Brazilian Ministry of Health that conducts biotechnology research, education, production, and quality control) to work for three months in the Office of Intellectual Property. The intern learned how intellectual properties are managed in the USA, which helps her refine the technology transfer procedures and IPR policy at the foundation.

## Building functional intellectual property policies, legislation, and their enforcement

As with all projects, there are certain management issues that are not under the direct control of the project. Training and education can affect national government policies, but an international collaborative project itself cannot change national policies. So while developing human capacity is essential, it does not guarantee that proprietary materials will be made available to developing countries. For example, in the case of Indonesia, lack of intellectual property protection and enforcement has kept one collaborator from transferring its genetically engineered maize to Indonesia for incorporation into their breeding program.

Change in government policies regarding IPR is underway. Countries are positioning themselves to respond to international treaties such as the North American Free Trade Agreement (NAFTA), the Convention on Biological Diversity (CBD), and the General Agreement on Tariffs and Trade (GATT). They are also trying to address the uncertainty of the types of intellectual property protection and the capacity for enforcement. Difficulties in complying with these agreements have kept many private companies and universities from transferring proprietary material to emerging economies. The ABSP project views these barriers as educational opportunities and provides expertise and guidance so that the policies that are developed are appropriate to the particular country, while still meeting their international obligations.

The ABSP project has used different mechanisms to influence national intellectual property policies, including one-on-one consultation, internships, workshops (Barton and Bedford 1993; Bedford & Maredia 1994; Maredia 1995). The ABSP project organized a plant variety protection and patents workshop in Jakarta, Indonesia, in 1995 and assisted in drafting a new plant variety protection law for Indonesia. This law is pending approval by the Indonesian parliament. In Egypt, ABSP-trained personnel have facilitated the inclusion of plant and food products into existing IPR legislation. ABSP partner country Morocco passed a new plant variety protection law in 1996. The ABSP project has provided one-on-one consultation and conducted a workshop in 1997 to assist Morocco in implementing this new law (Ives 1997).

In addition, the ABSP project has assisted in building intellectual property management capacity at the institutional level. In cooperation with the Office of Intellectual Property and the Institute of International Agriculture at Michigan State University, the project organizes annual internship programs in IPR and technology transfer to provide hand-on experience in day-to-day handling and management of intellectual property (Maredia et al. 1996).

## Conclusion

The ABSP project combines policy and research efforts within a single program. This unique feature has helped scientists, administrators, and policymakers clearly see the complex issues surrounding the access, use, and management of intellectual property. The successful management of agricultural biotechnology intellectual property is one of the central aims of the ABSP project. This goal presents a number of challenges for project management, including raising a general awareness and understanding of the importance of intellectual property; developing a system, both within the project and within partner countries and institutions to access, use, and manage intellectual property; and building functional national and institutional intellectual property policies, and their proper enforcement.

These challenges have been addressed through educational initiatives that vary depending upon the need of countries, institutions, and individuals. In addressing the need for an overall system within ABSP to handle intellectual property, the responsibility for intellectual property management was assigned to a technology transfer coordinator.

With the assistance of technology transfer personnel at MSU, a formalized system for accessing and exchanging materials was developed for the ABSP project. Partner countries and institutions in developing countries have been encouraged to develop similar focal points through training and consultations. The final management challenge, the lack of national policies and effective enforcement of policy, cannot be directly altered by the ABSP project. Here, the project's approach has been to provide consultations, training, and advice so that national policies will be changed for the benefit of the country.

The overall goal of the ABSP project is to create an integrated R&D project involving selected developing countries. This involves the latest agricultural biotechnology research to address selected aspects of these countries' needs along with building human capacity that is necessary to allow this research to be commercialized. It is the union of these two aspects of the project, along with the spirit of cooperation between the partner institutions and USAID through the MSU coordinator that makes the ABSP project an interesting model for intellectual property management for developing-country managers.

## References

Bayh-Dole Act, 35 USCA sections 200-212, and 37 CFR section 401

Barton, J.H. and B. Bedford. 1993. Intellectual Property Rights: the ABSP Intellectual Property/Patent Workshop. *BioLink* vol.1 no.4.

Bedford, B. and K. Maredia. 1994. Proceedings of AGERI and ABSP Workshop Series on Intellectual Property Rights, Biosafety and Project Evaluation. January 24-31, 1994, Cairo, Egypt. East Lansing, MI: Michigan State University.

Ives, C. 1997. Implementation of plant variety protection in Morocco. *Seed World* 135: 34-36.

Maredia, K. M. 1995. ABSP Workshop targets intellectual property rights in developing countries. *BioLink* vol. 2, nos. 2 & 3.

Maredia, K. M., F.E. Erbisch, and J.H. Dodds. 1996. Strengthening Technology Transfer Framework in Developing Countries: A Michigan State University Internship Program. *Journal of the Association of the University Technology Managers* 8: 21-32.

Youngers, J.A. 1994. Pertinent Laws: Compliance Requirements. *In* Association of University Technology Managers Technology Transfer Manual.

# 23 Industrial Research and Business Development: Experiences from the Singapore Institute of Molecular Agrobiology

*Tai-Sen Soong*

## Abstract

*The increasing importance of agricultural biotechnology is affecting traditional agricultural institutions in developing countries. In addition to maintaining sound scientific and academic standards, many of these institutions are tasked with building links with the private sector, as this becomes a key issue for agricultural research in these countries. This chapter reviews approaches for industrial research, drawing on experiences from the author and the Singapore-based Institute of Molecular Agrobiology. This institute has developed strategies for industrial research and development (R&D) and commercial development. Emphasis is placed on managing industrial R&D, technology transfer, business development mechanisms and implications for managers in national research organizations.*

## Introduction

Over the past decade, agricultural biotechnology has become a core technology for improving agricultural productivity, in particular in industrialized countries. Products such as multivalent animal vaccines, biological pesticides, transgenic plants, diagnostic kits, and enzymes used in feed and food testify to the applicability of molecular biology to modern agroindustry.

Traditionally, research results from public agricultural science were freely shared among the producers and users of the results. However, with the growing role of the commercial sector in research, intellectual property rights and laws are increasingly used to protect inventions and discoveries. Developing countries are therefore strengthening their abilities to negotiate and manage intellectual property as one way of increasing capacity in agricultural biotechnology research and development (R&D). These experiences help countries negotiate rights for access and use of emerging technologies.

Products derived from new biotechnologies are being adapted to local environments. National agricultural research organizations (NAROs) can contribute their knowledge of local varieties and collaborate in transforming or improving molecular-based crops, which, in turn, can address end-user needs, fulfill NARO objectives, and increase the commercial

© CAB *International*. 1999. *Managing Agricultural Biotechnology—Addressing Research Program Needs and Policy Implications* (ed. J.I. Cohen)

utility of inventions. R&D institutes that focus on biotechnology have new roles to play in advancing ongoing scientific programs and in considering new mechanisms for business development.

This chapter reviews means for implementing and planning research, with special emphasis on collaboration with the industrial sector. Suggestions by the author provide one means of comparing and contrasting current R&D practices with alternative practices highlighted in the examples provided.

## Institute of Molecular Agrobiology: Private-public project development

Located in Singapore, the Institute of Molecular Agrobiology (IMA) was established to enhance agricultural productivity in Asia through extensive research in and application of agricultural biotechnology (IMA 1999). The institute has the following objectives:

- to undertake innovative research in agrobiology at the genetic and molecular levels
- to provide training at the postgraduate level in the area of agrobiology
- to provide the critical mass and international standing to attract high-caliber researchers to undertake first-rate agrobiological research
- to facilitate the development of niche technologies and commercialization opportunities for agriculture in Asia
- to provide international standing in order to facilitate industry taking up of research and development, as well as manufacturing activities in agroindustry in Singapore

IMA's level of funding and facilities is consistent enough to provide high-quality, basic research. Complementing this basic research, the institute supported an extensive applied research program to address immediate and long-term needs of industry. These projects are developed in collaboration with multinational and regional companies.

Soon after IMA was established, and even before its own research program was fully developed, an alliance was built with Delta & Pine Land and Monsanto (now merged) and a number of Singapore parties including IMA. The alliance was established to conduct a cotton quality improvement project in China. Delta & Pine Land supported IMA at US $1.8 million to do the research. The Singapore government also co-funded the project. IMA recruited specialists from the USA and China to build the necessary laboratories. IMA and Delta & Pine Land jointly wrote the project specifications. This helped ensure that research results would have immediate commercial utility. For example, the transgenic cotton produced through the joint venture is already on the market in two provinces in China.

The following sections review some of the techniques and strategies that IMA and others have employed in developing research products that respond to commercial needs. They also provide ideas on how to interface with business partners such as multinationals.

## Managing systems for industrial research

Research managers who direct their activities towards the industrial development of biotechnology need to consider two key issues: (1) whether the research organization puts its R&D investment into areas of technology that are commercially attractive, and (2) what mechanisms or plans exist to commercialize these research programs. An organization may do research to address existing clients' needs with defined specifications, or it may do research on the organization's perceived research priorities in its "core capability programs." The first type of research is aimed at developing technologies for immediate commercial needs, such as the expansion of market shares of existing product lines, the second type for developing new products that can create new markets.

The success of commercial research programs also depends on managers' ability to interface with business partners. They can do this through a strategy of industrial integration (i.e., to locate industrial partners and first identify the needs and then design a research project). This strategy can be developed through R&D contracts, technology licensing, and joint ventures. New joint ventures between R&D organizations and business partners will be the most beneficial for the long-term interest of any research organization. For example, they affect R&D programs by providing market information for project development.

## Developing industrial R&D programs

### Market-oriented R&D

Industrial R&D, i.e., the development of commercially competitive processes and products, consists of a series of developmental phases: discovery, laboratory testing, pilot-scale testing, market-testing and refinement, manufacturing operations, regulatory registration (for certain products), and marketing. Few new discoveries, however, result in commercially valuable products. For R&D to be cost effective, projects should therefore be designed from the marketing phase to the upstream research phase. Using this approach, project developers can work with industrial partners and use important market information.

### Project management

Project management is the key to the success of commercial development. When developing the project, project managers benefit from adopting the basic principles of SMART, i.e., the objective of the project must be Specific, expected results must be clear and Measurable, the project members must be held Accountable, merits must be timely rewarded, and a Time frame should be specified for each activity (Randolph and Posner 1988). This way, the project can be monitored closely. Commercial partners in particular are keen to know what can be offered, when it can be delivered, who is responsible, and how each party will benefit once the project is completed. In the course of the project, marketing managers of the industrial partner should participate actively to provide feedback on market changes. If managers understand this process well and use it properly, then it can be a powerful tool for integrating R&D with business development.

## Key issues in technology transfer and industrial collaboration

### Technology identification

The identification of appropriate technology to suit commercial needs is normally the beginning of industrial collaboration. R&D can be categorized as "incremental," "radical," and "fundamental" in meeting the needs of business at various stages (Roussel et al. 1990). "Incremental" R&D represents low risk and moderate rewards to exploit existing technologies in new ways. It is highly acceptable to industry with few financial resources but eager to gain market shares (in other words, R&D could be represented as D only). "Radical" R&D may create new knowledge to industry and is often characterized by higher risk and increased reward (r&D). It is suitable for companies with strong financial resources and seeking opportunities for diversification. "Fundamental" R&D creates new knowledge for the world that broadens a company's understanding of scientific areas (R only). Because of its high-risk nature and uncertain application, such R&D is usually funded by the government.

After identifying the most suitable category of R&D, suitable partners should be identified. Even for fundamental R&D, some multinationals may be interested in collaborating if the programs are well planned. If the institute can assemble a team of highly qualified scientists to explore a given subject, it can generate impact and thus strengthen its reputation, thereby attracting the attention of multinational companies. For example, even though gene discovery and function are based on fundamental research, future impacts of commercial importance and interest can be envisioned.

## Intellectual property status of the technology

It is necessary first to define the intellectual property status of a given technology. This can be in the form of patent, copyright, trade secret, or technology licenses. It is also necessary to determine whether this position can be enforced if necessary or defended if challenged. This may involve working closely with a patent attorney to investigate if competing intellectual property or "prior art" already exists.

## Costs, royalties, and benefits

Project developers must consider the cost for downstream development of any technology, including investment in manufacturing, and for marketing. If the technology is outstanding but requires significant capital to bring to market, it will be less attractive and its value reduced. An estimate of the cost of development for the licensor to bring a technology to a stage where it is ready for licensing will often serve as baseline figure for calculating the value of any technology.

A measure that can help in this regard is the so-called 25% rule. According to the rule, a licensor should receive a royalty of at least 25% of a licensee's gross profit before tax (Parr and Sullivan 1996). This figure is estimated to be about 5% of the market price. For technologies that require larger capital inputs for the manufacturing facilities, the royalty can be reduced to 10% of gross profit before tax or 2% of the market price. Such rules are likely to be acceptable and serve at least as a starting point for negotiation.

Royalty income will be determined by industrial standards. Following the 25% rule this can fall between 1% and 5% of the sales. However, with certain technologies the royalty income can be calculated if the market value is clearly calculated and the return is clear. If the technology strengthens a company's reputation or helps achieve larger market shares for other related products of the same company, then these benefits should also be considered.

## Business development mechanisms

Traditionally, industrial research focuses on whether R&D results can be used to establish new industries. The research institute served its industrial partner by providing R&D services on a contract basis to upgrade its traditional business. While this has certain merits in advanced nations with a strong industrial basis, it does not help build new bioindustries in developing countries. While in the Asian Pacific region, public investments in agricultural biotechnology have risen over the past 10 years, the expected industrial development that should result from these investments has not materialized. Alternative approaches of setting up businesses may therefore be considered, such as technology assembly, technology transfer, and joint ventures with multinational corporations. These options count on spin-offs from research staff that can eventually result in building an industrial base to support national research organizations.

### Strategies for business development

There are several strategies for business development. Establishing a joint venture as a subsidiary of the research organization can serve as an efficient commercial vehicle for R&D results. The following models should be considered.

#### Team up with scientists and technologies to form a new company

Agricultural biotechnology companies require several years for them to break even financially. A start-up company will therefore face a shortage of operational funds to support the required downstream production, registration needs, field trials, and market testing. Therefore, it is more effective to invite a company with its own marketing channels, some knowledge of the products, and enough funds to be the main partner for driving the business. Scientists can team up with the new company to focus on production-related issues. A venture fund would help the new company's position in negotiating with its main business partner. In many countries, governments encourage public-sector staff to establish such ventures by offering capital investment and leave without pay for two to three years, so that start-up funds are guaranteed, as well as a job if the project fails.

#### Establish a new business by working with reputable multinational companies

Local circumstances in developing countries often prevent multinational companies from being as effective as they wish. National institutions can provide the necessary resources, i.e., local strains, varieties, connections to market and government agencies, regulatory issues, etc., to facilitate the new joint venture to grow.

#### Upgrade local companies

Purchasing a certain percentage of shares in local companies and assisting these companies in taking up new technologies licensed abroad is another approach to partnering with industry. Many local companies in agrobusiness may already have a sound sales network and a portfolio of products. These companies are looking for new or better products to compete in the market. If the research institution can identify the needed technologies, through direct licensing and capital investment, the partnership will be valuable to either party.

## Implications for managers in public institutions

Managers of public institutions need to be aware of changes and new trends in handling R&D results and their management. Some of the issues for public institutions include the following, with further details outlined by Crespi (1998):

### Notebook system

In many public institutions, laboratory notes are typically taken in a casual manner, with little attention to detail. The notebook system is an excellent tool to help protect the rights of the inventors. In the USA, for example, the detailed recorded data in a researcher's notebook is crucial in determining the inventorship of any new technology. Any patent application has to demonstrate where, how, and when the invention was done and recorded. Therefore, managers of public institutions need to establish a notebook system and require their scientists to follow the guidelines properly to secure any potential invention for the institutions in the future.

### Confidentiality agreements

Before entering into any discussions with clients or other institutions, a confidentiality agreement must be signed to protect the rights of the institutions. The content of such an agreement should include the following:

- definition of the proprietary information
- duration for both parties to keep the information confidential
- procedures for disclosing this information
- exceptions to the disclosure of information
- a brief framework for settling any possible legal disputes that may arise. It is important to specify the materials and information in the agreement. Vague or general technology descriptions should be avoided.

### Material transfer agreements

In the past, experimental materials were freely exchanged between institutions. Materials from agricultural biotechnology typically need to be covered by material transfer agreement before the material is released to the other party. In this agreement, the receiving party receives the material only for research purposes, and the release of the material to third parties is restricted. The material transfer agreement has become standardized in many institutions in advanced nations, but it is rarely used in the developing countries.

## Conclusion

Rapid developments resulting from increased use of modern agricultural biotechnology, including technologies and materials protected by some form of intellectual property, are affecting agricultural research and related industries. Public research institutes need mechanisms that would allow them to work, in a mutually beneficial manner, with these technologies and with industry. Mechanisms that can facilitate this collaboration include technology licensing, business assembly, direct investment, and technology development. Not all of these suggestions may be compatible or appropriate for meeting the mission and objectives of national research organizations. However, when such mechanisms are found to be useful, then structural/organizational changes of institutions, policies, investment tools/capital, and flexibility in providing business options may be adjusted accordingly. These adjustments would help ensure that national research organizations are equipped to be practical and beneficial to industry and the organization itself.

## References

Crespi, R.S. 1998. Patenting for the research scientist – bridging the cultural divide. TIBTECH 16 (November 1998): 450-55.
IMA. 1999. Institute of Molecular Agrobiology Home Page. http://www.ima.org.sg.
Parr, R. L. and P. H. Sullivan. 1996. Technology Licensing, Corporate Strategies for Maximizing Value. New York: John Wiley & Sons.
Randolph, W. A., and B. Z. Posner. 1988. Effective Project Planning & Management, Getting the Job Done. New Jersey: Prentice Hall, Englewood Cliffs.
Roussel, P.A., K.N. Saad and T.J. Erickson. 1990. Third Generation R&D, Managing the Link to Corporate Strategy. Boston: Harvard Business School Press.

# 24 Introducing Transgenic Crops in India: A Joint Venture Approach

*Ellora Mubashir*

## Abstract

*This chapter describes an India-based joint venture company, Proagro-PGS, which was established to bring to market the fruits of conventional and biotechnological breeding. It describes how the company was formed, shows the importance of negotiating agreements in detail, explains the significance of compatibility, and assesses success factors.*

## Introduction

It is estimated that by the year 2002, India will need around 230 million tons of foodgrain to feed its population. This means that the country needs to add 35 million tons to its current annual foodgrain production of 195 million tons. To achieve this, the Five Year Plan 1997–2002 pursues an ambitious annual agricultural growth rate of 5%, with technology as one of the key driving forces. Similar increases are required in all other food segments, such as pulses, oilseeds, fodder, vegetables, milk, and eggs.

The Government of India is encouraging the public sector and the private sector to collaborate to build capacity and to achieve the government's goal of self-sufficiency in foodgrain production. A joint venture between a public-sector institute from a developing country and a private company from a technologically more advanced country is a particularly useful mechanism for developing countries to build capacity and accesss technology. In turn, such a joint venture helps an industrialized country access new markets, provided that appropriate intellectual property regimes are in place to facilitate the protection and transfer of technologies. This "added value" for either partner is the driving force behind any joint venture.

Technology-based joint ventures are becoming better positioned in India as an intellectual property rights (IPR) regime is emerging, thanks, in part, to the Agreement on Trade-Related Aspect of Intellectual Property Rights (TRIPs) (Dhar and Rao 1999). Considering IPR issues for crop plants is a multidisciplinary activity in India (Sehgal 1999), involving the Ministry of Agriculture (for plant variety protection), the Ministry of Environment and Forests (for issues related to the Convention on Biological Diversity), the

---

This chapter, begun in late 1998, was authorized for publication by the management of Proagro-PGS. Proagro Seed Company, Ltd was acquired by Hoechst Schering AgrEvo GmbH of Germany in February 1999.

© CAB *International*. 1999. *Managing Agricultural Biotechnology—Addressing Research Program Needs and Policy Implications* (ed. J.I. Cohen)

Ministry of Commerce (for issues related to the General Agreement on Tariffs and Trade [GATT] and TRIPs), and the Ministry of Industry (for patents).

While IPR has become an important issue in worldwide trade, India faces many difficulties in this regard. A recent study of the country's position towards the year 2020 shows that India needs many new technologies to achieve its agricultural goals (Abdul and Rajan 1998). Better IPR protection will help India become a technology provider rather than a technology seeker. This is vital for a country that depends on technology imports; according to a study by the Chamber of Commerce and Industry, 95% of industrial production in India is based on imported technology. India curently spends less than 1% of its gross domestic product (GDP) on research and development. Of this, commercial contributions amount to only 15%, or less than 0.12% of GDP.

### *The Proagro-PGS joint venture company*

One example of such an international, technological joint venture is Proagro-PGS India Ltd—an enterprise between Proagro Seed Company Ltd (PSCL) of India and Plant Genetic Systems (PGS) of Belgium. Proagro Seed is a leading seed company in India with an established infrastructure, experienced and competent human resources, and a strong market presence since 1988. Proagro Seed and its affiliated companies own and develop germplasm of various crops and have expertise in traditional plant breeding. The company has an extensive production, marketing, and distribution network for seeds of various crops. Over the years, it has gained wide experience in, for example, brand image, and it developed a broad network of official, institutional, and banking contacts in India and neighboring countries.

Plant Genetic Systems is a biotechnology R&D company with nearly 20 years of experience and global contacts in several biotechnology business areas. It has developed novel technology to produce hybrid oilseed *Brassica*, hybrid vegetables, and other commercially valuable traits, which are conferred on plants through genetic engineering. Patent applications and patents are pending for part of their technology in several countries. Prior to participating in the joint venture, Plant Genetic Systems was strictly a research company with little business experience. It has several proprietary technologies that are either commercialized or are about to be used commercially. The company's transgenic *Brassica napus* hybrid was first commercialized in Canada in 1995 and subsequently in several other countries. The company's genetically engineered nuclear male sterility trait has proven to be efficient in oilseed rape, corn, tobacco and vegetable crops. The company has also developed new plant breeding know-how (for example, techniques such as double haploid and RFLP) for several crops. An additional strength of the company is the extensive experience gained in many countries in the application of laws and regulations related to transgenic plants.

Proagro-PGS was established in December 1993 as a limited company engaged in R&D, commercial production, and marketing of value-added hybrid seeds of oilseed *Brassica* and vegetable crops. The mission of the joint venture company is to develop, produce, and commercialize high-quality seeds of hybrid oilseed *Brassica* and vegetables.

The objectives of the company are also in line with the country's national agricultural objectives in the Five Year Plan 1997–2002 of fostering high-yielding, value-added hybrids, and excellence in high-technology strategic research. Proagro-PGS currently spends around 20% of its net sales on R&D; most Indian companies spend less than that on R&D. The partnership emphasizes rigorous product testing, area-specific product placement, access to niche markets, and human resource development. Several international seed companies have produced highly successful technologies, but as their breeding base is usually their home countries (e.g., USA, Japan, and Europe), their varieties are often not optimally adapted to India. As reported here, the joint venture helped address these local requirements.

## Motivation for the joint venture

Having complementary strengths, Proagro Seed and Plant Genetic Systems believed that joining forces would result in a uniquely competitive business to produce and market hybrid varieties. By teaming up with Proagro Seed, Plant Genetic Systems could use Proagro's strengths in commercializing its transgenic technology, while Proagro would use Plant Genetic System's technological expertise. Another important consideration was the reasonably favorable business opportunities for seed companies in India, helped by economic liberalization and the revision of the Seed Act, coupled with the relatively lower cost of production and availability of both skilled and unskilled manpower in the country.

Another motivation for acquiring and developing new products and technologies in a joint venture was the concern that the productivity of many crops had leveled off. New biotechnologies have the potential to overcome yield barriers as well as provide new sources of weed and pest resistance. International experience has demonstrated that elite varieties derived from such new technologies can capture substantial market share within three years of introduction.

### *Market opportunity*

Hybrid oilseed *Brassica* is not yet on the market in India, and the hybrid technology of Plant Genetic Systems would offer the first-ever economically viable production of such hybrids. The use of hybrid seed is increasing by 2–3% each year, and there is significant scope for further growth. By the year 2000, the Indian seed market is expected to be worth about US $600 million. India's seed industry was virtually built on value-added products such as hybrids, of which almost 90% comes from the private sector. The public sector is mainly engaged in high-volume, open-pollinated crops.

Converting the open-pollinated Indian oilseed *Brassica* market to hybrids is expected to increase the oilseed *Brassica* seed market. At present the total market value of oilseed *Brassica* seed is estimated at around $12 million. Profit margins to farmers for oilseed *Brassica* production are higher than for cereals. The margins can be influenced by government policies, introduction of hybrids, stability of production, demand for oilseed, and general end-user market trends. The joint-venture hybrids are expected to yield between 20–25% over the best current open-pollinated varieties available at the time they are released. Gross margins for open-pollinated varieties are less than 25%, whereas for hybrid corn and sunflower they are 40–60%. Oilseed *Brassica* hybrids are expected to earn similar margins.

Plant Genetic System's Bt insecticidal trait should also have high market value, as current losses in vegetables due to fruit worm range between 15% and 20% and can be as high as 30% in infected areas.

## Contribution of the companies to the joint venture

Proagro Seed and Plant Genetic Systems each owns half of the joint venture. The parent companies contribute know-how, material, infrastructure, and equity to the joint venture company to carry out its mission. Each company contributes equally in various ways (e.g., "germplasm" versus "technology"). Several intangible areas had to be quantified and equated. During negotiation, it was agreed that a number of contributions by each parent company were equivalent in nature and would be provided free of charge to the joint venture.

Proagro Seed transferred to the joint venture its existing breeding program on oilseed *Brassica* and hybrid vegetables, together with the elite germplasm base. The company also provided initial support in sharing production locations and its strong marketing network. It also permitted the joint venture to use the "Proagro" brand name and company logo, which is recognized all over the country.

Plant Genetic Systems transfers technology and grants of exclusive licenses (for markets in certain countries) to use Plant Genetic System's proprietary nuclear male sterility state-of-art hybridization technology, including future improvements (for 15 years). It grants nonexclusive right to use Plant Genetic Systems technology in the backcrossing, breeding, production, and commercialization of oilseed *Brassica* and vegetables directly or indirectly derived from the basic material of the agreed countries. It trains joint-venture personnel in double haploid technology, molecular characterization, and certain breeding and seed quality analysis aspects. It provides know-how in the field of compliance with regulatory requirements for transgenic crops (until the fifth anniversary of the signature date). Similarly, the joint venture will inform Plant Genetic Systems regularly of the laws and regulations related to transgenic products applicable in the agreed countries of operation.

## Salient features of the joint venture agreement

In the joint venture agreement, an exclusivity or noncompetition clause ensures that each partner will only undertake commercialization activities in the agreed-upon countries and will operate in other countries through the partner that is allocated that territory. The following are the main objectives of the joint venture:

1. to develop, grow, raise, process, buy, sell, export, import, and deal in oilseed *Brassica* and vegetable seeds
2. to develop, establish, and maintain seed farms, greenhouses, and processing and distribution centers and stores for all other kinds of seeds

In addition, the joint venture agreement also provides for the following:

1. breeding, production, and commercialization of nonhybrid oilseed *Brassica* seeds and vegetables in its territory, pending the commercialization of hybrid oilseed *Brassica* and vegetables seeds
2. breeding, production, and commercialization of vegetable seeds that are not based on any technology or germplasm proprietary of the parent companies

The joint venture will establish a full-fledged breeding operation and a supporting biotechnology laboratory that will develop enabling technologies, such as tissue culture (double haploid) and molecular marker technologies (PCR, RFLP). The joint venture will act as a socially responsible company, abiding by all regulations pertaining to the production of transgenic plants and ensuring good laboratory practices.

The primary marketing and distribution network of oilseed *Brassica* seed is that of the National Seed Corporation and private seed companies. In due course, the joint venture will enter into marketing and distribution agreements with selected groups or alternatively develop its own marketing and distribution network.

The five-year business plan provides for annual reviews and improvements. Any dividends will only be distributed once the joint venture has shown profits for three consecutive years. Part of the profits will be invested into the joint venture to strengthen its capacity.

### Present management structure

The joint venture is a professionally managed company, with an equal number of directors from each partner company on its board. The board appoints the managing director and the general manager for a fixed term. These directors, who have been provided by Proagro Seed thus far, manage the day-to-day operations. As there is a great deal of trust between the partners, this management structure has worked well.

## The joint venture's IPR agreement

The IPR agreement for the joint venture was drawn up in 1994, before India signed the TRIPs agreement. The joint venture IPR agreement may have to be modified once the various IPR laws have been passed by the Indian parliament.

However, there will be no change in the ownership or IPR that is related to any material of Plant Genetic Systems or Proagro Seed that is transferred to the joint venture. Ownership of any improvement or new technology developed by either company that is transferred to the joint venture remains in the parent companies or in the party from whom a license has been obtained.

Germplasm and products developed by the joint venture/Plant Genetic Systems, based on germplasm made available by either of the partners, shall be owned exclusively by the company that developed the new material.

Any patentable invention by the joint venture may be patented in any country of the world and will be owned exclusively by the joint venture. Plant Genetic Systems shall automatically be granted a nonexclusive, worldwide license, with the right to grant sublicenses.

While exclusive licenses have been granted to the joint venture, the parent companies shall have exclusive right to protect their own germplasm or technology. If a company lacks interest in pursuing such protection in any country, then the joint venture/Plant Genetic Systems has the right to do so in the name of the owner company, but at the interested party's own cost.

If one partner's inbred line is used by the other in production, the partner using the germplasm shall pay royalties to the developer company, based on the net sales. Also, any royalty for IPR in the country of sale will be borne by the partner who sells it there.

Currently there is no patent protection for plants in India and only limited plant variety protection. It will therefore be crucial to protect the joint venture's hybrids by controlling parental-line production.

## Strategy to survive the gestation period financially

A strategy was planned to sustain operations until the new transgenic products reached the commercialization phase. Proagro-PGS would create a brand image through the sale of value-added hybrids (own hybrids as well as licensed hybrids) and open-pollinated varieties. The product mix would be gradually changed by phasing in new hybrids as they were developed and phasing out the open-pollinated lines. The joint venture's capability in seed production would be developed for additional earnings in the initial years by producing customized seeds. This strategy would also enable Proagro-PGS to position itself for introducing transgenic varieties once they reached the commercialization stage.

In the first five years of its establishment, the joint venture has a market presence with a product range of 30 open-pollinated and 58 hybrid varieties in the area of vegetables and oilseed *Brassica*. The company's traditionally bred vegetables are exported to Bangladesh, Japan, and Syria. A research program is in progress to use modern biotechnology breeding techniques and develop transgenic crops. Transgenic crops are at different stages in the regulatory pipeline in India.

## Management issues

In a joint venture, each partner's objectives and business goals must be clear and understood, with no hidden agendas. The rationale behind the joint venture also needs to be understood, and when creating the new enterprise, the focus should be on "the marriage rather than the wedding." Companies that agree on a clear strategy and management structure (the so-called soft issues) before they tie the knot stand a better chance of success than those who do not. In a joint venture, both partners should be comparable in size and have complementary core competencies, with strong personal rapport at the top level. Proagro Seed and Plant Genetic Systems spent about two years developing a business plan for what was to become the joint venture company. Such negotiations resulted in a detailed joint venture agreement, including realistic forecasts for a seven-year period.

The Proagro-PGS alliance is marked by transparency, confidence in fair treatment, an atmosphere of friendly relations, and mutual respect for each partner's capabilities.[1]

The partners also recognize each other's problems. For example, obtaining regulatory permits in India was very time-consuming, in part because it was a new experience for the regulators and the regulatory setup was untested. Plant Genetic Systems was very sympathetic in accepting this delay, and it helped the joint venture company to establish an in-house biotechnology R&D center for genetic transformation work. Some of the molecular biology analyses required for Indian regulatory purposes are conducted by Plant Genetic Systems in Belgium. Similarly, Proagro Seed freely shares its marketing and production network and some of its infrastructure until the joint venture company becomes more self-sufficient. Even after Plant Genetic Systems became a subsidiary of AgrEvo in 1997 and after AgrEvo subsequently merged with Rhone Poulenc, the joint venture continued to work smoothly, with the greater global recognition and contacts proving advantageous.

---

[1] Much credit for this positive environment goes to Dr. S.M. Sehgal, currently chief advisor to the Board of AgrEvo. Dr. Sehgal was instrumental in forming the joint venture and, thanks to a long-standing professional and personal relationship, was trusted by both parties.

## Conclusion and success factors

Joint ventures work successfully when the partners have complementary core competencies, when they can construct sound contracts following extensive negotiations, and when they foster transparent relationships in an environment of cultural compatibility. Negotiations on management responsibilities play a key role. These initial negotiations must be held under the supervision of competent lawyers.

Factors that contributed to the success of the Proagro-PGS joint venture include the following:

- a productive analysis of objectives and expectations, producing an exhaustive memorandum of agreement
- the sprit of partnership, often extending beyond the written agreement
- mutual dependence of partners (not necessarily equality)
- complementarity of interests
- value creation
- recognition of each partner's strengths and weaknesses
- agreement on short-, medium-, and long-term goals
- a management team that focuses on what is good for the joint venture rather than what is good for one partner
- provision of sufficient start-up funding in the form of equity and loans

## Acknowledgment

The author acknowledges the following individuals for their help in writing this chapter: Dr S.M. Sehgal, Biogenetic Technologies, Inc.; Mr S.K. Kapoor, Managing Director, Proagro-PGS; Dr. A. Kapur, Director, Proagro-PGS; Ms H. Abichandani, Finance Manager / Company Secretary, Proagro-PGS.

## References

Abdul, A.P.J. and Y.S. Rajan. 1998. India 2020: A Vision for the New Millennium. New Delhi: Viking Press, Penguin India.

Dhar, B. and C.N. Rao. 1999. Plant breeders and farmers in the new intellectual property regime: conflict of interests? *In* Biotechnology, Biosafety, and Biodiversity: Scientific and Ethical Issues for Sustainable Development, edited by S. Shantharam and J.F. Montgomery. Enfield, NH: Science Publishers.

Seghal, S. 1999. IPR controversy and the Indian seed industry. *In* Biotechnology, Biosafety, and Biodiversity: Scientific and Ethical Issues for Sustainable Development, edited by S. Shantharam and J.F. Montgomery. Enfield, NH: Science Publishers.

# APPENDIX

# Accessing Electronic Information

*John Komen and Patricia Traynor*

## Introduction

The quality and success of most research depends on researchers' ability to access and use good-quality information. The increased use of biotechnology in agriculture makes research and development even more information intensive and makes research management more dependent on information. In recent years, electronic information in particular has become a major resource in research. Electronic information can often be retrieved more easily, at lower cost, and faster than traditional, printed information. This appendix discusses electronic resources for research such as genome databases and resources for research management (for example, patent and biosafety information). Rather than giving a comprehensive picture of this rapidly evolving area, the appendix provides an overview of resources; the technical (hardware and software) requirements for accessing electronic resources are therefore not addressed.

The increased availability and use of electronic information does not mean that the traditional carriers of information (books, journals, as well as people) have lost any of their value. The authors focus on electronic information sources because these are growing so rapidly in size and importance that some guidance for managers on what material is available and where it can be found is warranted.

The information requirements for biotechnology in agriculture and particularly molecular biology research have yielded large amounts of data and databases on the genetic make-up of major crops and livestock species. Today, much of the information never touches paper and is primarily available electronically, for example on CD-ROM and on-line on the World Wide Web. The diversity of the information itself is also growing. Patent literature, for example, has found a place next to scientific papers as a source of technical information. As a result, managers of biotechnology research programs need access to a much wider variety of information than do managers of "conventional" research. The following section provides a first look at the numerous electronic information resources on agricultural biotechnology.

© CAB *International*. 1999. *Managing Agricultural Biotechnology—Addressing Research Program Needs and Policy Implications* (ed. J.I. Cohen)

## Information resources for research

Research-related information forms most of the information that is electronically available. The possibility of combining text files with graphics and direct links to other sources of information makes the World Wide Web particularly suitable for exchanging information on research and specific techniques. The following are some examples of particularly useful services:

### R&D information

Thousands of national and international public and private research organizations have developed "sites" on the Web to publish past and current research activities. And the number is growing every day. Some services focus on one specific research area, such as the Plant Tissue Culture Information Exchange[1], or the ExPASy Molecular Biology Server. Containing information on the genomes of plants and animals, molecular databases are another growing electronic resource for research. A very useful, free, service is provided by the National Agricultural Library of the United States Department of Agriculture (USDA). The library manages the Agricultural Genome Information Service, which provides access to a range of genome databases of agriculturally important species. It uses a generic genome database management system, which allows the user to query multiple databases.

### Genetic resources data

Detailed documentation and evaluation of germplasm accessions, the "raw material" needed for agricultural research, is increasingly made available through the Web. Examples include national genetic-resource services such as those collaborating under the European Cooperative Programme for Crop Genetic Resources; USDA's Genetic Resources Information Network; the recently established Chinese Crop Germplasm Resource Information System (see box 1); and the System-wide Information Network for Genetic Resources (SINGER) of the Consultative Group on International Agricultural Research (CGIAR). SINGER is the world's largest, international germplasm information service, linking the genetic resources databases of the CGIAR centers. It enables the user to search for information on the identity, origin, characteristics, and distribution of the genetic resources in these collections and to access specific data, e.g., crop characterization data.

### On-line bibliographic information

The Biotechnology Information Center (BIC) of the US National Agricultural Library provides bibliographic on-line information on agricultural biotechnology, free of charge. BIC publishes bibliographies on different research topics, such as herbicide tolerance and virus resistance. These are regularly updated. BIC also provides addresses for other information resources on agricultural biotechnology, such as relevant Web sites, education resources, and newsletters.

---

[1] All information services discussed in this chapter are listed in more detail in appendix tables 1 and 2.

> **Box 1**. Establishing an Information Base for Genetic Resources in China
>
> Records of germplasm traits, data from identification and evaluation studies of those traits, and illustrations of the morphological features of germplasm forms the basic information that is needed for biotechnology research. Such data must be collected, managed, and used in an efficient manner.
>
> During the Seventh Five Year Plan and the Eighth Five Year Plan (together covering the period 1986–95), the Chinese Science and Technology Commission assigned a team of scientists to systematically identify and evaluate the agronomic characteristics, quality traits, and stress- and pest-resistance of China's crop germplasm resources. The data collected were organized in the Chinese Crop Germplasm Resource Information System (CGRIS). Covering 141 crop species and containing over 12 million records in 664 datafiles, CGRIS is currently one of the largest database systems of plant genetic resources in the world. CGRIS is managed by the Chinese Academy of Agricultural Sciences, which also publishes standard guidelines for collecting and processing information. The academy promotes the use of CGRIS software to harmonize the management of a countrywide germplasm information system.
>
> The system includes the following elements:
> 1. single-crop germplasm resource data for managing records on the characteristics of germplasm resources, based on identification and evaluation studies
> 2. multi-crop germplasm resource data with multiple searchable fields
> 3. image and text information on morphological features
>
> After collecting, classifying, and standardizing the data, crop specialists check and confirm the information before it is entered into the database. Using the CGRIS system, researchers can carry out comprehensive queries and assessments of traits such as fast growth, high yield, dwarfing, multiple resistance, and product quality of germplasm resources in China.
>
> Databases have to be updated continuously to guarantee the authenticity and reliability of the data. Databases also need to use up-to-date software and hardware to exchange the information easily with partners in other countries. Database administrators and professionals can do the day-to-day maintenance of the system, while scientists and germplasm specialists should be responsible for the system's scientific content. The accuracy of the data is extremely important; using national and international criteria, database managers need to ensure that only standardized data is entered into the database. CGRIS overhauls its entire database every six months.
>
> ---
>
> Based on Yeding Mao, Junfang Lin, and Shoucai Chen. 1998. "Managing Information for Biotechnology and Germplasm Resources." Unpublished paper prepared for the International Service for National Agricultural Research.

A good deal of bibliographic information on agricultural biotechnology is contained in the databases of organizations that specialize in supplying agricultural information. These include the Agricultural Information System (AGRIS) of the Food and Agriculture Organization of the United Nations (FAO). Operational since 1975, AGRIS has accumulated a database of more than 2.7 million references. AGRIS is available on CD-ROM and on-line through commercial electronic information providers such as

Dialog[2] and Silverplatter.[3] An on-line current awareness service maintained by FAO provides details on new additions. FAO also created the Current Agricultural Research Information System (CARIS), with information about agricultural research projects carried out in or on behalf of developing countries. CARIS is accessible on-line on FAO's Web site and is available on CD-ROM.

Another traditional provider of general agricultural information, CAB International (CABI), is well-known for its abstract journal *AgBiotech News and Information*. CABI produces CAB Abstracts, a comprehensive database of agricultural information containing records from CABI's abstract journals and 8,500 other journals in 37 languages. CAB Abstracts is available on CD-ROM and on-line through Dialog and Silverplatter. Specialized subsets of CAB Abstracts are available as well, such as PlantGeneCD for abstracts on plant breeding, genetic resources, and biotechnology.

*On-line journals and newsletters*

A growing number of scientific journals are becoming available on-line, including leading journals such as *Nature Biotechnology*. However, most of the on-line versions only provide tables of contents or abstracts of articles, and many charge for accessing the full text (printed or on-line). An interesting service in this area is provided by Bioline Publications. It collaborates with scientific publishers, newsletter editors, and authors of reports and makes scientific material available electronically at reduced prices. Journals available through this service include *Biocontrol Science and Technology* (published by Carfax) and AgBiotech News and Reviews (CABI). Bioline material can be accessed through annual subscription or by ordering single documents. A number of electronic newsletters are available free of charge. Table 2 lists selected Web sites that are most relevant to the topics covered in this book. In addition, it recommends a limited number of sites that contain information that is of particular relevance to the technical aspects of biotechnology.

*Services from international programs*

An increasing number of international initiatives in agricultural biotechnology[4] are developing their presence on the Web. International programs form an excellent entry point for developing-country institutes to gather information as these Web sites give easy access to sources relevant to specific commodities. They also provide on-line electronic information services specifically targeted at developing countries. Examples include the following:

- **International Centre for Genetic Engineering and Biotechnology (ICGEB).** ICGEB provides access to a bioinformatics service called ICGEBnet. ICGEBnet helps molecular biologists analyze nucleotide and protein sequences and access a large variety of databases in such fields as genetics and structural biology. Access to ICGEBnet is available free of charge to scientists in ICGEB member countries.

---

[2] www.dialog.com

[3] www.silverplatter.com

[4] See chapter 10 of this volume for more details on international initiatives in agricultural biotechnology.

Preference is given to those scientists whose research activities are directly related to the research goals of the center.

- **Applied Biotechnology Center at the International Maize and Wheat Improvement Center (CIMMYT).** This center provides free on-line access to maize linkage maps.
- **Technical Co-operation Network on Plant Biotechnology in Latin America and the Caribbean (REDBIO).** FAO's REDBIO supports its networking activities by managing an e-mail discussion list and by disseminating a database that contains profiles of some 400 research institutes included in the network. The database, CATBIO, is available on diskette (searchable by different fields) and on-line (organized by country).

## Information on management aspects

A small part of the electronic resources include information on management aspects of biotechnology research, such as biosafety and intellectual property rights. Information and links to expert organizations on biosafety aspects are available through a number of specialized services, for example:

- **Biosafety Information Network and Advisory Service (BINAS).** BINAS is supported by the United Nations Industrial Development Organization. It provides on-line access to biosafety-related documents from international organizations as well as guidelines and laws regulating biotechnology from a range of developing and developed countries.
- **Animal and Plant Health Inspection Service (APHIS).** USDA's APHIS provides detailed information on US regulations and field trials currently taking place in the USA. APHIS also provides on-line copies of forms and user guides for requesting field-trial permits or submitting notifications.
- **Biotechnology Information Center** and **CABI.** Bibliographical information on biosafety is available through USDA's Biotechnology Information Center and CABI. The Biotechnology Information Center publishes a regularly updated bibliography of *Biotechnology: Legislation & Regulation.* CABI publishes electronic bibliographies (E-bibs) on diskette, one of which deals with biosafety. The E-bibs are a subset of CAB Abstracts.

## Constraints in managing information resources

The section above points to the large number and wide variety of available electronic information resources that are relevant to biotechnology research managers. The fast-growing volume and variety of information relating to agricultural biotechnology also raises a number of issues in managing this resource. Despite these constraints, electronic information, especially that available through the Web, is a rich resource for research managers involved in agricultural biotechnology. It takes time, money, and knowledge to mine this resource, but it should be part of every research institute's infrastructure.

## Access

Most electronic information is provided on-line on the Web. Other electronic sources are CD-ROMs and diskettes. While the computers that are needed to access electronic information are usually standard equipment for most research organizations, a connection to the Web is essential for accessing on-line electronic information. Web access is growing rapidly in many developing countries. However, in many countries and institutions, outdated telecommunications policies and infrastructure still hamper access.

## Variety of formats

While the sheer magnitude of electronic information that is available presents a problem, the highly variable ways in which different providers organize and structure their information makes finding the right information even more difficult. Biotechnology research has many aspects and applications, and different databases extract different subsets of information. But even resources that are closely related in content are often incompatible in format.

## Quality

The quality of information provided through electronic resources is obviously of critical importance. However, as most documents and scientific data are not peer-reviewed before they are posted on the Web, it is difficult to assess the quality of the data provided. Procedures for error detection and correction are still under development. Maintaining and updating the information is also an issue. In a rapidly evolving field such as biotechnology, information quickly becomes obsolete. As long-term maintenance is often more time-consuming and costly than the initial set-up, it is sometimes neglected. Also, the quality of technical support and user manuals is highly uneven and sometimes non-existent. These obstacles are major challenges to using and integrating the information.

## Cost

A number of electronic information aspects incur costs. Web access involves initial costs such as additional hardware and software to get connected to the Web, as well as maintenance costs such as charges for using a telephone line and fees charged by the service provider. Costs that are often ignored or not recognized are those to train managers and users in accessing and using these newly available resources. Third, good information comes at a price. Many databases for agricultural biotechnology (scientific abstracts, patent information, etc.) that are sufficiently comprehensive and up-to-date, are usually distributed by commercial providers. On-line searchable databases usually charge an access fee for each password and a fee for each database record viewed. Also, very few CD-ROMs and diskettes are available free of charge.

# Bibliography

BAC. 1996. Data Needs Survey Report Prepared for the African Regional Biosafety Focal Point. Stockholm: Biotechnology Advisory Commission.

Frederick, R.J. 1996. Biosafety Information Exchange: Do we really know how to do it? Paper presented at the UNEP workshop on Follow-up to the International Technical Guidelines for Biosafety. Buenos Aires, Argentina, October 30– November 2, 1996.

King, D. 1996. Developing an Information System. Appendix III in BAC, 1996.

Kornhauser, A. and B. Boh. 1992. Information Support for Research and Development in Biotechnological Applications. *In* Biotechnology Economic and Social Aspects: Issues for Developing Countries, edited by E.J. DaSilva, C. Ratledge and A. Sasson. Cambridge: Cambridge University Press.

Saracevic, T. and M. Kesselman. 1993. Trends in Biotechnology Information and Networks: Implications for Policy. *In* Biotechnology R&D Trends: Science Policy for Development, edited by G.T. Tzotzos. *Annals of the New York Academy of Sciences* Vol. 700: 135-144.

Appendix Table 1. Biotechnology-Related World Wide Web Sites

| Source of information | Web address |
|---|---|
| **Research** | |
| The **Agricultural Genome Information Server (AGIS)** is a cooperative effort between the Department of Plant Biology, University of Maryland, and USDA's National Agricultural Library Genome Informatics Group (GIG). The server provides genome information for agriculturally important organisms. It also includes a number of databases with related information, e.g. on germplasm and plant gene nomenclature data. | http://probe.nalusda.gov:8300 |
| FAO's **Agricultural Information System (AGRIS)** and the **Current Agricultural Research Information System (CARIS)** can be found at FAO's website. AGRIS has collected 3 million bibliographic references to either "official" publications (e.g., journal articles, books) or "unofficial" materials (grey literature, e.g., theses and reports) that are not available through the normal commercial channels. CARIS identifies specific projects on all aspects of agriculture. | http://www.fao.org/agris/default32.htm |
| **CAB International** is a not-for-profit organization with three principal divisions: (a) Publishing, which publishes and markets a wide range of printed and electronic products, (b) Information for Development, which assists developing countries in acquiring and managing scientific information, and (c) Bioscience, which undertakes research and training in biological pest management, biodiversity, biosystematics, and the environment. | http://www.cabi.org/ |
| The **ExPASy Molecular Biology Server** of the Swiss Institute of Bioinformatics is dedicated to analyses of protein sequences and structures. It contains many links to databases, analytical tools, and software and many other resources and servers on molecular biology. | http://expasy.hcuge.ch/ |
| The **Plant Tissue Culture Information Exchange**, maintained by Texas A&M University, provides access to technical information about plant tissue culture. It includes links to relevant literature, supplier companies, and micropropagation companies. | http://aggie-horticulture.tamu.edu/tisscult/tcintro.html |
| **Management issues** | |
| **AgBiotechNet** publishes current information about biotechnology and biosafety for researchers and policymakers worldwide. The site provides access to research developments in genetic engineering and updates on economic and social issues. | http://agbio.cabweb.org/index.htm |

| Source of information | Web address |
|---|---|
| **Biotechnology Industry Organization** is the largest trade organization representing the biotechnology industry in the world. The site includes a media guide to biotechnology; a biotech food products list; a citizen's guide to biotechnology; laws and policies; and a guide to bioethics. | http://www.bio.org |
| The **Biotechnology Information Center (BIC)**, one of 10 information centers of the National Agricultural Library of the US Department of Agriculture, provides access to a range of information services and publications covering many aspects of agricultural biotechnology. | http://www.nal.usda.gov/bic/ |
| The **European Association for Bioindustries (EuropaBio)** represents the interests of 39 multinational members and 14 national associations involved in the research, development, testing, manufacturing, and distribution of biotechnology products | http://www.europa-bio.be/index.html |
| **ISNAR's Biotechnology Service (IBS)** is an independent advisor to developing countries on matters of biotechnology policy and management and on socioeconomic and technical issues. The IBS site is linked to ISNAR's Information and Discussion Forum on Policy and Management Issues in Agricultural Biotechnology. | http://www.cgiar.org/isnar/projects/ibs/index.htm |
| The US **National Agricultural Biotechnology Council (NABC)** is a neutral forum where people can freely exchange ideas on issues of agricultural biotechnology. Leading US and Canadian agricultural research and educational institutions are members of NABC. | http://www.cals.cornell.edu/extension/nabc |
| The **Union of Concerned Scientists (UCS)** monitors and evaluates the agricultural biotechnology and sustainable agriculture policies and regulations of the US Department of Agriculture, the US Food and Drug Administration, and the US Environmental Protection Agency. | http://www.ucsusa.org |

## *Priority setting*

| | |
|---|---|
| Additional details on strategic planning and the Analytic Hierarchy Process (AHP) are found on the homepage of **Expert Choice**. The company's site also provides access to free trial downloads of decision-support software. | http://www.expertchoice.com/ |
| **ISNAR's Discussion Forum on Priority Setting** provides a good introduction to and relevant material on priority setting. Various methods are explained, with relevant cases in which they have been applied. | http://www.cgiar.org/isnar/fora/priority/index.htm |

| Source of information | Web address |
|---|---|
| **Genetic resources and biodiversity** | |
| The objectives of the **Convention on Biological Diversity (CBD)** are "the conservation of biological diversity, the sustainable use of its components and the fair and equitable sharing of the benefits arising out of the utilization of genetic resources." | http://www.biodiv.org/ |
| The **European Cooperative Programme for Crop Genetic Resources (ECP/GR)** is a collaborative program of European countries that is aimed at ensuring the long-term conservation of plant genetic resources in Europe and facilitating the use of these resources. | http://www.cgiar.org/ecpgr/Index.htm |
| The **Genetic Resources Action International (GRAIN)** is an international nongovernmental organization, established in 1990, to help further a global movement of popular action against genetic erosion, one of the most pervasive threats to world food and livelihood security. | http://www.grain.org/index.htm |
| The **Germplasm Resources Information Network (GRIN)** web server provides information on plant, animal, microbe, and invertebrate germplasm available within the National Genetic Resources Program of the Agricultural Research Service of the US Department of Agriculture. | http://www.ars-grin.gov |
| The **Rural Advancement Foundation International (RAFI)** is an international nongovernmental organization dedicated to the conservation and sustainable improvement of agricultural biodiversity and to the socially responsible development of technology useful to rural societies. | http://www.rafi.ca/ |
| The **System-wide Information Network for Genetic Resources (SINGER)** of the CGIAR gives access to information on the germplasm collection held by the CGIAR centers. These collections comprise over half a million samples of crop, forage, and tree germplasm. | http://singer.cgiar.org/ |
| **International collaboration** | |
| The **Agricultural Biotechnology Support Project (ABSP)** builds linkages between developing-country public and private sectors, the US public sector, and the US private sector, which holds much of the technology. It focuses on specific product-oriented research activities in the context of an integrated management scheme emphasizing human resource development in policy as well as technical areas. It gives access to research and policy information for developing-country scientists through a global approach to networking. | http://www.iia.msu.edu/absp |

| Source of information | Web address |
|---|---|
| The **Bean/Cowpea Collaborative Research Support Program (B/C CRSP)** is a research and training program that supports international research partnerships to increase the availability of beans and cowpeas. | http://www.isp.msu.edu/scripts/CRSP.pl |
| The **Center for Tropical Agricultural Research and Higher Education (CATIE)** conducts research and provides postgraduate training in the fields of agricultural research and renewable natural resources in Central America. | http://www.catie.ac.cr/catie/ |
| The **Centre de coopération internationale en recherche agronomique pour le développement (CIRAD)** is a French organization specialized in tropical agricultural research. CIRAD's program covers research, training, and scientific information. | http://www.cirad.fr/ |
| The 16 international research centers of the **Consultative Group on International Agricultural Research (CGIAR)** conduct research and training programs on a range of agricultural commodities in collaboration with a large number of national institutes in developing countries and advanced research institutes worldwide. | http://www.cgiar.org/ |
| The objective of the **Applied Biotechnology Center** of the International Center for Maize and Wheat Improvement (CIMMYT) is to make plant breeding more effective through DNA marker techniques and genetic transformation of maize and wheat and to transfer useful technology to developing countries through training and consulting. | http://www.cgiar.org/cimmyt/biotechnology/biotechnology.htm |
| The **International Centre for Genetic Engineering and Biotechnology (ICGEB)** is an international organization established "to promote the safe use of biotechnology worldwide and with special regard to the needs of the developing world." It includes a database containing over 1700 abstracts and references to biosafety-related research papers. | http://www.icgeb.trieste.it/biosafety/ |
| The **International Centre of Insect Pest Physiology and Ecology (ICIPE)** develops technologies for managing and controlling both harmful and useful arthropods. ICIPE's activities focus on improving and promoting human, animal, plant, and environmental health by interdisciplinary teams working in ecosystems science, behavioral biology, chemical ecology, molecular biology and biotechnology, and social sciences. | http://nbo.icipe.org/ |
| The **International Laboratory of Molecular Biology (ILMB)** conducts and coordinates a research program that brings together experts in molecular biology to facilitate work on the pathogenetic mechanisms of human and animal tropical diseases. It does research on the molecular biology of disease agents and develops vaccines and rapid diagnostic kits to assist in tropical disease control. | http://www.vetmed.ucdavis.edu/centers/ilmbtd.htm |

| Source of information | Web address |
|---|---|
| The **International Service for the Acquisition of Agri-biotech Applications (ISAAA)** transfers and delivers appropriate biotechnology applications to developing countries and builds partnerships between institutions in the South and the private sector in the North and strengthens South-South collaboration. | http://www.isaaa.org/ |
| The **Network on Plant Biotechnology in Latin America and the Caribbean (REDBIO/FAO)** is a regional network established to accelerate the process of adaptation, generation, transfer, and application of plant biotechnology to address crop production constraints and genetic resources conservation in the countries of the region. | http://www.cenargen.embrapa.br/~redbio |
| **Biosafety** | |
| The Web site of the UK **Advisory Committee on Releases to the Environment (ACRE)** includes summary records for experimental releases in the UK and for marketing genetically modified organisms in Europe. It publishes press releases and advice to the secretary of state and ministers. | http://www.environment.detr.gov.uk/acre/ |
| The Belgian **Biosafety Server** covers regulatory information for Europe and other countries, and it provides news about biosafety-related meetings, conferences, and courses. | http://biosafety.ihe.be |
| The **Biosafety Information Network and Advisory Service (BINAS)** is a service of the United Nations Industrial Development Organization (UNIDO), covering global developments in regulatory issues in biotechnology. It has full-text regulations and guidelines from many countries, a library of publications on regulatory policy and biological risk assessment issues. | http://binas.unido.or.at/binas/binas.html |
| The **Biosafety Research and Assessment of Technology Impacts** project is a core activity of the Priority Programme Biotechnology of the Swiss National Science Foundation. It holds scientific reports on safety assessments of recombinant rabies vaccine, foods derived from genetically modified organisms, genetic engineering for plant protection, and environmental and agricultural safety considerations of transgenic crops. | http://www.eurospider.ch/bats/index.html |
| **Biotrack Online** of the Organization for Economic Cooperation and Development (OECD) helps OECD member-country governments and industries notify and assess biotechnology products. It makes the information available to non-OECD countries. | http://www.oecd.org/ehs/bioabout.htm |
| The Australian **Genetic Manipulation Advisory Committee (GMAC)** oversees the development and use of innovative genetic manipulation techniques in Australia to identify and manage safety risks for workers, the community, and the environment. | http://www.dist.gov.au/science/gmac/gmachome.htm |

| Source of information | Web address |
|---|---|
| **Information Systems for Biotechnology (ISB)** provides information to support the environmentally responsible use of agricultural biotechnology products. It posts documents and searchable databases on the development, testing, and regulatory review of genetically modified plants, animals, and microorganisms within the USA. | http://www.isb.vt.edu/ |
| Japan's **Innovative Technology Homepage** of the Ministry of Agriculture, Forestry, and Fisheries gives guidelines and information about the current status of field tests, releases, and commercialization of transgenic plants in Japan. | http://ss.s.affrc.go.jp/docs/sentan |
| The **Joint Research Centre** of the European Commission includes summaries of field trials held in European Union member countries. | http://biotech.jrc.it/ |
| The USDA **Biotechnology Risk Assessment Research Grants** program assists US federal regulatory agencies in making science-based decisions about the safety of introducing into the environment genetically modified organisms in plants, microorganisms, fungi, bacteria, viruses, arthropods, fish, birds, and mammals and other animals. | http://www.reeusda.gov/crgam/biotechrisk/biotech.htm |
| USDA's **APHIS Biotechnology and Scientific Services**, consisting of USDA's Biotechnology Evaluation (BE) and Coordination and Technical Assistance (CTA) divisions and the Animal and Plant Health Inspection Service, and the Biotechnology and Scientific Services, regulate the importation, interstate movement, and environmental release of certain genetically engineered plants and microorganisms. | http://www.aphis.usda.gov/bbep/bp/ |

*Public acceptance*

| | |
|---|---|
| The **European Federation of Biotechnology Task Group on Public Perceptions of Biotechnology** raises public awareness and understanding of biotechnology and encourages public debate. | http://www.kluyver.stm.tudelft.nl/efb/tgppb/home.htm |

*Intellectual property rights*

| | |
|---|---|
| The **European Patent Organisation (EPO)** administers a uniform patent system in Europe and a centralized patent grant system administered by the European Patent Office on behalf of the contracting states. The EPO Web site gives access to the "esp@cenet" patent information service that covers European, Japanese, and worldwide patents. | http://www.european-patent-office.org/index.htm |

| Source of information | Web address |
|---|---|
| On the **IBM Intellectual Property Network**, one can search for US and European patent documents as well as patent applications published by the World Intellectual Property Organization (WIPO). | http://www.patents.ibm.com/ |
| The Swiss-based **International Union for the Protection of New Varieties of Plants (UPOV)** is an intergovernmental organization based on the International Convention for the Protection of New Varieties of Plants, as revised in 1961. The objective of the convention is to protect new varieties of plants by intellectual property rights. | http://www.upov.org/eng/index.htm |
| The US **Patent and Trademark Office (PTO)** promotes intellectual property rights systems in the USA and advises the government on trade-related aspects of intellectual property. Their Web site gives access to bibliographic and full-text databases of US patents. The databases cover the period from 1 January 1976 to the most recent weekly issue date. | http://www.uspto.gov/ |
| The **World Intellectual Property Organization (WIPO)** is an intergovernmental organization headquartered in Switzerland responsible for promoting the protection of intellectual property throughout the world through cooperation among states. It also administers various multilateral treaties on the legal and administrative aspects of intellectual property. | http://www.wipo.int/eng/newindex/index.htm |

## Appendix Table 2. Electronic Journals and Newsletters

| Title | Description | Web or e-mail address |
|---|---|---|
| AgBioForum | Publishes articles on the on-going dialogue on the economics and management of agricultural biotechnology. The purpose of AgBioForum is to provide unbiased, timely information, and new ideas leading to socially responsible and economically efficient decisions in science, public policy, and private strategies pertaining to agricultural biotechnology. | http://www.agbioforum.missouri.edu/ |
| BioLink | Quarterly newsletter of the Agricultural Biotechnology Support Project (ABSP) is available electronically on the Web and by e-mail. | http://www.iia.msu.edu/absp/biolink.html To subscribe by e-mail, contact gibbons3@pilot.msu.edu |
| **Biosafety** | On-line journal published by Bioline prints original papers on the effects of genetically modified organisms and introduced species on people and the environment; the containment of potentially hazardous organisms; biosafety reviews; accounts of risk assessment and hazard analysis; and follow-up monitoring. | http://www.bdt.org.br/bioline/by |
| Biotechnology and Development Monitor | Joint publication of the Department of Political Science of the University of Amsterdam and the Netherlands' Ministry of Foreign Affairs. It reports on local, national, and international applications of biotechnology that are of special interest to developing countries. It has editorials, articles on policies, research and new technology developments, and book reviews. | http://www.pscw.uva.nl/monitor/index.html |
| Electronic Journal of Biotechnology | On-line, international scientific journal publishes papers on all areas related to biotechnology, ranging from molecular biology and the chemistry of biological process to aquatic and environmental aspects to computational applications and policy issues related to biotechnology. | http://www.ejb.org/ |
| Gene Exchange | Covers US government policy regarding genetic engineering, federal applications for field tests, and legislative initiatives for environmental release. The primary focus is on raising awareness about current or potential problems and negative consequences of genetic engineering. | http://www.ucsusa.org/publications/index.html For a text-only version, send message SUBSCRIBE GENEX <your E-mail address> to: genex@ucsusa.org. |
| Intellectual Property & Biodiversity News | Covers items related to biotechnology and biodiversity. Topics include patenting, regulation, legislation and finance; both domestic and international news is covered. Each issue includes a list of resources (publications, centers, etc.), and a calendar of events. | http://www.newsbulletin.org/bulletins/ Send email to: iatp@iatp.org |

| Title | Description | Web or e-mail address |
|---|---|---|
| ISB News Report | Covers research news and information, US regulatory news, international developments, business and industry news, meetings, and so on. A free monthly print and email publication from Information Systems for Biotechnology. | http://www.isb.vt.edu<br>To subscribe, send an email to: news@nbiap.biochem.vt.edu; type *subscribe newsreport [your-name]* in the message field. |
| Plant Breeding News | Published by FAO. Promotes the exchange of information among all interested persons including plant breeders, researchers, policymakers, crop extension experts, NGOs, students, etc. The newsletter is meant to be an informal forum on plant breeding technologies and related issues including announcements, inquiries, technical discussions, un-refereed articles, etc. | To subscribe, send e-mail to mailserv@mailserv.fao.org, leaving the subject blank, and write in the first line of the e-mail message: SUBSCRIBE PBN-L. |

# Glossary

**Abiotic stress**
Nonbiological factors that may cause a damaging effect on plants. Examples include pesticides, temperature and moisture extremes, and soil and water salinity or acidity.

**Accession**
Plant or seed sample, strain or population held in a genebank or breeding program for conservation and use.

**Amplified Restriction Fragment Polymorphism (AFLP)**
A DNA fingerprinting technique that detects distinguishing restriction fragments by means of PCR amplification; can be applied in genetic studies, such as the analysis of germplasm collections or the construction of genetic marker maps.

**Analytic hierarchy process (AHP)**
A multicriteria decision-making approach that uses pairwise comparisons to determine preferences among a set of projects. AHP systematically structures a complex decision problem and analyzes only two elements at a time. The approach incorporates subjective judgments and combines them with hard data.

***Bacillus thuringiensis* (Bt)**
A soil bacterium that produces a protein highly toxic to specific classes of insects. Bacterial spores containing the toxin are used as an environmentally benign pesticide. When ingested, Bt toxin kills certain insect larvae, but it is regarded as harmless to humans, animals, and most beneficial insects, such as bees.

**Biodiversity**
The total variability with and among species of living organisms and their habitats.

**Bioinformatics**
The application of computer technology to the management of biological and genetic information.

**Biological control**
A means of managing pest populations by purposefully introducing natural enemy species, e.g., release of insects that are predatory or parasitic on pest insects, or use of microbial species that cause insect disease.

**Biopesticide**
A pesticide made from biological sources, such as bacterial toxin proteins, plant extracts, compounds derived from animal by-products, etc.

---

© CAB *International. 1999. Managing Agricultural Biotechnology—Addressing Research Program Needs and Policy Implications* (ed. J.I. Cohen)

# Glossary

**Bioprospecting**
The systematic search for and development of new sources of chemical compounds, genes, micro- and macroorganisms, and other valuable products from nature. Managed bioprospecting fosters the sustainable use and conservation of biological resources and promotes the scientific and socioeconomic development of source countries and local communities.

**Bioremediation**
The use of living organisms to remove pollutants from the environment through biological processes, e.g., use of a fungus to detoxify chemically contaminated soil at sawmills; use of a hydrocarbon-metabolizing bacterium to help clean up oil spills.

**Biosafety**
The goal of ensuring that the development and use of transgenic plants and other organisms does not negatively affect plant, animal or human health, genetic resources, or the environment.

**Biotechnology**
A variety of techniques that use living organisms or substances from them to make or modify products, improve plants, animals, and microorganisms, or develop novel organisms for specific uses.

**Biotic stress**
Physiological stress imposed by one organism on another, causing deleterious effects. Examples include physical damage or disease caused by pests and pathogens, crowding and competition for limited resources, and predation.

**Center of diversity**
The geographic region in which the greatest variability of a crop occurs. A primary center of diversity often coincides with the species' center of origin; secondary centers of diversity are regions of subsequent spread and diversification of the species.

**Center of origin**
The geographical area in which a species or taxon first arose.

**Conservation**
The judicious use and management of nature and natural resources in order to preserve them for the benefit of human society.

**Cost-benefit analysis**
The appraisal of an investment project which includes all social and financial costs and benefits accruing to the project.

**Cryopreservation**
The long-term storage of living cells at ultra-low temperatures, usually in liquid nitrogen. This conservation method is increasingly used in the management of crop plant genetic resources.

**Deoxyribonucleic acid (DNA)**
The chemical name of the linear molecule that encodes genetic information in living organisms.

**DNA delivery**
Introducing isolated DNA directly into cells or tissues by various chemical, physical, or biological methods.

**DNA sequencing**
Determination of the order of individual subunits comprising a DNA molecule.

**Economic surplus analysis**
A refinement of cost-benefit analysis, measuring the economic benefits associated with research alternatives. Economic surplus analysis shows to what extent research-induced reductions in production costs may reduce market prices and thus change the distribution of benefits between consumers and/or producers of a commodity. It can also be used to show how economic policy interventions, such as commodity price ceilings or subsidies, distort the welfare gains obtained from research.

**Ecosystem (agricultural)**
The organisms and abiotic factors with which they interact in a field or other portion of an agricultural enterprise.

**Electrophoresis**
A laboratory technique for separating molecules in a matrix such as agarose or polyacrilamide gels, according to their electrical charge and size.

**Enzyme**
A protein that accelerates a specific biochemical reaction, without itself being destroyed.

**Ex ante evaluation**
An assessment of alternatives before any are undertaken.

**Ex post evaluation**
An assessment of alternatives after they are completed.

**Ex situ conservation**
Literally "out of place;" conservation outside the original or natural habitat, e.g., in a genebank.

**Familiarity**
Having enough information and experience to predict behavior, judge safety, or make decisions. For GMOs, a relatively low degree of familiarity may be compensated for by appropriate management practices. Familiarity can be increased as a result of a trial or experiment. This increased familiarity can then form a basis for future risk assessment

**Farmer's rights**
Rights arising from the past, present, and future contributions of farmers in conserving, improving, and making available plant genetic resources, particularly those in centers of origin/diversity.

**Gene**
The fundamental physical and functional unit of heredity; the segment of a DNA molecule made up of ordered subunits (nucleotide base pairs), which encode information for producing a specific product or have an assigned function.

**Genebank**
The storage facility where germplasm is stored in the form of seeds, pollen or tissues cultured in vitro. A field genebank is an organized collection of plants growing in the field.

**Geneflow**
The exchange of genes between populations of related and sexually compatible organisms; in plants, usually via transfer of pollen.

**Genetic engineering**
The process of isolating genes from an organism, manipulating them in the laboratory, and inserting them into the same or another organism; the use of recombinant DNA methods to alter or improve plants, animals, and microorganisms.

**Genome**
All the genetic material in the chromosomes of a particular organism; the full complement of functional and nonfunctional genes, regulatory elements, and other DNA sequences.

**Genomics**
The study of genes and their functions. Genomics allows genetic information from one organism to be directly compared with that of other organisms

**Germplasm**
The total genetic variability, represented by germ cells or seeds, available to a particular population of organisms.

**Green Revolution**
The technological advancements in developing-country agriculture after 1960; usually refers to the development and spread of high-yielding plant varieties, increased application of chemical pesticides and herbicides, and use of fertilizers and irrigation techniques.

**Haploid**
Cell or organism that contains one set of chromosomes.

**Hybrid**
The offspring of a cross between two individuals that are not genetically identical.

**Innovation**
A new or improved product, process or service; may concern high technology and elements of common life.

**In situ conservation**
Preservation of genetic resources within ecosystems and natural habitats; for domesticated or cultivated species, in the surroundings where they developed distinctive properties.

**Integrated pest management**
An ecologically based multipart strategy to control pests that combines such tactics as crop rotation, tillage methods, judicious and targeted application of chemical controls, deployment of biological control agents, etc.

**Intellectual property rights**
A broad term for the various rights provided by law for the protection of economic investment in creative effort, such as patents, plant variety protection, and trademarks.

**Internet**
A global network connecting millions of computers. The Internet has hundreds of millions of users worldwide, and the number is growing rapidly. Nearly all countries are now linked to the Internet in some way and exchange information, data, knowledge, news, and opinions.

**Introgression**
Stable incorporation of genes transferred from one genome into another.

**In vitro**
Literally "in glass;" loosely applied to biological procedures carried out in a laboratory setting, such as protein synthesis by in vitro translation; also cultural practices carried out under sterile conditions.

**Isozymes**
Multiple, physically distinct forms of an enzyme, each encoded by one of a family of highly similar genes.

**Landrace**
A crop cultivar or animal breed that evolved with and has been genetically improved by traditional agriculturalists, not influenced by modern breeding practices.

**Management**
The judicious use of means and resources that together help achieve a defined goal.

**Marker**
An identifiable physical location on a chromosome (e.g., restriction enzyme cutting site or gene) whose inheritance can be monitored; used to track the inheritance of specific traits whose genes are physically linked to the identifiable site.

**Material transfer agreement (MTA)**
A form of intellectual property protection applied to material not generally protected by patents; used to set out the terms or conditions for using the material obtained under such an agreement. MTAs are used by most international agricultural research centers for the exchange of genetic resources they hold in trust.

**Mathematical programming**
A problem-solving technique aimed at maximizing the expected benefits from a research program. Instead of using simple scoring, this technique helps identify an optimal research portfolio based on multiple goals and resource constraints.

**Molecular biology**
The branch of science that studies the molecular structure and function of cell components.

**Monitoring**
Surveillance to ascertain the status of an experiment, the extent of compliance with pre-established requirements, or the degree of deviation from an expected norm.

**Mutation**
Any change in the genotype of an organism occurring at the gene, chromosome, or genome level.

**National agricultural research organization (NARO)**
The aggregation of national agricultural research institutes into one formal organization with an apex authority governing the whole organization.

**National agricultural research system (NARS)**
The research system comprised of all entities responsible for organizing, coordinating, or executing agricultural research within a country to contribute to the development of agriculture and maintenance of natural resources.

**Native (plants, etc.)**
Plants, animals, fungi, and microorganisms that occur naturally in a given area or region.

**Net present value**
The discounted value of the net benefits of use of a resource.

**Nitrogen fixation**
Biological assimilation of atmospheric nitrogen to form organic nitrogen-containing compounds.

**Parasite**
An organism that derives all its needed nutrients from a host organism, usually without killing the host.

**Patent**
A legal instrument granting an inventor the exclusive right, for a limited time, to exploit the invention in exchange for disclosure about it.

**Pathogen**
Any organism able to cause disease in a specified host.

**Plant breeders' right (plant variety right)**
A legal instrument granting the developer of a plant variety the exclusive right to market it for a limited time. Varieties protected by such legislation may be used by others in the development of new varieties.

**Plant genetic resources**
The genetic material of plants, including modern cultivars, landraces, and wild relatives of crop species, that serves as a resource for present and future generations of people.

**Plasmid**
A circular piece of DNA not part of a chromosome, found in bacteria. Plasmids are the principal tool for inserting new genetic information into microbes or plants.

**Polymerase chain reaction (PCR)**
Multiplying a particular DNA segment in repeated cycles. The "copies" made in a previous cycle are used as "originals" or templates in the next cycle. For example, PCR enables forensics experts to do DNA testing on very small blood samples.

**Priority setting**
The process of choosing among different sets of research activities to make the most effective use of available resources.

**Random Amplified Polymorphic DNA (RAPD)**
A technique for amplifying anonymous stretches of DNA using PCR.

**Recalcitrant seed**
Seed that cannot be dried and so cannot be stored at low temperatures without damage.

**Recombinant DNA technologies**
Procedures used to combine DNA segments that previously were separate; the technical basis for genetic engineering, cloning, and some types genetic analysis.

**Refuge area**
Area within a field of a transgenic Bt crop, which is planted to a non-Bt variety of the same crop; part of a strategy to delay the emergence of pests resistant to the Bt toxin by providing an area for habitation and immigration of susceptible insects.

**Resistance management**
Use of agronomic and integrated-pest-management practices to delay the emergence of resistant individuals within a population of pests targeted for control by a pesticide; essential to the proper use of chemical pesticides, as well as biological pest controls such as Bt.

**Restriction fragment length polymorphism (RFLP)**
Differing patterns of DNA fragments that distinguish individuals, produced by cutting DNA with restriction enzymes and analyzing the size of the fragments; can be used as a tool in breeding programs to monitor the inheritance of genes associated with a particular fragment.

**Ribonucleic acid (RNA)**
A nucleic acid synthesized from a DNA template and usually having one of three functions: (1) encoding "instructions" for the synthesis of a protein, (2) forming a structural component in a ribosome (the protein synthesizing machinery of a cell), or (3) acting as a shuttle vehicle for delivering protein subunits to a ribosome.

**Risk assessment**
The evaluation of a proposed activity or event to identify potential hazards, the probability of their occurring, and the estimated degree of resulting harm.

**Risk management**
The measures used to minimize a potential risk or reduce it to an acceptable level; steps taken to ensure the desired degree of safety.

**Safe**
Found, with reasonable certainty, to pose acceptable or negligible risk to human health or to managed or natural ecosystems.

**Simulation model**
A tool to generate a set of research priorities that best meet the objectives of a research system; in it, mathematical relationships among variables are exposed to different scenarios to assess the best outcome.

**Species**
A category of biological classification comprising related organisms or populations potentially capable of interbreeding.

**Strain**
A group of individuals from a common origin; generally a more narrowly defined group than a variety.

## Substantial equivalence
The concept that existing organisms used as food, or as a source of food, can be used as the basis for comparison when assessing the safety of a food or food component that has been modified or is new.

## *Sui generis* system
A Latin term literally meaning "of its own kind of class." WTO member states that do not allow patents on plant varieties or plants in general have to provide for the protection of plant varieties by an effective *sui generis* system. Although the TRIPs agreement does not give any details on what elements this effective *sui generis* system would have to include, certain minimum requirements are indicated that such a system would have to fulfil.

## Sustainable use
The use of biological and nonbiological resources in a way and at a rate that does not lead to the long-term decline of biological diversity.

## Taxonomy
A branch of science dealing with the classification of organisms.

## "Terminator technology"
The name given by critics to a technical method—officially named "Technology Protection System"—that incorporates into seeds a mechanism for making second-generation seeds nonviable.

## Tissue culture
The propagation of tissues in a laboratory environment under strict sterility, temperature, and nutrient requirements.

## Transformation
Introduction and assimilation of DNA from one organism to another via uptake of naked DNA.

## Transgenic animals or plants
Animals or plants whose hereditary DNA has been modified in a laboratory by the addition of DNA from a source other than parental germplasm, using recombinant DNA techniques.

## Variety
A subdivision of a species below subspecies.

## Vector
A carrier or transmission agent. In the context of recombinant DNA technology, a vector is the DNA molecule used to introduce foreign DNA into host cells. Recombinant DNA vectors include plasmids, bacteriophages and other forms of DNA.

## Virus
Microscopic particle that contains genetic information, but must invade a cell to reproduce.

## Weediness
Characteristic of a plant that allows it to invade, persist, and become established in a new place; traits that make plants a pest or nuisance in an agronomic setting.

**World Wide Web**
A subset of the Internet; a collection of millions of electronic documents that are linked together on many thousands of "servers" worldwide and that can be read on computers.

# Index

## A

AARD
    *See* Agency for Agricultural Research and Development
abiotic  70, 74, 103, 129
    stress  17
ABSP
    *See* Agricultural Biotechnology for Sustainable Productivity
accession  105-106, 287
accountability  51, 144, 150, 203, 225, 228-229
added value  279
adoption rate  46, 58
Advisory Committee on Novel Foods and Processes (UK)  18
AFLP  106
Africa  116, 156, 269
Agency for Agricultural Research and Development (AARD)  26, 37, 66
Agreement on Trade-Related Aspects of Intellectual Property Rights  18, 209, 214, 220, 222-224, 240-241, 279-280, 283
Agricultural Biotechnology for Sustainable Productivity project  72-73, 110, 117, 135-136, 262-270
agricultural biotechnology policy seminars  8, 12, 40, 83, 85, 115, 155
agricultural development  19, 21, 27, 41-42, 66, 223, 258
Agricultural Genetic Engineering Research Institute (AGERI) (Egypt)  110, 263, 266, 268
agricultural gross domestic product
    vs research expenditure  31
agricultural industry  93
agricultural productivity  1, 47, 66, 68, 80, 164, 272
agricultural research policy  20

*Agrobacterium*-mediated transformation  185
agrochemicals  93
agroindustry  70, 272
alien genes  16
allergens  18
American Association for the Advancement of Science  133
analytic hierarchy process (AHP)  9, 48-51, 54-55, 63-64
    in private-sector reseach  54
antibiotechnology groups  174
antibiotic resistance  18, 181, 185, 188
applicant  156-157, 159-160, 162, 178, 215-218
appropriation  224, 247
ARBN
    *See* Asian Rice Biotechnology Network
Asia  xvii, 84, 116-117, 186, 190-191, 269, 273
Asian Development Bank  87
Asian Pacific region  276
Asian Rice Biotechnology Network  73, 89-90, 117
Association of Biotechnology Companies  135
Association of South-East Asian Nations  224
auditing  83
Australia  89, 157, 166, 172, 215, 229, 231
Australian Centre for International Agricultural Research  73
Autonomous National University of Mexico  26, 35
Autonomous University of Chapingo (Mexico)  35
Autonomous University of Morelos State (Mexico)  35

## B

*B. sphaericus* 116
*Bacillus thuringiensis* 12, 17, 69, 74, 116, 132, 141, 157, 169, 185-186, 191, 195-204, 215, 263, 268, 281
Bangladesh 89, 284
Bayh-Dole Act 264, 266-267
bbiosafety committee
    National Committee on Biosafety of the Philippines 167
Berne Convention for the Protection of Literary and Artistic Works 222
bilateral initiative 128
biodiversity 17, 22, 89, 92-93, 103, 151, 160, 168, 187, 189, 194-195, 259, 261, 269
biofertilizer 25
biofora 178
bioinformatics 1, 90, 289
bioinsecticides 25, 35
biological control 43, 93, 168
biological diversity
    sustainable use of 92
biopesticide 12, 35, 116, 142
bioprospecting 92
bioprospecting framework 94-95
bioprospector 98
bioremediation 164
bioresources 93
biosafety 2-3, 8, 10, 27, 45, 64, 75, 85, 111, 115, 129, 132-133, 142, 145, 149, 161, 166, 168, 177, 195, 247, 259, 262, 290
    education and training 162
    guidelines 10, 156-157, 168, 195
    information 286
    management of 155
    policy 156-157
    procedures 262
    protocol 130, 158
    regulation 130, 158
    regulators 170
    research 161
    review 10, 115, 117, 130, 156
    reviewers 161, 173
    system 156, 159, 167
    training 162
    *See also* biosafety committee
Biosafety Commission (Indonesia) 75
biosafety committee 134, 158
    institutional 158-159, 162, 167, 169-170, 173
    Kenya Biosafety Committee 27
    Mexican Biosafety Committee 27
    national 159-160, 166
    national and institutional 163
    National Biosafety Committee (Zimbabwe) 27
    National Technical Biosafety Committee (Brazil) 247
BioServe 12
Biotechnology in Europe, Manpower, Education & Training 80
biotechnology research indicators 24
biotic 70, 74, 103
Bogor Agricultural University 26, 37
bottom-up 42, 68, 70, 74
brand name 282
*Brassica* oilseed 280
Brazil 116, 229, 240, 250, 255, 269
breeder's exemption 210, 213
breeder's privilege 211
breeders 16, 21, 64, 93, 103, 132, 181, 190, 192, 210-213, 223, 245
breeders' rights 26-27, 213-214, 229-230, 251
Budapest Treaty 216
budget cycle 70
business development 75, 94-97, 272-273, 276
    integrating R&D with 274
    strategies for 276
business partners 96, 246, 273-274

## C

CAB International 290, 293
Canada 182, 280
capacity building 138, 141, 170
    human 54
    institutional 54
case studies 2, 8, 10
cassava 113
CBD
    *See* Convention on Biological Diversity
center
    of diversity 10, 160, 223
    of origin 160, 189, 223

Center for Research and Advanced Studies
	(INVESTAV) (Mexico)  26, 35,
	110, 117
Central America  269
central control of seed  202
Central Research Institute for Food Crops
	(CRIFC)  67-68, 72-73, 263
centralization  44
CGIAR
	See Consultative Group on International
		Agricultural Research
Chile  54, 250, 255
China  89, 273, 288
Chulalongkorn University  231
CIMMYT
	See International Maize and Wheat
		Center
Coconut Research Institute (Sri Lanka)  87
collaboration  12, 27, 32, 72, 85, 94, 140,
	178
	between public and private sector  130,
		145, 229
	between researchers  61
	contracts relating to  232
	in bioprospecting  93, 95
	in managing genetic resources  105-106
	industrial  274
	initiatives in international  120-127
	international  8, 11-12, 88-89, 110-119,
		130, 257, 261
	maximizing benefits of  115
	with industrial sector  273
	with industry  274
	with international biotechnology
		programs  85
	with multinational and regional
		companies  273
	See also Indo-Swiss Collaboration in
		Biotechnology
collaborative opportunities  110
collaborative research  12, 136
Colombia  250, 255
Colorado State University  129-130
commercial  25, 99, 133, 210
	assessment  219
	Bt transgenic plants  200
	collaborators  232
	companies  170
	contracts  245
	crops  21
	cultivar development  133
	development  85, 274
	electronic information providers  288
	exploitation  212, 231
	GMO crops  21
	growing of GMOs  173
	institutions  135
	marketing  211, 213
	opportunities  130
	organizations  88
	partners  274
	partnerships  262
	potential  267
	potential of inventions  232
	product development  117
	research partnerships  129
	returns  229
	sector  13, 29, 132, 134, 138, 272
	transgenic crops in Japan  181
	unions  224
	use of first GMO  15
	use of GMOs  158
	utility  273
	utility of inventions  272
	varieties  102-103
commercialization  71-72, 130, 134, 136,
	141, 182, 232-233, 268, 273, 282,
	284
	of GMOs  181
	of oilseed *Brassica*  282
commercializing  262
comparative judgments  56
competitive financing mechanisms  12
conduit of skills  85
confidential information  170, 221-222,
	235-236, 265
confidentiality  136, 235, 243, 245, 259, 277
	agreement  237, 277
	from visitors  243
	obligations  259
conservation  225
	ex situ  102
	in genebanks  102
	in vitro  103
	of biodiversity  22
	of biological diversity  92
	of food crops gene bank  74
	of germplasm  189
	of root crops germplasm  74
	of soil  16
	of the environment  240, 245
	priorities  104

Consultative Group on International
    Agricultural Research 31, 111, 114,
    129, 131-133, 136, 229-230,
    249-250, 252-254, 257, 259-260,
    287
consumer activism 196
consumer awareness 179
consumers 40, 180
consumers' organizations 158, 181
containment 75, 159, 169, 189-190, 262
contract research 12
Convention on Biological Diversity (CBD)
    10, 92, 95, 99, 155, 158, 168, 209,
    223, 241, 270
conventional (plant) breeding 16, 45, 180
conventional breeding 10, 102, 185
    vs genetic engineering 185
copyright 210, 222, 235-236, 242, 275
Cornell University 73, 244, 263
corporate control 18
cost 21, 44-46, 50, 54, 63, 87, 102, 105,
    136, 157, 163, 178, 190, 198-199,
    203, 216, 219, 233, 244, 275, 283,
    286
    of electronic information 291
    of human resource development 162
    of maintaining patents 232
Costa Rica 94, 116, 138, 250, 255, 269
cost-benefit analysis 93, 225
Côte d'Ivoire 116
cotton 15
Crawford Fund xvii, 89
critical mass 11, 82, 86-87, 163, 273
Crop Germplasm Resource Information
    System (CGRIS) (China) 288
cross-pollination 189
cryopreservation 103

# D

decision makers 7-8, 12, 31, 42-43, 54-55,
    86, 209
decision making 8, 40-42, 55, 156
    for patents 217
    framework for 7
decision-making framework
    phases 9
defensive patenting 250
Delta & Pine Land 273
Department for International Development
    (DfID) 105, 134

Department of Biotechnology (DBT) (India)
    140
Development Planning Agency (Bappenas)
    (Indonesia) 70
disclosure of inventions 216
discovery 44, 97, 196, 214, 232, 267-269,
    272, 274-275
DNA 93, 132, 179, 185
    delivery method 187
    marker technology 90
    markers 106
    sequencing 188
drought 16, 66, 103, 129

# E

Earth Summit 241
economic impact 64, 100, 135
economic surplus method 46-50
ecosystem 10, 96, 155, 161, 184, 195
    diversity 93
education 96, 162, 178, 198, 202, 269-270,
    287
educational activities 178
efficiency 40, 43-44, 46-47, 70, 81, 129,
    184-185
Egypt 110, 116-117, 138, 264, 266
electrophoresis 106
Embrapa 240, 258
Empresa La Moderna 27
end users 8-9, 11, 58, 133, 198-199,
    201-203, 272
environment 39, 46, 75, 96, 155, 159, 168,
    173, 182, 194, 198, 245, 261
environmental
    activism 196
    activists 15, 20
    analysis 135
    contamination 197
    criteria 56
    diversity 66
    groups 158, 179
    implications of GMOs 19
    management 22
    organizations 174
    protection 17, 19
    risks 43, 156, 197
    safeguards 158
    safety of GMOs 177
    stresses 16
    sustainability 1, 43

Index                                                                                                                                                   315

enzyme   97, 188, 272
equity   43, 46
ETH
    *See* Federal Institute of Technology
Europe   114, 161, 182, 219, 281
European Patent Convention   216, 219
European Union   224
evaluation criteria   43
ex ante
    analysis of research benefits   46
    assessment of biotechnology products   10
    evaluation   45, 63
    rate of return   47
    socioeconomic analysis   10
ex post
    evaluation   45
    socioeconomic analysis   10
ex situ   96, 98, 103, 106
Expert Choice©   55
extension   22
extensionists   27, 158, 202

## F

familiarity   157, 177
FAO   293
    *See* Food and Agriculture Organization of the United Nations
farmer's exemption   213
farmer's privilege   211, 213
farmers   1, 9, 13, 16, 18, 21, 42-43, 102, 116, 128, 133, 187, 189, 191, 194, 198, 213, 247, 252
farmers' rights   223
Federal Institute of Technology (ETH) (Switzerland) 54, 140
feedback   145, 173, 204
    mechanisms   156, 160
    on biosafety guidelines   169
    on consumer seminars   179
    on market changes   274
    procedural   162
    scientific   161
field-testing   17, 19, 129, 133, 135, 158, 185, 189, 266
field tests   135, 158, 161, 262
field trials   15, 110, 117, 178, 212, 276
financial
    constraints   12
    resources   25

    rewards   81
food
    prices   21
    safety   18-20, 157, 178, 181, 198
    security   41, 43, 46, 66, 68, 118, 171
    self-sufficiency   66
Food and Agriculture Organization of the United Nations (FAO)   54, 104, 118, 180, 223, 288
Frankenfoods   19
fungicides   194

## G

gene transfer   19
genebanks   103, 229, 250
geneflow   10, 160, 171
General Agreement on Tariffs and Trade   241, 270, 280
genetic
    building blocks   102
    code   16
    diversity   10, 93, 105
    engineering   1, 29, 32, 73, 103, 129, 166, 280
    engineering of rice   184
    interchange   17
    mapping   132
    markers   250
    recombination   10
    stability tests   262
    structure   105
    transfers   16
    transformation   42, 102, 179, 284
    variance   93
genetic resources   73
    managing   102
    preserving and managing   200
genetically modified food   177
genetically modified organisms (GMOs)   10, 15-17, 154-155, 157, 168, 174, 195
    evaluation of   159
genetics
    molecular   87
genomic
    databases   67
    research   1
genomics   97
geographic marks   221

banks 98
long-term conservation 105
global insecticide market 197
Global Plan of Action for the Conservation and Sustainable Utilization of Plant Genetic Resources for Food and Agriculture 103
GMO
   *See* genetically modified organisms
Green Revolution 19, 180
gross domestic product 25
group decision making 55
guiding principles 250

# H

haploid 132, 280
herbicide tolerance 16
herbicides 194
human
   capacity building 54, 61, 262, 271
   health 45, 141, 155, 162, 181, 197, 261
   health benefits of Bt transgenic plants 198
   nutrition 66, 184
   resource development 11, 64, 73, 112, 114-115, 132, 140, 143, 162, 281
   resource development and management 80
   resources 7, 11, 25, 67-68, 72, 74, 81, 129, 250, 267, 280
hybridization 19
hybrids 18, 45, 280, 284

# I

IBS
   *See* ISNAR Biotechnology Service
impact 10, 12, 24, 41, 43-44, 58, 75, 86, 99, 119, 131, 159, 161, 169, 177, 194, 275
   of IPR 225
   analyses 135
in situ 96, 106
   collections 98
   germplasm conservatories 189
in vitro 105
   conservation methods 103
   culture 103
   germplasm preservation 74
   propagation 74
   selection 129

in vitro culture 105
INBio 99
incentive 33, 85, 95, 98, 202-203, 225, 229
   to researchers 233
   to the private sector 13
India 89, 116, 140, 213, 223, 231, 279
Indian Institute of Technology 140
indicators 83
individual judgments 45
Indonesia 25, 37, 66, 83, 89, 116, 138, 229, 264, 269-270
Indo-Swiss Collaboration in Biotechnology (ISCB) 114, 140
industrial applicability 215
industrial sector 273
industry 229
information
   need for 22
information management 96
infrastructure 7, 74, 80, 89, 97, 104, 111, 116, 118, 140, 143, 170, 282, 284
   regulatory 242
innovation 1, 11, 19, 21, 83, 131, 209, 214, 241, 250, 265
   system 143
   third-party 258
insect resistance 16, 113, 181, 197, 199
insecticide 25, 194
insecticide use 197
Institute of Molecular Agrobiology 272
institutional 7
   accountability 228
   and policy development 26
   biosafety training 162
   capacity 116, 134
   decision criteria 56
   development 54, 89
   intellectual property committee 243
   intellectual property policies 271
   linkages 64
   policy for managing IP 242
   priorities 116, 131
   priority setting 41
   restructuring 72
   *See also* biosafety committee
integrated pest management 199, 204
integrated value chain 148
intellectual property 75, 85, 87, 135, 272
   awareness 242
   coordinator 231
   infrastructure 229

management of 149, 228, 240, 261
ownership by a NARO 229
policies 269
policy 231
protection 2, 134
statute 234
strategy 246
third party 251
intellectual property rights (IPR) 104, 240, 272
  accountability of NAROs for 228
  affirm ownership of 230
  as an IBS activity 115
  as an incentive for collaborative research 95
  as part of legal framework 45
  costs and benefits of 225
  in ABSP project 262
  in commercial product development 117
  in decision-making framework 8
  in joint ventures 279
  in negotiating biodiversity prospecting contracts 99
  integration with technical dimensions 129
  internships 73
  involvement of USAID 132, 136
  management of 111, 145, 231, 290
  regimes in biodiversity-rich countries 98
  -related technology transfer 75
  scientists' "perceived need" for 11
  statutory 238
  study of inputs protected by 249
  system 209
  training in 133
  TRIPs vs CBD 209
  use by industry 229
  use of internships 134
  *See also* Agreement on Trade-Related Aspects of Intellectual Property Rights
interbreeding 17
interdisciplinary teams 94, 98
interest groups 174
International Center for Agricultural Research in the Dry Areas (ICARDA) 230
international collaboration
  *See* collaboration

International Crops Research Institute for the Semi-Arid Tropics (ICRISAT) 230
International Food Policy Research Institute (IFPRI) 13
International Livestock Research Institute (ILRI) 114, 123
International Maize and Wheat Improvement Center (CIMMYT) 290
International Plant Genetic Resources Institute (IPGRI) 87
International Potato Center (CIP) 229
International Rice Research Institute (IRRI) 73, 89-90, 102, 117, 166, 184, 221
International Service for National Agricultural Research (ISNAR) 2, 7-9, 12, 25, 40, 54, 73, 83, 112, 147, 155, 160-161, 249, 260
International Union for the Protection of New Varieties of Plants (UPOV) 210-214, 224, 241, 243
  member states 211
Internet 87, 164, 180
internships 73, 129, 134
introgression 189
inventions 27, 75, 117, 210, 214-215, 217-219, 222, 233, 254-255, 265-268, 272, 283
  disclosure of 216, 232
inventiveness 214
inventors 245, 264
IPR 118
  continuing debate on 223
IRRI
  *See* International Rice Research Institute
ISNAR
  *See* International Service for National Agricultural Reseach
ISNAR Biotechnology Service (IBS) 25, 40, 73, 112, 115-116, 147, 155, 161
isozymes 106

# J

Japan xvii, 73, 115, 166, 174, 219, 281, 284
joint programs 21
joint research activities 141
joint venture 12-13, 274, 276, 279
  agreement 265
  management issues 284
Jomo Kenyatta University of Agriculture & Technology (Kenya) 36

## K

Kasetsart University 117
Kenya 25, 36, 83, 116, 138
Kenya Agricultural Research Institute
(KARI) 27, 263

## L

labeling 20, 203
landrace varieties 102
Latin America 12, 116, 156, 249, 269
lawyers 94, 98-99, 244, 246, 285
legal framework 45, 54, 64, 223, 242, 246, 255, 264
level of biotechnology development 32
license 251-252, 275
    agreement 268
licensing 202, 245, 247
    arrangement 18
linkages 64
    between public and private sector 68
    international 82
    with commercial organizations 85, 88
    with other research institutes 233
    with regulatory bodies 134
livestock research 32, 112, 114
local research capacity 111

## M

macropolicies 95
maize 15, 17, 72-73, 113, 188, 229, 266
Malaysia 89, 230
management
    challenges 12, 94, 143, 156, 186, 262
    challenges in formulating and implementing biosafety 169
    of natural resources 148
    seminars 135
    structure 283
    tasks 86
marker gene 18, 160, 187, 250, 255, 265
market shares 275
marketing 275, 280
    analyses 135
    managers 274
    network 282
Master's Program in Biotechnology 85
material transfer agreement 104, 210, 222, 224, 230, 245, 251-252, 257, 259, 267-268, 277
mathematical programming 47

Mercosur 224, 247
metabolism 16
Metropolitan Autonomous University (Mexico) 35
Mexico 25, 35, 83, 116, 250, 255
Michigan State University (MSU) 72, 138, 262-263, 270
micropropagation 12, 43, 113, 116, 132, 262-263
midcareer scientists 88
Middle East 266
molecular biology 1, 19, 25, 73, 84, 87-88, 90, 92, 103, 170, 179, 250, 272
molecular markers 42, 45, 73, 93, 103, 105, 113
Monsanto 110, 202, 273
Morocco 264, 270
MSU
    See Michigan State University
MTA
    See material transfer agreement
multidisciplinary 279
    approach 87
    teams 94, 98
multinational 21, 158, 247, 254, 265, 275-276
    companies 16, 158, 273
multinationals 20-21
mungbean 72
Murdoch University 231
mutation 16, 190

## N

NAFTA
    See North American Free Trade Agreement
national agricultural research organizations (NAROs) 82, 131, 228-232, 249-250, 272
    Latin American 255
    patents 257
national agricultural research systems (NARS) 40-41, 81, 105, 133, 192
    leaders 51
    scientists 186
National Biodiversity Institute (INBio) 94
national capacity 110
national funding 87
National Institute of Agricultural Research (INIA) 54

national policies  8, 21, 168
National University of Science and
    Technology (Zimbabwe)  36
net present value  43
Netherlands  27, 85
network  111, 129, 135
networking  132, 144-145
    international  88
NGOs
    *See* nongovernmental organizations
nitrogen fixation  93
nongovernmental organizations (NGOs)
    149, 169, 172, 247
nontarget organisms  186
nontransgenic plants  199
nontransgenic rice  189
North American Free Trade Agreement
    (NAFTA)  27, 224, 270
notebook  277
novel  214
novelty  212, 215

## O

Ohio State University  187
open-pollinated crops  281
option to license agreement  268
Organisation for Economic Co-operation and
    Development (OECD)  177
ownership  183, 264
    at the project level  150
    of biological resources  95
    of current and future technologies  266
    of genes and processes  265
    of intellectual property  228-230, 267
    of IPR  283
    of priority-setting decisions  63
    of research  86
    of research tools  265
    of technologies  252

## P

Pakistan  89
parasite  114
parastatals  25
Paris Convention  217, 240
participation  42
patent  11, 27, 81, 117, 130, 209, 242, 251,
    267, 275, 280
    attorney  219, 275

    covering genes and transformed plants
        217
    law  75, 217
    pipeline  246
    protection  211, 214
    rights  18
Patent Cooperation Treaty  216, 219
patenting  245
pathogen  17, 132, 160, 168, 194, 296
PCR  114, 143, 188, 283
peanut  72
peer review  58, 128, 135-136, 145
pest  103, 195
    adaptation  194
    control  132
    resistance  11, 194
    resistance management  195
    resistant strains  17
pesticides  16
pharmaceutical industry  93
Philippine Biosafety Guidelines  169, 189
Philippines  89, 166
physical resources  25
pilot project  99
planning cycle  70
plant
    breeding  16, 45, 180, 280
    disease  132
Plant Genetic Systems  280
plant variety rights (PVR)  11, 210, 228
    duration  212
    registration process  212
plasmid  187-188, 217
policies  8
policy seminars
    *See* agricultural biotechnology policy
        seminars
policymakers  1, 8, 27, 156, 199, 201, 264,
    269
political
    considerations  42
    objectives  158
    pressure  198
pollution  194
postgraduate programs  84
potato  113
poverty  21
    alleviation  144, 148
priority date  217-218
priority setting  9, 27, 53, 68, 128
    appropriate method for  49

basic steps  42
levels  41
methods  40
structure  41
private
   companies  9
   industry  95, 131
   investment  12
   seed industry  242
private sector  8, 12-13, 21, 24, 27-28, 54, 56, 68, 82, 92, 95-96, 100, 111, 117, 132, 136, 145, 158, 178, 196, 209, 225, 248-249, 261, 279, 281
   agreements with  259
   participation  82
   partnerships  115
   scientists  159
   transferring technologies to  269
privatization  133
Proagro Seed Company Ltd  280
Proagro-PGS India Ltd  280
productivity  22, 43, 81, 131, 134, 186-187, 189, 192, 262, 281
   loss of  162
   of biotechnology research  41
   of crops  103
   of research  67
products  10, 12, 16, 26, 68, 74-75, 81, 92, 94, 98, 115, 133, 142, 149, 157, 255, 266, 272, 275
   building public acceptance of  158
   commercialization of  72
   delivery  8, 13
   development  8, 141, 144-145
   distribution to end users  138
   impact of  10
   industrial  17-18
   public acceptance of  45
   public perception in introducing  53
   sale of  31
   value-added  281
profits  283
program management  11
project
   impact  58
   management  89
   manager  86, 135
propagation  74, 212
proprietary
   genes  247
   information  261

rights  249
science  249
technology  117, 245, 249, 267
protoplast transformation  185
public
   acceptance  1-2, 21, 43, 45, 53, 62, 158, 174, 177, 186, 192, 247
   awareness  1
   comment  172
   concern  174
   confidence in review procedures  177
   consultation  172
   good  223, 228, 254, 258
   hearing  172
   interest  20
   interest organizations  181
   opposition  181
   perception  8, 53, 186, 192
   relations strategies  21
   welfare  171
public sector  8, 12, 21, 24, 29, 82, 92, 132, 192, 261, 279
   researchers  20
   scientists  159
Purdue University  73
PVR
   *See* plant variety rights

## Q

questionnaires  181

## R

R&D planning  86
RAPD  106-107
rapeseed  15
ratio
   of researchers to managers  83
   of researchers to technical support staff  83
   of technical support to researchers  29
   research intensity  31
rDNA  12, 111, 116-117
recalcitrant seeds  103
recombinant DNA  69, 74, 88, 157, 178
recombinant livestock vaccines  10
refuge area  17
registrable marks  220
regulation  20, 202
   of GMOs  19

regulatory
    agencies  173, 174
    authorities  157
    framework  32
    mechanisms  85
    policy  134
research
    exemption  211, 224
    grants  86
    indicators  25
    managers  82
    partnerships  129, 133
    priorities  8
    success  58
    teams  86
Research and Development Center for
    Biotechnology (Indonesia)  37
research exemption  213
Research Institute for Food Crops
    Biotechnology (RIFCB) (Indonesia)
    37, 73, 74-75
research liaison office (RLO)  231, 232
resistance cultivars  194
resistance management strategies  199
resource allocation  41
resource manager  86
RFLP  57, 69, 106, 280
rice  66, 69, 73-74, 89-90, 104-105, 107,
        113, 129, 179, 184, 196, 247
    basmati  106
    biotechnology  117
    breeding  106-107
    Bt-  169, 186, 191
    center of origin of Asian  189
    cultivars  185, 189
    farmers and consumers  191
    field-testing of  186
    genome  107
    germplasm  105, 107
    germplasm management  106
    indica  106
    javanica  106
    landraces  107
    nontransgenic  189
Rice Biotechnology Network  72
risk
    analysis  10, 75, 172
    assessment  20, 75, 157, 161, 163, 170,
        177, 181
    management  75, 142, 157, 159-161,
        163-164, 168

RLO
    *See* research liaison office
RNA  188
Rockefeller Foundation  72-73, 118, 131
royalties  95, 98-99, 214, 242, 275

## S

safety assessment of food and feed  178
scientific capacity  7
Scientific Research Center of Yucatan  35
scoring method  9, 46, 56
scoring technique  131
SDC
    *See* Swiss Agency for Development
        Cooperation
selectable markers  187
self-sufficiency  70
simulation models  47
Singapore  272
skill conduit  82
skills  83, 86, 89, 96, 116
social criteria  64
socioeconomic analysis  10
sorghum  72, 113
Southeast Asia  12, 83, 89, 156, 189
soybean  15, 72, 215
species diversity  93
Sri Lanka  89
stakeholders  8, 48, 133, 157, 163, 172, 268
Stanford University  244, 264
strategic
    alliances  12, 44
    decisions  2
    planning  8, 93
    research  280
strategy  68, 83, 104, 128, 209, 258, 273
    for dealing with biotechnology  19
substantial equivalence  177
*sui generis*  214, 224
super pests  171
sustainability  43
    of funding  31
sustainable development  24
Sweden  85
sweet potato  113
Swiss Agency for Development Cooperation
    (SDC)  73, 140
Switzerland  140
Syria  284

System-wide Information Network for
        Genetic Resources (SINGER)   287

# T

Taiwan   89
target gene   18
taxonomic   96, 107
teamwork   71
technology gap   111
technology transfer   9-10, 12, 90, 98-99,
        111, 135, 141, 143, 145, 158, 209,
        243, 245, 255, 262, 274, 276
    agreements   267
    from public- to private-sector
        organizations   13
    from US universities to private industry
        264
    international   2
    IPR-related   75
    system   269
    within ABSP   267
terminator technology   1, 19
Thailand   89, 116
thematic sections   2
third party   231, 243, 245, 251, 258, 277
tissue culture   1, 25-26, 29, 32, 35-37, 43,
        72, 103, 129, 132, 180, 282
tomatoes
    genetically engineered   160
top-down   42, 68, 70, 74
total resource management   83
toxicity   160
trade   247
trade secret   275
trademark   11, 220, 228
traditional breeding   210
traditional carriers of information   286
traditional plant breeding   98
training   3, 81, 87, 97-98, 105, 111, 114,
        129, 133, 141, 143, 162, 178, 202,
        270, 273
    intensive technical   163
    midcareer   88
    module   3
    program for HRD   73
transdisciplinary teams   98
transformation   42, 73, 102, 132, 187-188,
        217, 251-252, 255
    of crops   113, 179
    of scientists to managers   82

protocols   90, 192
systems   250
technology   174
vectors for   265
transgenic
    approach   191
    Bt potato   202
    Bt variety   195
    cotton   273
    crops   16, 130, 134, 160, 171, 177-178,
        279, 284
    cultivars   246
    greenhouse   185
    material   265
    melon   177
    melons   177
    organisms   10, 177
    plant deployment   3, 194
    plant varieties   13, 117
    plants   10, 93, 159, 161, 185, 217, 262,
        272
    potatoes   110, 197
    products   10, 27, 63, 186
    protein   186
    regulatory policy   201
    rice   189-190
    seed   197, 203
    soybean and corn seeds   247
    technology   180, 197, 281
    tomatoes   117
    traits   153, 187, 247
    vs nontransgenic seed   203
transgenic plant deployment
    policy for   201
transparency   50
TRIPs
    *See* Agreement on Trade-Related
        Aspects of Intellectual Property
        Rights

# U

UK   15, 18, 20, 134, 212-213, 229
uncertainty   45, 53, 58, 217, 251, 275
United Nations Conference on the
    Environment and Development
        130, 155, 241
United Nations Development Programme
    (UNDP)   118
United Nations Environment Programme
    (UNEP)   162

United Nations Industrial Development Organisation (UNIDO)  162
United States Agency for International Development  72, 128-136, 162, 261-262, 264-267, 269
United States Department of Agriculture (USDA)  19, 131, 161, 215
United States Patent Office  215
Universiti Putra Malaysia  230
    intellectual property statute  234
universities  13, 16, 25-26, 28, 31, 68, 80, 88, 95, 99, 114, 117, 129-131, 134, 149, 219, 231, 233, 245, 264, 266
University of Birmingham  105-106
University of Florida  117
University of Indonesia  67
University of Nairobi  36
University of the Philippines in Los Baños  166
University of Viçosa (Brazil)  244
University of Zimbabwe  36, 85
university research  130
upstream research  274
USA  80, 114, 157, 161, 166, 195, 197, 201, 215, 219, 229, 262, 264, 273, 281

## V

vaccines  12
vector  19, 159, 177, 255
Vietnam  89
visitors  230-231, 243, 246, 259

## W

weediness  160, 171, 177, 187, 189
weeds  10, 181
wild germplasm  189
World Bank  27, 29, 72
world hunger  21
World Trade Organization (WTO)  18, 214
World Wide Web  180, 286
WTO  224

## Y

yam  113

## Z

Zimbabwe  25, 36, 83, 116